DESIGN OF EXPERIMENTS

A Realistic Approach

STATISTICS

Textbooks and Monographs

A SERIES EDITED BY

D. B. OWEN, *Coordinating Editor*

Department of Statistics
Southern Methodist University
Dallas, Texas

PETER LEWIS
Naval Postgraduate School
Monterey, California

PAUL D. MINTON
Virginia Commonwealth University
Richmond, Virginia

JOHN W. PRATT
Harvard University
Boston, Massachusetts

OTHER VOLUMES IN PREPARATION

DESIGN OF EXPERIMENTS

A Realistic Approach

Virgil L. Anderson and Robert A. McLean

Department of Statistics
Purdue University
Lafayette, Indiana

Department of Statistics
University of Tennessee
Knoxville, Tennessee

MARCEL DEKKER, INC.   New York and Basel

MARCEL DEKKER, INC.

270 Madison Avenue, New York, New York 10016

LIBRARY OF CONGRESS CATALOG CARD NUMBER: 73-90767

ISBN: 0-8247-7493-0

Current printing (last digit):

20 19 18 17 16 15 14 13

PRINTED IN THE UNITED STATES OF AMERICA

To Avis and Marjorie

For when fear is overcome, curiosity and constructiveness are free, and man passes by natural impulse towards the understanding and embellishment of life.

Will Durant: Our Oriental Heritage

CONTENTS

v

PROBLEMS

PREFACE

In this book we have tried to express rather complex ideas on how and why scientific investigators should design experiments. It is assumed that the reader knows only high school algebra and basic statistical concepts such as normal, t, χ^2, and F distributions; sampling distributions of the mean and variance; and has been exposed to simple linear regression and one way analysis of variance. Many of the ideas have been tried out on students at Purdue University, the University of Tennessee and in numerous industrial type classrooms.

The authors have both taught these concepts at this comparatively low mathematical level to an extremely diversified audience. The students have ranged in mathematical competence from Ph.D. candidates in mathematics to students who had only high school algebra. The experimental backgrounds of the students have ranged from undergraduates with no experience to a professor with a Ph.D. (who took the course for credit). The latter had many years of experience but no formal training in design of experiments. In all cases, if the student had no fear of statistics, wanted to learn and possessed a certain amount of mathematical maturity, the varied backgrounds did not detract from the success of the course. In fact, many times this diversity made students recognize similarities among problems from different fields of applications. This gave them a broader scientific approach to experimental problems than a similar course taught only with familiar examples. This book is meant to be read by people from any field who work, or will possibly work, on problems dealing with scientific experiments. Some of the areas from which students have come are Agriculture (all departments), Biology, Bionucleonics, Chemistry, Computer Sciences, Education,

Engineering (most of the schools), Genetics, Home Economics,
Industrial Management, Mathematics, Medicine, Pharmacy, Social
Sciences, Statistics and Veterinary Science. The varied background
of the students makes it impossible to present examples for
everyone. The mature students, however, appear to have little
difficulty in visualizing similar examples in their own fields.

This book is unusual in regard to the emphasis on the
usefulness of the results of the experiment. We want the reader to
appreciate the inferences he can make once he has the results. This
means he must understand how the samples were taken, or in design of
experiments terminology he must know what restrictions on
randomization he has imposed on his observations and how these
restrictions affect his inferences. The usual concepts of blocking
and efficiency of designs are given, but we wish to emphasize the
real dangers the experimenter runs when blocking and sometimes how
little information can be obtained when he inadvertently or
purposely makes restrictions on randomization. We use a restriction
error in a pedagogical sense to make the reader grasp certain
deficiencies that happen so often in experimentation. It is not
uncommon for a so called "expert" in design of experiments to ruin
an experiment because the emphasis in so many design of experiments
books has been the straight application of the designs set up for
agricultural experiments. These people do not understand the
assumptions made in this particular subject matter area and apply
the techniques without careful thought. One of our main purposes,
then, in this book is to try to give a broad, general base for
designing experiments in any subject matter area and have the
reader thoroughly understand what and why he is doing certain
things at the design stage.

The book is arranged so that the reader may go from the simple
to the complex designs and grasp the appropriate analyses from the
resulting data. In Chapter 1 we review analyses for one way
classified data and in Chapter 2 we introduce two way analyses.
Random sampling concepts are used in these two chapters, assuming

the design of each experiment is completely randomized although the design is not fully defined until Chapter 4. The method of review that is utilized is to present some non-standard, as well as standard, material. The most controversial subject that is presented in this review is that of the use of multiple comparison tests. The reason for this controversy is that there is no one correct answer as to whether one should always strive to have small Type I or Type II errors for either the individual comparisons among treatment means or for the entire experiment. Obviously, the selection of the most appropriate multiple comparison test is dependent upon the individual situation and usually the experimenter does not have sufficient information available to make a good selection. Consequently, since we have experienced and observed larger economic losses whenever Type I errors are made, a modified version of the Student-Newman-Keuls test, which has excellent control over the Type I error rate, is used throughout this text. The modification arises by requiring a significant F test in the analysis before the multiple comparison test is utilized. This appears to be a very realistic approach when one is involved with experiments containing several factors and associated interactions.

Chapter 3 is the real beginning of this design book. In this chapter we wish to show the reader where design of experiments fits in the scientific method and how most experiments come about.

Chapters 4 through 7 cover our concepts of design, the thorough usage of the restriction error concept and the unusual analyses that result. This is the body of the book, the basic designs that set the stage for a most realistic look at the problems encountered by an experimenter when he decides to design an experiment.

Chapter 8 expresses a view on Latin square type designs and Chapters 9 and 10 attempt to provide a practitioner's approach to the 2^n completely randomized factorials. Chapters 11 and 12 have useful designs in them but if time were a factor in using this book for a quarter or a semester, it would do little harm to go to

Chapter 13 directly from Chapter 10. We believe it would be bad
for a student not to be exposed to the designs related to response
surface exploration. Hence, we recommend that, even though Chapter
13 is not as thorough as Myers, R. H., Response Surface Metholology,
Allyn and Bacon, Inc. (1971), it should be covered in a design of
experiments course.

The authors chose to leave out the use of concomitant variables
(including covariance) in designing experiments in this book because
of the possible background of the readers and the experience of the
authors in teaching courses with and without this additional
information. It has been our experience that students do appreciate
the difference between concomitant information and blocking, or
additional terms in models expressing the effect of concomitantly
measured variables; however, to cover this material in the desired
depth to make it understandable would require more time in the
analysis phase (Chapters 1 and 2) than we believe is warranted for
this text. For those readers who would like to know something
about these concepts we recommend they read Chapter 4 of Planning
Experiments, Cox, D. R., Wiley and Sons, Inc., 1958.

We are indebted to Oliver and Boyd Publishers for allowing us
to reprint most of pages 609 and 610 and Section 11.4 from their
book edited by Davies, O. L., entitled Design and Analysis of
Industrial Experiments (1963). These appear in Appendix 4 and
Section 13.2.4, respectively. We also wish to thank S. S. Shapiro
and M. B. Wilk for allowing us to use their tables in Appendices
9 and 10, and H. Leon Harter for allowing us to reproduce in
Appendix 6 part of the tables in Order Statistics and their Use in
Testing and Estimation, Volume 1: Tests Based on Range and
Studentized Range of Samples from a Normal Population (Gov't.
Printing Office, Washington, D. C., 1970). We are indebted to the
Biometrika Trustees for giving us permission to use Tables 8, 12,
and 18 from Biometrika Tables for Statisticians, Volume I, third
edition (1966). These tables appear in Appendices 7, 3, and 5 in
this book.

We gratefully acknowledge the help and criticism from students and colleagues who gave freely of their time through the many revisions of this book.

Most of all we wish to thank Mrs. Eddie Hicks of the Statistics Department, Purdue University, for her patience through the endless changes and expert typing of the final manuscript.

Robert A. McLean   Virgil L. Anderson
Knoxville, Tennessee   West Lafayette, Indiana

DESIGN OF EXPERIMENTS

A Realistic Approach

Chapter 1

REVIEW OF SOME BASIC STATISTICAL CONCEPTS

Before the ideas of designing experiments can be fully
appreciated certain standard statistical techniques must be
presented. The ideas developed here are not intended to encompass
all of the elementary procedures, but merely to display enough
examples to introduce the terminology required to structure the
material on designing experiments.

Our methodology throughout the book will be to introduce and
explain ideas through examples and provide references for details.

In the first two chapters the design of the experiment is
assumed to be completely randomized. That is, the structure of the
experiment is assumed to be such that the treatments were placed on
the experimental units completely at random.

Details for constructing a completely randomized design are
given in Chapter 4. Suffice it to say here that for the first two
chapters the structure of the experimental arrangement of the
material used in the investigations is such that no restriction on
randomization of the treatments onto the experimental units is
assumed.

1.1 TESTING HYPOTHESES AND SAMPLE SIZE

In a Highway Research Organization one aspect of a most
complicated experiment had to do with comparing the effect of
percent cement and curing time on the strength of cement-stabilized
base material. The dependent variable, y, was pounds per square

inch (psi) measured by an indirect tensile strength method. For
this part of the study, previous investigations showed that the
measured variable responded well enough for the usual t-test
assumptions (normal distribution and equal variances) to hold.
Hence the measured variable, psi, could be used as the variable to
be analyzed.

In order to get an indication of the number of observations
that would be necessary to show a significant increase in psi using
8% cement and 7 days of curing time over using 4% cement and 21 days
of curing time if there truly was an increase, the experimenter
decided to run 10 cylindrical specimens, each of which may be
considered as an experimental unit, measuring 4 inches in diameter
and 2 inches in height for each treatment. This would allow an
estimate of the variance from the 20 specimens and then one could
estimate the additional number of specimens needed to show
significance if it really existed. Stein (1945) shows a
theoretically sound procedure for handling this type of sequential
sampling; however, an approximation of this technique will be
utilized here.

For this particular experiment the investigator was willing to
run a 5% risk of saying there was a significant difference when
there really was not.

As pointed out in this study, most experimenters know quite a
bit about their product and reflect this knowledge by using a one-
tail test. In this case the investigator is only interested in
obtaining a less expensive product (4% cement) which will perform
the same task as the more expensive product (8% cement). Most
investigators ignore the fact that they have two factors, i.e.,
percent cement and curing time, confounded and merely compare
"Method 1" to "Method 2." This is a satisfactory procedure in this
case, but, as we will see later, when one wants to write a
mathematical model which relates percent cement and curing time to
tensile strength one will not be able to use an experiment which

confounds the two factors. "Confounds" merely means that the effects
of the two factors are not separable.

Assume that the experiment was carried out appropriately at the
mixing and curing stages and the specimens were obtained by a
completely randomized procedure. (By completely randomized, we mean
that a table of random numbers, Appendix 1, or some procedure, was
used to decode the order of obtaining the 20 specimens.) The data
obtained are given in Table 1.1.

TABLE 1.1

4% cement and 21 days curing time (psi)	8% cement and 7 days curing time (psi)
70	89
81	79
82	85
68	76
74	74
76	81
65	70
74	79
60	65
70	72
$\sum_{i=1}^{10} y_{1i} = 720$	$\sum_{i=1}^{10} y_{2i} = 770$
$\bar{y}_1 = 72$	$\bar{y}_2 = 77$
$\sum_{i=1}^{10} (y_{1i} - \bar{y}_1)^2 = 422$	$\sum_{i=1}^{10} (y_{2i} - \bar{y}_2)^2 = 460$

The inference from this experiment is that the results apply to
stabilized base material in which all other factors are held constant.
Such factors are water at 3%, wrapping specimen during curing,

limestone soil, curing temperature at 40°F, high compactive effort, and gyratory shear compaction.

The analysis of the data follows:

H_0: $\mu_1 = \mu_2$ where μ_1 = population mean psi of 4% cement at 21 days curing

H_1: $\mu_1 < \mu_2$ (one tail) μ_2 = population mean psi of 8% cement at 7 days curing

In this case the experimenter is only interested in finding out whether or not there is a significant decrease in the mean psi when he uses the 4% cement cured for 21 days as compared to the 8% cement cured for 7 days. Under the assumption that the psi observations are normally distributed, the test statistic is

$$t_{df} = \frac{\bar{y}_1 - \bar{y}_2}{s_{\bar{y}_1 - \bar{y}_2}} = \frac{72 - 77}{\sqrt{49(0.2)}} \cong -\frac{5}{3.13} = -1.6$$

where t_{df} may be compared to t in Appendix 3 (df = $n_1 + n_2 - 2 = 18$), $\bar{y}_1$ and $\bar{y}_2$ are sample means estimating μ_1 and μ_2, respectively,

$$s_{\bar{y}_1 - \bar{y}_2} = \left[\frac{\sum\limits_{i=1}^{10} (y_{1i} - \bar{y}_1)^2 + \sum\limits_{i=1}^{10} (y_{2i} - \bar{y}_2)^2}{n_1 + n_2 - 2} \left(\frac{1}{n_1} + \frac{1}{n_2} \right) \right]^{\frac{1}{2}} .$$

Since $n_1 = n_2 = 10$ and $\alpha = 0.05$ the critical value for t is -1.734. The value α is the probability of making a Type I error, i.e., rejecting the null hypothesis (H_0) when in fact it should have been accepted. The selection of the magnitude of α is done by the experimenter and is possibly based on the economics of the problem or in many cases merely selected at a value, say 0.05, which is consistent with previous studies of this type. Note that

$$\frac{\sum\limits_{i=1}^{10} (y_{1i} - \bar{y}_1)^2 + \sum\limits_{i=1}^{10} (y_{2i} - \bar{y}_2)^2}{n_1 + n_2 - 2} = \frac{882}{18} = 49$$

is a pooled estimate of the within-treatment variability which will be denoted as s_{pooled}^2 = 49. This calculation is the one that requires the assumption of equal variances. The method of testing for homogeneous variances is reviewed in Section 1.3. This pooled estimate of the within variance is the best estimate of the true variance, σ^2, and will be used to determine the approximate sample size required.

Based on the numerical results above some practitioners would say the inference must be made that there is no difference in using 8% cement for 7 days curing and 4% cement for 21 days curing. It is true that the means are not statistically significant at the 0.05 probability level, but what are the consequences of accepting H_0 in this case?

An error of accepting H_0 when in fact it should have been rejected is referred to as a Type II error. The probability of making an error of this kind and consequently accepting a false hypothesis is referred to as β. Naturally one would like to have β as small as possible because a decision error of this type would result in an economic loss.

The next step in this experimental procedure is to decide on the number of additional specimens to take to reduce β to a reasonable size, say, 0.10. This decision on the size of β and α, for that matter, must be made before the experiment is run or the sequential procedure recommended here or by Stein (1945) must be followed. In either case the Type I and II errors must be understood by the experimenter and be decided upon.

If for example, there was an actual difference of 5 psi between the two populations, i.e., $\delta = \mu_2 - \mu_1 = 5$ and using $s_{pooled} = 7$ as the σ in Appendix 4, β can be estimated.   In this case D = δ/σ = 0.71 and for a one-tailed test with α = 0.05 and n_1 = n_2 = 10 we find that β > 0.5.   This is extremely bad from a practitioner's view because he wants to make the right decision as frequently as he can. Note that we have utilized the information obtained in our experiment as if these were true values, rather than just estimates of these values.   One will resort to this type of reasoning quite frequently and hence conduct the experiment in a sequential fashion. The practitioner must recognize that the β values in Appendix 4 are approximate, as a result of estimating σ, but they do provide excellent guidelines for obtaining sample sizes.

Continuing our example, let us assume that α = 0.05 (one tail), δ = 5, σ = 7 and β = 0.10, what should the sample size be for each treatment?   Using D = δ/σ = 5/7 $\simeq$ 0.7 in Appendix 4 we find n_1 = n_2 = n = 36.   Hence we should take 26 more specimens of each (percent of cement, curing time) treatment combination to force β to be small and allow a more clear-cut decision.

>   Problem 1.1.1.   If the economics of the problem caused the
>   experimenter to let δ = 7 and he could take only 13 more
>   observations per treatment after the first 10, what combination
>   of α and β would you recommend?   Explain your reasoning.

1.2   ONE WAY CLASSIFICATION, ASSUMPTIONS MET

In Section 1.1 general ideas on testing hypotheses and determining sample size in experimentation were discussed.   The example in that section had only two treatments and utilized the t-test to bring out these general ideas.

In this section the generalization is to set up a testing procedure to handle any number of treatments, namely, the analysis of variance (ANOVA).   The example will have only four treatments but it should be understood that the technique is easily extended to any number of treatments.   To allow the reader to review the

ANOVA method without other disturbances, we will say that all
assumptions necessary to make the F-test have been met. In the
following sections discussions will cover cases in which some
assumptions have not been met.

A research worker in pharmacy was interested in finding out
whether or not there were differences in disintegration times of a
narrow group of four types of tablets; A, B, C, and D. He was
interested in only these four types which make the factor fixed.
Hence the mean time (in seconds) for disintegration should be
analyzed. If he had been interested in the population of all tablets
from which these four were drawn and he had drawn these four at
random from the possible types, he could have called the factor
random and analyzed the variance of the means of disintegration time.
The inference space, that space within which the investigator may
apply the results of his experiment, is much broader and has wider
implication for the random model than for the fixed model. This is
a general concept for any number of factors in a large experiment.
Two references on fixed and random factors are Eisenhart (1947) and
Dixon and Massey (1969, Chapter 10).

This experiment on tablet manufacturing was set up so that the
manufacturing and testing procedures were carried out in a completely
randomized manner. In addition, previous studies of tablets of these
general types indicated that if there were differences among types,
at $\alpha = 0.05$ level, that the number of tablets per type had to be at
least six in order for the experiment to detect this difference.
References on the subject of determining sample size in ANOVA
problems include Kastenbaum, Hoel and Bowman (1970a), (1970b) and
Bratcher, Moran and Zimmer (1970).

This experiment had six tablets selected at random from each
type and the disintegration time (in seconds) was measured. It is
assumed that the response variable in this case is normally and
independently distributed, that the terms in the model are linear
and that the variance among the tablets within types are all equal.

The model used as the basis for the analysis is

$$y_{ij} = \mu + T_i + \varepsilon_{(i)j} \qquad \begin{array}{l} i = 1,\ 2,\ \ldots,\ t = 4 \\ j = 1,\ 2,\ \ldots,\ n = 6 \end{array} \qquad (1.2.1)$$

where

y_{ij} = the disintegration time of the jth tablet in i^{th} type

μ = overall mean

T_i = the effect of the ith type of tablet (fixed)

$\varepsilon_{(i)j}$ = the random error of the jth tablet in the ith type, NID $(0,\ \sigma^2)$ or normally and independently distributed with zero mean and variance σ^2.

The data are presented in Table 1.2.1.

TABLE 1.2.1

Disintegration Times (seconds) for Types of Tablets

Type of Tablets (i)	Observations (j)	Totals	Means
A	6, 2, 5, 4, 6, 7	30	5.0
B	10, 8, 11, 7, 7, 9	52	8.7
C	3, 7, 6, 4, 8, 6	34	5.7
D	10, 4, 6, 6, 7, 8	41	6.8

Sums of squares (SS):

Given that $\quad \sum_i \sum_j y_{ij} = 157$

and $\quad \sum_i \sum_j y_{ij}^2 = 1141$

(a) Correction term to allow for deviation around mean (CT)

$$CT = \frac{1}{nt} \left(\sum_i \sum_j y_{ij} \right)^2 = \frac{1}{24} (157)^2 = 1027.0$$

(b) SS types $= \sum_i \frac{1}{n} \left(\sum_j y_{ij} \right)^2 - CT = 1073.5 - 1027.0 = 46.5$

(c)   SS total = $\sum_i \sum_j y_{ij}^2$ - CT = 1141.0 - 1027.0 = 114.0

(d)   SS Within Types (error) = SS Total - SS Types

$$= 114.0 - 46.5 = 67.5$$

From these results we can set up the ANOVA in Table 1.2.2. An abbreviated form of notation for the above material and the intuitive reasoning which supports the F-test shown in Table 1.2.2 is given in Section 1.2.1. The student who is truly interested in understanding the philosophy behind the ANOVA should study this section in detail.

TABLE 1.2.2

ANOVA for Tablet Manufacturing Data in Table 1.2.1[a]

Source	df	SS	MS	F	$F_{crit}^{(0.05)}$
Types of tablets	3	46.5	15.5	4.6	3.1
Within types (error)	20	67.5	3.4		
Total	23	114.0			

[a]df = degrees of freedom; SS = sums of squares, deviations around means; MS = SS/df = mean square; F = MS types/MS error calculated from data; $F_{crit}^{(0.05)}$ = the critical value of F for $\alpha = 0.05$ and df are 3 and 20 found in Appendix 5.

Conclusions:

Since the calculated F from the data is greater than the critical value of F at the 0.05 level, the experimenter concludes that there is a significant difference among the mean disintegration times of the four types of tablets. This does not allow him to say anything about which type has the shorter or longer disintegration time, however. Since there is an overall significant effect of types, the investigator would ordinarily examine the individual means to

find this out. One test that allows investigation of all possible
pairs of means in a sequential manner, has very good power $(1 - \beta)$
and keeps the α level constant for investigation of all pairs of
means is the Newman-Keuls test. The two basic references for this
test (See the preface for a philosophical discussion of multiple
comparison tests) are Newman (1939) and Keuls (1952). In addition
to the Newman-Keuls test the researcher is apt to encounter other
multiple comparison procedures such as: the least significant
difference, Scheffé's method, Tukey's honestly significant difference,
and Duncan's multiple range test. Two excellent references which
compare these various methods are Bancroft (1968, Chapter 8), and
Carmer and Swanson (1973).

To run the Newman-Keuls test, arrange the means in rank position

$$\bar{y}_1 > \bar{y}_2 > \bar{y}_3 > \bar{y}_4 .$$

In our example we rank the means as follows:

1	2	3	4
B	D	C	A
8.7	6.8	5.7	5.0

Next prepare a table of differences of means from the largest
difference to the smallest in all possible pairs forming a triangular
arrangement as follows:

Number of Means Spanned

	k	k-1	k-2
	4	3	2
1	B-A 3.7	B-C 3.0	B-D 1.9
2	D-A 1.8	D-C 1.1	
3	C-A 0.7		

where column headings indicate the number of means spanned for each test on that diagonal.

Prepare a list of the least significant ranges as follows:

$R_k = q_\alpha(k,df)\ s_{\bar{y}}$; where $q_\alpha(k,df)$ means the entry in Appendix 6 for the desired α and k = 4, number of means compared in the experiment, and $s_{\bar{y}}$ = the standard error of the mean

$$s_{\bar{y}} = \sqrt{\frac{MS\ error}{n}} = \sqrt{\frac{3.4}{6}} = 0.75$$

where n = number of observations in that type.

$R_{k-1} = q_\alpha(k-1,df)\ s_{\bar{y}}$
$\vdots \qquad \vdots$
$R_2 = q_\alpha(2,df)\ s_{\bar{y}}$

and for our example using α = 0.05,

$R_k = R_4 = q_{0.05}\ (4,20)\ s_{\bar{y}} = (3.96)(0.75) = 2.97$

$R_{k-1} = R_3 = q_{0.05}\ (3,20)\ s_{\bar{y}} = (3.58)(0.75) = 2.68$

$R_{k-2} = R_2 = q_{0.05}\ (2,20)\ s_{\bar{y}} = (2.95)(0.75) = 2.21$

Referring back to the triangular table, we look at each diagonal element and compare in order from k to 2

B-A = 3.7 vs 2.97 $\therefore$ significant
B-C = 3.0 vs 2.68 $\therefore$ significant
D-A = 1.8 vs 2.68 $\therefore$ not significant
B-D = 1.9 vs 2.21 $\therefore$ not significant
D-C = 1.1 vs 2.21 $\therefore$ not significant
C-A = 0.7 vs 2.21 $\therefore$ not significant

Hence the final conclusions for comparing individual means are:
(a) Type B has a significantly longer disintegration time than types A and C.
(b) All other types are not distinguishable.

In general, when making individual comparisons of this type one has one additional rule to follow. That is, one continues across the first row until the first nonsignificant difference is encountered. At this point one must stop and declare all remaining differences in the first row nonsignificant.

In some instances it will be possible to continue across the row and find another value which is greater than its respective least significant range. This value would still be declared not significant in order to avoid any logical contradictions. In addition, all differences below any nonsignificant difference are declared nonsignificant. Next, one repeats this procedure for row 2, and continues until finished. Note that our example satisfies this particular rule.

To complete this investigation the research worker may want the longest disintegration type for some reason and in this case he should choose B. Next he would almost undoubtedly like to know the 95% confidence interval on the mean disintegration time. To obtain this confidence interval he merely needs to set up

$$(\bar{y}_B - t_{20\ df}^{(0.025)}\ s_{\bar{y}}) < \mu < (\bar{y}_B + t_{20\ df}^{(0.025)}\ s_{\bar{y}})$$

and use Appendix 3 to obtain

$$t_{20\ df}^{(0.025)} = 2.086$$

Hence he has the interval

$$[8.7 - (2.086)(0.75) < \mu < 8.7 + (2.086)(0.75)]$$

or

$$(7.1 < \mu < 10.3)$$

as the 95% confidence interval on the mean disintegration time (in seconds) of the type B tablet.

1.2.1   ANOVA Notation and Rationale

The type of notation that is used in most statistical books on the ANOVA is the so called "dot" notation. To introduce this notation let us consider the general form of the model which is given in Eq. (1.2.1). The symbols representing the data are displayed in an array for n observations for each of t treatments as shown in the following tabulation.

| Treatment | \multicolumn{6}{c}{Observation} | Total | Mean |
	1	2	3 ...	j ...	n		
1	y_{11}	y_{12}	y_{13} $\cdots$	y_{1j} $\cdots$	y_{1n}	$T_{1\cdot}$	$\bar{y}_{1\cdot}$
2	y_{21}	y_{22}	y_{23} $\cdots$	y_{2j} $\cdots$	y_{2n}	$T_{2\cdot}$	$\bar{y}_{2\cdot}$
3	y_{31}	y_{32}	y_{33} $\cdots$	y_{3j} $\cdots$	y_{3n}	$T_{3\cdot}$	$\bar{y}_{3\cdot}$
.							
.							
.							
i	y_{i1}	y_{i2}	y_{i3} $\cdots$	y_{ij} $\cdots$	y_{in}	$T_{i\cdot}$	$\bar{y}_{i\cdot}$
.							
.							
.							
t	y_{t1}	y_{t2}	y_{t3} $\cdots$	y_{tj} $\cdots$	y_{tn}	$T_{t\cdot}$	$\bar{y}_{t\cdot}$

The symbol $T_{i\cdot}$ represents the total of all observations for the ith treatment and the "dot" indicates a summation over the subscript that has been replaced by the dot, that is $T_{2\cdot} = \sum_{j=1}^{n} y_{2j}$. In addition, $\bar{y}_{i\cdot} = T_{i\cdot}/n$, the sample mean of the ith treatments. Note that the grand total of all observations is $T_{\cdot\cdot}$ and the grand mean is $\bar{y}_{\cdot\cdot} = T_{\cdot\cdot}/tn$.

The method of calculation of the sums of squares for types of tablets and the total sum of squares was indicated in the previous section. The error sum of squares shown in Table 1.2.2 was then obtained by subtracting SS types from SS total. In order to understand the reasoning behind this calculation, observe that the basic definition for the total sum of squares is

$$SS \text{ total} = \sum_{i=1}^{t} \sum_{j=1}^{n} (y_{ij} - \bar{y}_{..})^2$$

By expanding $\sum \sum (y_{ij} - \bar{y}_{..})^2$ this expression can be seen to be equal to the formula for calculation, namely,

$$SS \text{ total} = \sum_{i=1}^{t} \sum_{j=1}^{n} y_{ij}^2 - CT$$

In addition, by writing

$$(y_{ij} - \bar{y}_{..})^2 = [(y_{ij} - \bar{y}_{i.}) + (\bar{y}_{i.} - \bar{y}_{..})]^2$$

it can be seen that

$$SS \text{ total} = \sum_{i=1}^{t} \sum_{j=1}^{n} (y_{ij} - \bar{y}_{i.})^2 + \sum_{i=1}^{t} \sum_{j=1}^{n} (y_{i.} - \bar{y}_{..})^2$$

$$+ \sum_{i=1}^{t} \sum_{j=1}^{n} 2(y_{ij} - \bar{y}_{i.})(\bar{y}_{i.} - \bar{y}_{..})$$

Note that in the last term of the above equation that $2(\bar{y}_{i.} - \bar{y}_{..})$ is a constant with respect to the inner summation over j and that

$$\sum_{j=1}^{n} (y_{ij} - \bar{y}_{i.}) = 0$$

consequently

$$SS \text{ total} = \sum_{i=1}^{t} \sum_{j=1}^{n} (y_{ij} - \bar{y}_{i.})^2 + \sum_{i=1}^{t} \sum_{j=1}^{n} (\bar{y}_{i.} - \bar{y}_{..})^2$$

Observing that

$$\sum_{i=1}^{t} \sum_{j=1}^{n} (\bar{y}_{i.} - \bar{y}_{..})^2 = n \sum_{i=1}^{t} (\bar{y}_{i.} - \bar{y}_{..})^2 \qquad (1.2.2)$$

$$= \frac{1}{n} \sum_{i=1}^{t} T_{i\cdot}^2 - \frac{T_{\cdot\cdot}^2}{nt}$$

$$= \frac{1}{n} \sum T_{i\cdot}^2 - CT$$

$$= \text{SS treatments}$$

we see that

$$\text{SS total} = \sum_{i=1}^{t} \sum_{j=1}^{n} (y_{ij} - \bar{y}_{i\cdot})^2 + \text{SS treatments}$$

so that

$$\text{SS error} = \sum_{i=1}^{t} \sum_{j=1}^{n} (y_{ij} - \bar{y}_{i\cdot})^2$$

which is intuitively correct since $\sum_{j=1}^{n} (y_{ij} - \bar{y}_{i\cdot})^2$ represents the sum of squares within treatment i and if divided by its degrees of freedom, n - 1, would yield an estimate of σ^2.

This fact should be intuitive since all the variation within treatment i is the result of the n ε_{ij} random variables each of which has been assumed to have variance σ^2. Thus the outer summation over i produces a pooled sum of squares which divided by the total degrees of freedom t(n - 1) yields the best estimate of σ^2.

It should be noted that the above data array and the algebraic relation among the SS total, SS treatments, and SS error was presented assuming an equal number of observations per cell. This assumption was only made for ease of notation and is not a mathematical requirement. This statement remains valid for the following discussion on the F-test utilized to test the hypothesis of no significant differences among treatment means.

The F-test is well known for its use in testing for equality of two variances. This test is carried out by computing the ratio of two independent variance estimates. If this ratio is greater than

some tabular value it is concluded that the population variance
estimated by the numerator variance estimate is greater than the
other population variance. Suffice it to say that it must be
demonstrated that under the null hypothesis of no treatment effects
that MS treatments and MS error are estimating the same population
variance and hence (MS treatments)/(MS error) is distributed as F.
The fact that these estimates are independent will be assumed.

It was shown above that (SS error)/t(n - 1) = MS error is an
estimate of σ^2. Thus it remains to show that MS treatments is also
an estimate of σ^2 under the hypothesis of no differences among
treatment means. Under the assumption of the null hypothesis
$(H_o : T_i = 0)$, Eq. (1.2.1) can be rewritten as

$$y_{ij} = \mu + \varepsilon_{(i)j} \quad \begin{Bmatrix} i = 1, \ldots, t \\ j = 1, \ldots, n \end{Bmatrix}$$

Thus, each of the treatment means, $\bar{y}_{i.}$, $i = 1, \ldots, t$, is merely a
random observation from the distribution of means which, by assumption,
must have mean μ and variance σ^2/n. It follows from Eq. (1.2.2) that

$$\frac{1}{t - 1} \sum_{i=1}^{t} (\bar{y}_{i.} - \bar{y}_{..})^2 = \frac{1}{n(t - 1)} \text{ (SS treatments)}$$

$$= \frac{1}{n} \text{ (MS treatments)}$$

and hence must be an estimate of σ^2/n so that MS treatments is an
estimate of σ^2 under the null hypothesis. Intuitively, it has been
demonstrated that the ratio

$$\frac{\text{MS treatments}}{\text{MS error}}$$

is distributed as F with t - 1 and t(n - 1) degrees of freedom.

1.3 UNEQUAL VARIANCES AND TRANSFORMATIONS IN ANOVA

In most ANOVA problems the assumption which was historically

thought to have been most critical was homogeneous variances. Box
(1954), however, demonstrated that the F-test in the ANOVA is most
robust for α^* while working with a fixed model with equal sample
sizes. Box (1954) showed that for relatively large (one variance was
up to 9 times larger than another) departures from homogeneity that
the α level may only change from 0.05 to about 0.06 which is not
considered to be of any practical importance. (It should be pointed
out that the only time the α level increased drastically in Box's
study was when the sample size was negatively correlated with the
size of the variance.) Whenever larger departures from homogeneity
occur it is felt that the data should be transformed in order for the
ANOVA to produce meaningful results. The intent of this section is
to discuss tests for homogeneity of variances, and the various
transformations that may be utilized in the case of "too much"
heterogeneity.

No one to our knowledge, has come up with an α-level on
homogeneity tests which will indicate when the experimenter should
become concerned about making a transformation. A set of working
rules which seem to be effective for the practitioner are as follows:

i.   If the homogeneity test is accepted at $\alpha = 0.01$ level,
       do not transform.

ii.   If the homogeneity test is rejected at $\alpha = 0.001$ level,
       transform.

iii.   If the result of the homogeneity test is somewhere
       between $\alpha = 0.01$ and 0.001, try very hard to find
       out from the investigator the theoretical distribution
       of the data. If there is a practical reason to transform
       go ahead and transform; otherwise we recommend not to
       transform. With the availability of transgeneration
       options in statistical programs for digital computers

---

*Robust for α in this situation means that even if the variances
are heterogeneous the α-level for the F-test on treatment means is not
changed materially.

there is a tendency to try various transformations, make the homogeneity test in each case and select the transformation which yields the test statistic which is most favorably located in the acceptance region. While this practice is not all bad, there is danger in losing "touch" with the meaning of the variable being investigated. Hence use as much theory about the variable as possible and select the transformation which makes sense from a physical point of view. An example of this type of selection is presented in Section 1.3.4.

An example in manufacturing tablets demonstrates tests of homogeneity of variances and the uses of transformations. Four types of magnesium trisilicate tablets:

(A) magnesium stearate   16 mesh granule size

(B) talc powder   16 mesh granule size

(C) liquid petrolatum   16 mesh granule size

(D) magnesium stearate   20 mesh granule size

were completely randomized in manufacturing and testing.

Ten tablets were measured for disintegration time in seconds for each of the four types. The data are presented in Table 1.3.1.

TABLE 1.3.1

Time (seconds) of Disintegration of Tablets for Four Types of Tablets

	Types of tablets						
	A		B		C		D
	20	42	8	12	50	124	151	178
	28	25	10	24	67	72	125	151
	36	24	12	10	90	78	180	152
	16	31	16	19	103	70	149	161
	25	33	9	10	90	76	175	118
Mean	28		13		82		154	
Variance	59.55		26.22		430.89		436.22	

On looking at the data one can see that there tends to be a
definite positive relationship between the means and the variance.
Ordinarily when one finds a relationship of this type between the
means and variances one also finds a lack of normality in the data.
It is our experience that transformations which improve the
heterogeneity situation also improve the lack of normality problem.
The relationship between the assumption of homogeneity of variances
and that of normality is discussed by Cochran (1947).

This visual relationship provides a warning that tests of the
homogeneity of variances should be considered. Three of these tests
follow in the subsections. The Bartlett (1937) test is an old,
well-established test that has been programmed for most computers,
but it is a bit sensitive to nonnormality, especially if the tails
of the distribution are too long it tends to show significance too
often. Using our working rules if the investigator has no
theoretical knowledge of his variable, we would suggest using
$\alpha = 0.001$ when the distribution of y seems to have excessively long
tails. This will tend to prevent too many unnecessary transformations.

Another test, suggested by Bartlett and Kendall (1946), utilizes
the $\log s^2$ as a variable. They suggest running ANOVA on the $\log s^2$,
assuming the variances of $\log s^2$ are homogeneous, in order to find
out which treatments cause this heterogeneity. The basis for using
$\log s^2$ as the variable is that it is approximately normally distributed
if $n \geq 5$. This test is not used as much as we think it should be
because too many practitioners do not know about it. One drawback,
of course, is that to obtain a minimum of two observations of $\log s^2$
per cell there should be at least 10 observations of y in each cell.
The group of 10 observations may be broken randomly into two groups
of 5 observations each and $\log s^2$ calculated for each group. Not
only does this test furnish the experimenter with another means of
investigating the assumption of homogeneity of variance but also
provides a procedure for determining the causal effects that
influence the variability of a process. Ordinarily the ANOVA is
used to study the effects of the factors on the response variable

but through the log s^2 technique one can also study the relationship between the factors and variability of the process.

The only other test of homogeneity we will describe is the Burr-Foster Q test, Burr and Foster (1972). This test has not been published and we believe it should be available to the practitioner because the test statistic is easy to calculate and does not have the sensitivity of normality departures encountered with the Bartlett test.

1.3.1   Bartlett Test (Equal Subclass Numbers)

Table 1.3.2 shows the procedure to test for homogeneity of the variances using Bartlett's test for equal subclass numbers on data in Table 1.3.1. Snedecor and Cochran (1967, p. 296 and 297), give a procedure for unequal subclass numbers.

TABLE 1.3.2

Bartlett's Test of Homogeneity

k	Type	s^2	$\log s^2$
1	A	59.55	1.77488
2	B	26.22	1.41863
3	C	430.89	2.63437
4	D	436.22	2.63971
	Total	952.88	8.46759

$$\overline{s^2} = \frac{\Sigma s^2}{k} = \frac{952.88}{4} = 238.22$$

$$\log \overline{s^2} = 2.37698$$

$$M = 2.3026 \ (df) \ [k \log \overline{s^2} - \Sigma \log s^2]$$

where: df = degrees of freedom per variance = 9

k = number of treatments or samples (type) = 4

$$M = 2.3026 \ (9) \ [4(2.37698) - 8.46759] = 21.56$$

$$C = 1 + \frac{k+1}{3(df)(k)} = 1 + \frac{5}{3(9)(4)} = 1.046$$

$$\chi^2_{3df} = \frac{M}{C} = \frac{21.56}{1.046} = 20.6$$

and

$$\chi^2_{crit} \; 3df(0.001) = 16.3$$

$\therefore$ Reject homogeneity of variances.

1.3.2   BARTLETT and KENDALL log s^2 ANOVA

The data in Table 1.3.1 were randomly assigned to two groups of 5 for each type of tablet. Table 1.3.3 shows the ANOVA of the log s^2.

TABLE 1.3.3

Log s^2 Analysis of Homogeneity

	Type of tablet						
A		B		C		D	
s^2	$\log s^2$	s^2	$\log s^2$	s^2	$\log s^2$	s^2	$\log s^2$
59	1.7709	10	1.0000	449.5	2.6527	493.0	2.6928
52.5	1.7202	39	1.5911	510.0	2.7076	478.5	2.6799

ANOVA of log s^2

Source	df	MS	F_{calc}	$F_{crit(3,4)}$ (0.01)
Types of tablet	3	0.96876	21.8	16.7
Within error	4	0.04439		
Total	7			

Since F_{calc} (calculated from data) $> F_{crit}$ we reject the hypothesis that the variances are homogeneous at $\alpha = 0.01$. A further refinement of the analysis would be to run a Newman-Keuls test on the type of tablets means of log s^2 and show which ones are different.

Problem 1.3.1.   Run the Newman-Keuls test on means of the (log s^2)'s of Table 1.3.3 and explain the results.

1.3.3 BURR-FOSTER Q-Test of Homogeneity

The Q-test for equality of variances is based on a statistic
which is a monotone function of the coefficient of variation of the
sample variances. As such it offers promise as a preliminary test
for the assumption of homogeneity of population variances which is
needed in the ANOVA technique. Although the Q-test is not a so-called
quick test, the test statistic is sufficiently simple to permit
calculation of its value on a desk calculator. A sample variance
taking the value zero does not disrupt this test (as it does to
Bartlett's test).

For equal sample sizes, n, from each of p parent populations,
let s_i^2 (for i = 1, ..., p) denote the ith sample variance. Denoting
the value of the test statistic by q, we have:

$$q = (s_1^4 + \ldots + s_p^4)/(s_1^2 + \ldots + s_p^2)^2$$

For unequal sample sizes we specify that each sample variance, s_i^2,
be calculated by dividing by the degrees of freedom, ν_i, rather than
by the sample size, n_i (where $\nu_i = n_i - 1$, for i = 1, ..., p). Let
$\bar{\nu}$ denote the arithmetic average of the degrees of freedom. In this
case we have:

$$q = \bar{\nu} (\nu_1 s_1^4 + \ldots + \nu_p s_p^4)/(\nu_1 s_1^2 + \ldots + \nu_p s_p^2)^2.$$

Large values of q lead to rejection of the hypothesis of equal
population variances. The critical values are given in Appendix 8
for various numbers of parent populations, p, and various possibilities
for equal degrees of freedom ν (where $\nu = n - 1$). This table can be
used directly for equal degrees of freedom. For unequal degrees of
freedom, q is calculated as indicated above, but $\bar{\nu}$ is to be substituted
for ν, provided that $\bar{\nu}$ and the harmonic mean of the ν's do not differ
greatly. For large sample sizes ($\nu > 60$), we note that $p \bar{\nu}(pq - 1)/2$
is asymptotically chi-square with (p - 1) degrees of freedom.

For the tablet example, the Burr-Foster test is carried out as
follows:

Types	s^2	s^4
A	59.55	3,546.20
B	26.22	687.49
C	430.89	185,666.19
D	436.22	190,287.89
Total	952.88	380,187.77

$$q = \frac{\sum\limits_{i=1}^{4} s_i^4}{\left(\sum\limits_{i=1}^{4} s_i^2\right)^2} = \frac{380,187.77}{(952.88)^2} = 0.419$$

where: $\nu = 9$; $p = 4$ and using
linear interpolation one gets

$$q_{crit}^{(0.01)} = 0.398$$

$$q_{crit}^{(0.001)} = 0.468$$

Since $q(0.01) <$ calculated $q < q(0.001)$ and theory indicates
that disintegration time is not normally distributed, we reject the
hypothesis and transform the variable (note paragraph iii of Section
1.3).

1.3.4   Transformation of y

The earlier results of this section indicated that if Bartlett's
test were used in analyzing the data in Table 1.3.1 the hypothesis of
homogeneity of variances would be rejected at the 0.001 level.   For
both the log s^2 and Q-test the hypothesis was rejected at the 0.01
level.   Based on this evidence, along with information from the
experimenter that this type of data usually appeared to have the
sample mean proportional to the sample variance (Dixon and Massey,
1969 p. 324) it was decided that the analysis of these data should
proceed by utilizing the square-root transformation.   The results of

this transformation on the data of Table 1.3.1 are shown in Table
1.3.4. In addition the results of performing Bartlett's test on the
transformed data, $\sqrt{y}$, is also included in this table.

TABLE 1.3.4

Analysis of Homogeneity of the Variances Using $\sqrt{y}$

Type	$s^2_{(\sqrt{y})}$	$\log [s^2_{(\sqrt{y})}]$
A	0.538	-0.2692
B	0.441	-0.3566
C	1.279	+0.1069
D	0.734	-0.1343
Total	2.992	-0.6522

$$M = 2.3026(9)[4(-0.1261) + 0.6522]$$
$$= 3.063$$
$$C = 1.046$$
$$\chi^2_{3df} = 2.93 : \chi^2_{3df}(0.01) = 11.34, \chi^2_{3df}(0.05) = 7.81$$

$\therefore$ Accept the hypothesis of the homogeneity of the
variances using $\sqrt{y}$ (Using Bartlett's test)

It is seen here that the hypothesis of homogeneity of variances is
now accepted even at the 0.05 level. This result is also substantiated
by the Q-test. At this point one is now ready to carry out the ANOVA
on $\sqrt{y}$ as described in Section 1.2.

Problem 1.3.2. (a) Carry out the Q-test on the data furnished
in Table 1.3.4. (b) Run the ANOVA on $\sqrt{y}$ in Table 1.3.1, p. 18
and explain your results. (c) Determine the 95% confidence
interval for the mean of the Type D population.

1.3.5 Other Transformations

As mentioned previously, transformations are utilized to obtain
a variate which when analyzed via ANOVA techniques will conform to
the basic assumptions of the analysis of variance. The most familiar

paper on this subject is that by Bartlett (1947). This paper
contains a listing of several often used transformations along with
the relationship which should exist between the mean and a function
of the variance of population.

The most commonly used transformation for "growth"-type data,
for example, changes in weight of animals caused by diets, is log y.
Log y is also used if the mean is proportioned to the standard
deviation.

In percentage data or proportions (p) dealing with the binomial
(0 or 1 response), the variance is proportional to [mean (1 - mean)]
and the correct transformation is arc sin $\sqrt{p}$. Usually one does not
carry out this transformation if the data are all between p = 0.30
to 0.70 because the variance is fairly well stabilized in this region.

If the standard deviation is proportional to the square of the
mean the reciprocal (1/y) is quite often used.

Ostle (1963, p. 340), provides a summary table of the four
transformations given here.

In addition, recently we have encountered data in which the mean
is negatively correlated with the variance. In this situation the
transformation, $\sqrt{B} - \sqrt{B - y}$, where B is the upper limit of all y's
in the data, has worked quite well.

1.3.6   Test for Normality

There are numerous tests for normality, however, we recommend
the W test developed by Shapiro and Wilk (1965). This test is such
that it is appropriate for a test of the composite hypothesis of
normality, i.e., one does not have to include the mean and variance
as part of the hypothesis as is common in some other tests for
normality such as the Kolmogorov-Smirnov and chi-squared tests.
These tests as well as others were compared by Shapiro, Wilk, and
Chen (1968) and they demonstrated that the W test was generally
superior in detecting nonnormality when evaluated on various

symmetric, asymmetric, short- and long-tailed alternatives over sample sizes ranging from 10 to 50.

Tables of coefficients required to calculate W are given in Appendix 9 and the percentiles of W appear in Appendix 10. When using these tables one must keep in mind that small values of W indicate nonnormality.

As an example to illustrate the use of this test consider the data of tablet Type B which appears in Table 1.3.1. The following steps must be carried out:

1. Order the n observations as $y_1 \leq y_2 \leq \ldots \leq y_n$.

2. Compute $\Sigma(y_i - \bar{y})^2$.

3. If n is even, n = 2k, compute

$$b = \sum_{i=1}^{k} a_{n-i+1} \, (y_{n-i+1} - y_i)$$

where the values of a_{n-i+1} appear in Appendix 9. If n is odd, n = 2k + 1 then one omits the sample median, y_{k+1}, and calculates

$$b = \sum_{i=1}^{k} a_{n-i+1} \, (y_{n-i+1} - y_i)$$

4. Compute $W = b^2/\Sigma(y_i - \bar{y})^2$.

5. Compare W to the percentage points given in Appendix 10. Again, small values of W indicate nonnormality.

In our example we have the following for each step:

1. Ordered observations: 8, 9, 10, 10, 10, 12, 12, 16, 19, 24.

2. $\Sigma(y_i - \bar{y})^2 = 236$.

3. b = (0.5739)(24-8) + (0.3291)(19-9) + ... + (0.0399)(12-10)

    = 14.0826.

4. $W = (14.08)^2/236 = 0.840$ (calculated).

5. W(0.05, 10) = 0.842, thus we would reject the hypothesis of
    normality at the α = 0.05 level (from Appendix 10).

The square root transformation on these data will tend to normalize
them. A reference on transformations is Bartlett (1947).

Problem 1.3.3. Use the W test to test the data utilized in the
above example for normality after carrying out the square root
transformation.

1.4 CURVE FITTING IN ONE WAY CLASSIFICATION

In most design of experiments problems involving quantitative
factors one usually becomes interested in describing the relationship
between the dependent variable (response or yield) and the independent
factors by means of a mathematical equation. In this section we will
describe the procedure of using orthogonal polynomials to obtain an
equation which estimates the response surface for the case of one
quantitative factor. The next chapter will generalize this procedure
for the case of more than one quantitative factor.

The procedure of determining estimates of coefficients in the
mathematical model and determining their relative importance in
explaining the total variation (sum of squares) in the response
variable is normally referred to as regression analysis. For
detailed procedures on this subject see Draper and Smith (1966),
Anderson and Bancroft (1952), Ostle (1963), and Myers (1971).
Regression analysis as described in these references is based on the
method of least squares. The use of orthogonal polynomials for
estimating the response surface associated with a "well"-designed
experiment is equivalent to using the method of least squares. In
this context, a well-designed experiment implies that the same number
of observations are taken at each level of the independent variable
(X) and that the levels of X are equally spaced. It is quite
feasible that one would be interested in designing an experiment
where some function of X was equally spaced rather than the values
of X. For example, the effect of the logarithm of time on the

electrical characteristics of electronic components is frequently
studied. In this case all calculations are carried out using the
logarithm of time instead of time itself. For the case of unequally
spaced X values one is referred to Anderson and Bancroft (1952) and
Robson (1959).

It should be pointed out to the practitioner that the use of
orthogonal polynomials merely estimates the response surface with a
certain degree polynomial. In most cases this polynomial will be
an approximation of some other mathematical function. If the true
functional relationship between the dependent and independent
variable is known it should be used in the analysis of the problem.
The use of the true function may possibly involve the use of nonlinear
models (see Draper and Smith, 1966, Chapter 10).

1.4.1 Orthogonal Polynomials

To introduce the topic of orthogonal polynomials we first
consider the case with one observation per treatment combination
(X level). We are also only considering one independent variable
where the responses are observed at equally spaced values of X. An
example might be hardness readings (y) taken at equal intervals across
a block of cured synthetic material. In this example the distance
from one edge would be the X value. These X values will be assumed
to be the values 1, 2, 3, ..., n where n is the number of X levels.
If this is not the case and we have a factor X' which has levels
2, 9, 16, and 23, then the levels of X will become 1, 2, 3, and 4 if
we code the X' values with the formula

X = [X' + (the spacing) - (the smallest X' value)]/(the spacing)

The practitioner immediately becomes interested in finding the
smallest degree of the polynomial in X which best describes his data.
The general form of the model is

$$y = \beta_0 + \beta_1 X + \beta_{11} X^2 + \ldots + \underbrace{\beta_{11\ldots1}}_{k} X^k + \varepsilon \qquad (1.4.1)$$

which can be written as an orthogonal polynomial model

$$y = \alpha_0 + \alpha_1 Z_1 + \alpha_{11} Z_{11} + \ldots + \underbrace{\alpha_{11\ldots 1}}_{k} \underbrace{Z_{11\ldots 1}}_{k} + \varepsilon \quad (1.4.2)$$

where the Z's are functions of X and the α's are the corresponding coefficients.

The Z(X) expressions with the various subscripts are the orthogonal polynomials which implies that $\Sigma Z_r Z_s = 0$ for $r \neq s$. The formulas for the first three orthogonal polynomials are given below where the λ's are scale factors and make the Z values become integers for easy calculations. References for this material are Anderson and Bancroft (1952) p. 210, Ostle (1963) p. 194, and Myers (1971) p. 35.

$$Z_1 = Z_{\text{linear}} = \lambda_1 (X - \overline{X})$$

$$Z_{11} = Z_{\text{quadratic}} = \lambda_2 [(X - \overline{X})^2 - \frac{(n^2 - 1)}{12}]$$

$$Z_{111} = Z_{\text{cubic}} = \lambda_3 [(X - \overline{X})^3 - (\frac{3n^2 - 7}{20})(X - \overline{X})]$$

In order to obtain the estimate of the response surface one must also determine estimates of the coefficients of the polynomials. For example,

$$\hat{\alpha}_0 = \frac{\Sigma y}{n} = \overline{y}$$

$$\hat{\alpha}_1 = \frac{\Sigma y Z_1}{\Sigma [Z_1]^2} \quad \text{and} \quad \hat{\alpha}_{11} = \frac{\Sigma y Z_{11}}{\Sigma [Z_{11}]^2}$$

The numerator of the $\hat{\alpha}_1$ and $\hat{\alpha}_{11}$ are sometimes called contrasts since ΣZ_1 and ΣZ_{11} in each case is zero. In general, if T_i represents a total of r observations and the c_i's are constants where $\Sigma c_i = 0$, then $C = \Sigma c_i T_i$ is a contrast and

$$SS\ C = \frac{[\Sigma c_i T_i]^2}{r\ \Sigma c_i^2} \quad .$$

When one recalls that we are going to assume that the X values take on the values 1, 2, ..., n, it is readily seen that one is able to create a table of Z values in order to simplify calculations. Appendix 11 gives an abbreviated table of Z's along with the corresponding λ's and $\Sigma(Z)^2$. The case where n = 6 is reproduced in Table 1.4.1 for illustrative purposes.

TABLE 1.4.1

Orthogonal Polynomials for n = 6

Z \ X	1	2	3	4	5	6	ΣZ^2	λ
Z_1	-5	-3	-1	1	3	5	70	2
Z_{11}	5	-1	-4	-4	-1	5	84	3/2
Z_{111}	-5	7	4	-4	-7	5	180	5/3
Z_{1111}	1	-3	2	2	-3	1	28	7/12

Notice for example that these values are orthogonal, e.g., $\Sigma Z_1 Z_{111}$ = (-5)(-5) + (-3)(7) + ... + (5)(5) = 0. In analyzing a set of data for trends of various degrees it becomes necessary to be able to compute the sums of squares which is explained by each of the terms included in the model of Eq. (1.4.2). For example, the formula required for the sum of squares (SS) for the quadratic term is

$$SS_{quadratic} = SS(\hat{\alpha}_{11}) = \hat{\alpha}_{11}(\Sigma y Z_{11})$$

The same procedure is used for any degree terms given in Eq. (1.4.2). The following example concerns the observation of hardness as measured every 2 centimeters across a block of synthetic material. The data and the method of calculating the $\hat{\alpha}$'s are shown in Table 1.4.2.

TABLE 1.4.2

Calculation Method for Determining the $\hat{\alpha}$'s

X'	y	X	Z_1	yZ_1	Z_{11}	yZ_{11}	Z_{111}	yZ_{111}
2	16	1	-5	-80	5	80	-5	-80
4	14	2	-3	-42	-1	-14	7	98
6	10	3	-1	-10	-4	-40	4	40
8	12	4	1	12	-4	-48	-4	-48
10	13	5	3	39	-1	-13	-7	-91
12	15	6	5	75	5	75	5	75
	80			-6		40		-6

$$\hat{\alpha}_0 = \frac{80}{6} = 13.33 \qquad \hat{\alpha}_1 = \frac{-6}{70} = -0.0857$$

$$\hat{\alpha}_{11} = \frac{40}{84} = 0.4762 \qquad \hat{\alpha}_{111} = \frac{-6}{180} = -0.0333$$

The resulting linear equation for the response surface would be

$$y = \hat{\alpha}_0 + \hat{\alpha}_1 Z_1 = 13.33 - 0.0857 \, [\lambda_1(X-\overline{X})]$$
$$= 13.33 - 0.0857 \, [2(X-3.5)]$$
$$= 13.93 - 0.1714 \, X$$
$$= 13.93 - 0.1714 \, (\frac{X'-0}{2})$$
$$= 13.93 - 0.0857 \, X'$$

The analysis of variance for this part of the calculations is given in Table 1.4.3. The sum of squares due to linear is basically that portion of the total sum of squares in y which can be explained by the first term in the polynomial. In regression analysis terms this is often referred to as the sum of squares due to regression. The error sum of squares is thought of as the sum of squares about the regression line. The three sums of squares in this ANOVA were computed as follows

$$SS_{linear} = SS(\hat{\alpha}_1) = \hat{\alpha}_1 (\Sigma \, y \, Z_1) \qquad = (-0.0857)(-6) = 0.514$$

$$SS_{total} = \Sigma (y - \bar{y})^2 = \Sigma \, y^2 - \frac{(\Sigma \, y)^2}{n} = 1090 - \frac{(80)^2}{6} = 23.333$$

$$SS_{error} = \Sigma (y - (13.93 - 0.0857 \, X'))^2 = SS_{total} - SS_{linear} = 22.819$$

TABLE 1.4.3

ANOVA for the Linear Model Using Data of Table 1.4.2

Source	df	SS	MS	R^2
Linear regression	1	0.514	0.514	$0.022 = \left\{ \dfrac{0.514}{23.333} \right\}$
Error	4	22.819	5.705	
Total	5	23.333		

The proportion of the total sum of squares in y which is explained by regression is given the name R^2.

In general the degrees of freedom for regression is one less than the number of coefficients estimated which would be k when one uses the models given in Eq. (1.4.1) and (1.4.2). As in the ANOVA section the degrees of freedom for total is the number of observations minus one. The coefficient, R^2, has a major weakness in that as the degrees of freedom for error becomes small, R^2 may be quite large (near one) but the prediction value may not be good. One may test the hypothesis that ρ^2, the parameter estimated by R^2, is zero by the test

$$F_{k,n-k-1} = \frac{MS \text{ due to regression}}{MS \text{ error}} = \frac{0.514}{5.705} = 0.090$$

which is quite non-significant. Thus we would conclude that this model does not contribute significantly towards explaining the variation in y.

If a plot of the data given in Table 1.4.2 were made it would easily be seen that a linear model would not be expected to explain

much of the variation in y. However, we continue this exercise as
this method will carry over to the case of several independent
variables where data plots are not as easily interpreted. Thus, on
the basis of the small R^2 and the appearance of a data plot, one
would want to include the quadratic polynomial in the response surface
model. In ordinary regression analysis a completely new set of
calculations would have to be performed. With orthogonal polynomials
this is not necessary as one is permitted to merely add on the
quadratic term in the model and add together the two sums of squares,
i.e., linear and quadratic.

The best fitting quadratic model for this example becomes

$$y = \bar{y} + \hat{\alpha}_1 \, Z_1 + \hat{\alpha}_{11} \, Z_{11}$$

$$= 13.33 - 0.0857 \; [\lambda_1 (X-\bar{X})] + 0.4762 \; \lambda_2 [(X-\bar{X})^2 - \frac{n^2-1}{12}]$$

$$= 20.60 - 5.1715 \; X + 0.7143 \; X^2$$

$$= 20.60 - 2.5857 \; X' + 0.1786 \; X'^2$$

The sum of squares contributed by the quadratic term is

$$\hat{\alpha}_{11}(\Sigma \, y \, Z_{11}) = 0.4762 \times 40 = 19.048$$

and the sum of squares due to regression becomes

$$SS_{regression} = 0.514 + 19.048 = 19.562$$

The ANOVA for this model is given in Table 1.4.4.

TABLE 1.4.4

ANOVA for the Quadratic Model Using Data of Table 1.4.2

Source	df	SS	MS	F	R^2
Linear	1	0.514			
Quadratic	1	19.048			
Regression	2	19.562	9.781	7.78	$0.838 = \frac{19.562}{23.333}$
Error	3	3.771	1.257		
Total	5	23.333			

The figures in Table 1.4.4 still do not indicate a significant R^2 when one uses the F-test; however, the relatively large jump in R^2 from 0.022 for the linear model to 0.838 for the quadratic model would probably lead one to consider the quadratic model as the appropriate model. If one continued this example and calculated the cubic sum of squares one would not find much of an increase in the regression sum of squares. In order to obtain any further information on the fit of this model we need additional observations at the same X values which is the topic of the next section.

1.4.2 Lack of Fit Principle

When one has repeated observations at each X value (when using orthogonal polynomials it is desirable to have an equal number of observations at each X value) the degrees of freedom for regression remains at k, assuming the model given in Eq. (1.4.1). Consequently, the degrees of freedom for error is increased by the number of additional observations. At this point we can consider partitioning the error sum of squares and the degrees of freedom into two parts. One part is that associated with "pure error" or the within cell sum of squares which is pooled across all X values. To calculate the pooled error one would calculate the sum of squares about the mean response at each X level and sum over all X, thus

$$SS_{\text{pooled error}} = \sum_{\text{all } X} \left(\sum_{j=1}^{m_i} y_j^2 - \frac{\left(\sum_{j=1}^{m_i} y_j \right)^2}{m_i} \right)$$

where m_i = number of observations at the ith X value and the summation over j takes place only for the y values associated with the ith X value. The degrees of freedom for pure error is $\sum_{i=1}^{n}(m_i - 1)$ where there are n X values. The "pure error" mean square thus becomes an estimate of the variance of y. The second part of the initial error sum of squares is now attributed to "lack of fit" and the degrees of freedom are calculated by subtraction. Under the null hypothesis

that the selected model is adequate to account for all possible
variation in y we see that the initial error mean square is also an
estimate of the variance of y. Thus when we partition the error sum
of squares the second part (the lack of fit sum of squares) divided
by its degrees of freedom will also be an estimate of the variance
of y. Thus the ratio of the lack of fit mean square to the pure
error mean square under the null hypothesis will be distributed as
F with n-k-1 and $\sum_{i=1}^{n}$ $(m_i - 1)$ degrees of freedom. The reader should
be able to picture the situation that occurs when the null hypothesis
is false. The initial error sum of squares is large due to the poor
fit. The pure error sum of squares will remain the same thus forcing
the lack of fit sum of squares to be large. Hence, the lack of fit
mean square becomes larger than it should be if the null hypothesis
were true.

The above procedure is illustrated by means of an example. This
example is merely an extension of the example given in the previous
section. Consider that two sets of hardness readings were taken
completely at random over the same set of X values. Thus we have
two observations at each X value. The calculations are essentially
the same except that one uses the mean of the y's for each X value
rather than the individual y's which were used in Section 1.4.1,
where there was only one y for each X value. For example,

$$\text{SS due to the quadratic term} = \hat{\alpha}_{11}(m \sum_{i=1}^{n} \bar{y}_i Z_{11})$$

where m is the number of observations at each X value and $\bar{y}_i$ is the
mean of the y's for the ith X value. The data and the calculations
for this example through the quadratic term appear in Table 1.4.5.

The ANOVA for this example appears in Table 1.4.6. Note that
the lack of fit is carried out twice and in general it would be
carried out on successive terms until the hypothesis of an adequate
model is accepted or there are no degrees of freedom remaining for
lack of fit. One must recognize that this testing procedure involves

TABLE 1.4.5

Calculation Method Using Two Observations per Level

X'	y	X	$\bar{y}$	$\bar{y}Z_1$	$\bar{y}Z_{11}$	$\bar{y}Z_{111}$
2	16,16	1	16.0	-80.0	80.0	-80.0
4	14,13	2	13.5	-40.5	-13.5	94.5
6	10,12	3	11.0	-11.0	-44.0	44.0
8	12,11	4	11.5	11.5	-46.0	-46.0
10	13,12	5	12.5	37.5	-12.5	-87.5
12	15,14	6	14.5	72.5	72.5	72.5
			79.0	-10.0	36.5	-2.5

$$\hat{\alpha}_0 = \frac{79}{6} = 13.17$$

$$\hat{\alpha}_1 = \frac{-10}{70} = -0.14$$

$$\hat{\alpha}_{11} = \frac{36.5}{84} = 0.43$$

$$SS(\hat{\alpha}_1) = SSL = \frac{2 \times (-10)^2}{70} = 2.86$$

$$SS(\hat{\alpha}_{11}) = SSQ = \frac{2 \times (36.5)^2}{84} = 31.72$$

contingent tests (Bozivich et al., 1956) and the α level will increase. Hence the β-error will decrease and if the hypothesis is accepted one may have more terms in the model than required, but there should be no more terms required in the model than that which is determined. This is a strong reason to use this method to evolve a model.

The appropriate quadratic models for this example in terms of the coded and uncoded variable are as follows:

$$y = 20.25 - 4.85X + 0.65X^2$$
$$= 20.25 - 2.42X' + 0.16X'^2$$

TABLE 1.4.6

ANOVA for the Data of Table 1.4.5

Source	df	SS	MS	F	R^2
Linear	1	2.86			0.072
Lack of Fit	4	32.81	8.20	12.30	
Quadratic	1	31.72			0.872
Lack of Fit	3	1.09	0.36	0.54	
Error	6	4.00	0.67		
Total	11	39.67			

Problem 1.4.1. Verify all calculations and results of the trend analysis example above.

Problem 1.4.2. An engineer conducted an experiment on a jet aircraft engine in which he was interested in finding out the polynomial that best described the temperature conditions around the engine. Once he obtained the equation he could keep the temperature as even as possible by designing the appropriate cooling system.

At each of seven equally spaced positions around the engine, the investigator took five readings at random over a specified time. The results were as follows:

y = °F

Positions

1	2	3	4	5	6	7
1750	1700	1730	1710	1630	1665	1775
1775	1750	1700	1665	1660	1690	1710
1730	1720	1690	1705	1645	1710	1735
1725	1725	1700	1680	1600	1705	1740
1770	1705	1680	1690	1615	1690	1740

Determine the best-fitting polynomial for these data.

Problem 1.4.3. Taking the data

X	y
1	4
2	5
3	7
4	8

(a) Demonstrate that for the nonorthogonal variable X, in the model $y = \beta_0 + \beta_1 X + \epsilon$ the estimates of β_0 and β_1 will change if either is deleted (assumed zero) from the model (by actual calculations) and plot the results to show this.

(b) Show, by calculations, that the estimates of α_0 and α_1 do not change for the orthogonal variable, Z, in the model

$$y = \alpha_0 + \alpha_1 Z + \epsilon$$

when either α_0 or α_1 is deleted (assumed zero) from the model. Plot the appropriate results.

(c) What have you discovered about the estimates of the coefficients obtained in (a) and (b) above?

(d) For each Z model, transform back to X.

(e) Do you prefer the X or Z model for predictive purposes? Explain your reasoning.

1.5 REFERENCES

Anderson, R. L. and Bancroft, T. A. Statistical Theory in Research McGraw-Hill, New York, 1952.

Bancroft, T. A. Topics in Intermediate Statistical Methods, Iowa State University Press, Ames, Iowa, 1968.

Bartlett, M. S. Proc. Roy. Soc. A, 160:268 (1937).

Bartlett, M. S. Biometrics 3:39 (1947).

Bartlett, M. S. and Kendall, D. G. JRSS Suppl. 8:128 (1946).

Box, G. E. P. Ann. Math. Stat. 25:290 (1954).

Bozivich, H., Bancroft, T. A., and Hartley, H. O. Ann. Math. Stat. 27:1017 (1956).

Bratcher, T. L., Moran, M. A., and Zimmer, W. J. J. Quality Tech. 2:156 (1970).

Burr, I. W. and Foster, L. A. A Test for Equality of Variances, Department of Statistics Mimeo Series No. 282, Purdue University, Lafayette, Indiana, 1972.

Carmer, S. G. and Swanson, M. R. JASA 68:66 (1973).

Cochran, W. G. Biometrics 3:22 (1947).

Dixon, W. J. and Massey, F. J., Jr. Introduction to Statistical Analysis, 3rd ed., McGraw-Hill, New York, 1969.

Draper, N. R. and Smith, H. Applied Regression Analysis, Wiley, New York, 1966.

Eisenhart, C. Biometrics 3:1 (1947).

Kastenbaum, M. A., Hoel, D. G., and Bowman, K. O. _Biometrika_ 57:421 (1970a).

Kastenbaum, M. A., Hoel, D. G., and Bowman, K. O. _Biometrika_ 57:573 (1970b).

Keuls, M. _Euphytica_ 1:112 (1952).

Myers, R. H. _Response Surface Methodology_, Allyn and Bacon, Boston, Mass., 1971.

Newman, D. _Biometrika_ 31:20 (1939).

Ostle, B. _Statistics in Research_, 3rd ed., Iowa State University Press, Ames, Iowa, 1963.

Robson, D. S. _Biometrics_ 15:187 (1959).

Shapiro, S. S. and Wilk, M. B. _Biometrika_ 52:591 (1965).

Shapiro, S. S., Wilk, M. B., and Chen, H. J. _JASA_ 63:1343 (1968).

Snedecor, G. W. and Cochran, W. G. _Statistical Methods_, 6th ed., Iowa State University Press, Ames, Iowa, 1967.

Stein, C. _Ann. Math. Stat._ 16:243 (1945).

Chapter 2

SOME INTERMEDIATE DATA ANALYSIS CONCEPTS

In Chapter 1 we described a few ideas in data analysis using the one-way classification from a completely randomized design. In this chapter we again utilize only the completely randomized design and discuss the two-way classification and its advantages over the one-way. The concepts used in analyzing two-way classified data are more involved than those encountered in the one-way.

For instance, in most research problems of any consequence, the information on the combined effects of two factors controlled in the experiment (two-factor interactions) on the response variable usually contributes more to solving the problem than the information on the individual effects of the factors themselves (main effects).

By extending the analytical procedures to many controlled factors in an experiment, the investigator is able to obtain the information on all the two-factor interactions simultaneously and additional information on the three-factor interactions also. Only recently (to our knowledge) have some research workers, for example, highway research engineers Hadley et al. (1969), interpreted three-factor interactions thoroughly enough to want to use them in the mathematical models to help to explain the responses obtained. As a general rule, however, the four-factor interactions and higher tend to be small contributors to the underlying ANOVA models used to explain the mathematical phenomenon. These high order interaction effects are frequently assumed zero to allow an estimate of the error if only one experimental unit has been used for each combination of the factor levels, often called the treatment combination. Details of this extended type experiment with many factors are given in Chapter 4 on factorial experiments.

In this chapter we are especially interested in expressing the interaction concept. In the first section two factors are investigated thoroughly, using only one experimental unit per treatment combination, or one observation per cell, and in the following section, concepts on more than one observation per cell are covered.

2.1 TWO FACTOR EXPERIMENTS WITH ONE OBSERVATION PER CELL

When experiments are run with only two factors (which is certainly more informative than running only one factor at a time because no information on the two-factor interaction can be obtained for one factor at a time experiments) and one observation per cell, there is not a clear-cut way to separate the effect of the interaction of the two factors from the experimental error.

In an experiment on stress-rupture life of material used to make turbine blades in fanjet aircraft engines, two factors; blade temperatures which might be encountered at takeoff "of the aircraft" and alloys or materials that are of interest to the research engineer, are to be investigated. Let us assume that for part of the experiment 12 treatment combinations (4 materials x 3 temperatures) are the only ones of interest and that these combinations can be run completely at random in the experiment. The experimenter can then analyze the data as if he has a fixed two-way ANOVA with only one observation per cell. The model for such an analysis is:

$$y_{ij} = \mu + M_i + T_j + MT_{ij} + \varepsilon_{(ij)} \qquad (2.1.1)$$

$$i = 1, 2, 3, 4 \quad j = 1, 2, 3$$

where

y_{ij} = hours to stress-rupture of the piece of metal made of the ith material at the jth temperature

μ = overall mean

M_i = the effect of the ith material (fixed)

T_j = the effect of the jth temperature (fixed)

MT_{ij} = the effect of the interaction of the ith material with the jth temperature

$\varepsilon_{(ij)}$ = the experimental error (random), NID $(0, \sigma^2)$

In this model the subscripts for MT and ε are the same, thus indicating that the effects due to the interaction and the error effects will be confounded. The brackets on the subscripts of ε indicate there is only one observation per cell or only one experimental unit (piece of metal) for each material-temperature combination. We could write the error symbol $\varepsilon_{(ij)1}$ indicating one observation per cell but the $\varepsilon_{(ij)}$ is sufficient if understood.

The data for this part of the experiment are given in Table 2.1.1.

TABLE 2.1.1

Stress-Rupture Life (hours) for Four Materials
and Three Temperatures

Materials	Temperatures			Total
	1	2	3	
a	185	182	182	549
b	175	183	184	542
c	171	184	189	544
d	165	191	189	545
Total	696	740	744	2180

Using $\alpha = 0.05$ the ANOVA for these data is given in Table 2.1.2.

TABLE 2.1.2

ANOVA of Data in Table 2.1.1[a]

Source	df	SS	MS	F	$F_{crit}^{(0.05)}$
Material (M)	3	8.67	2.89	<1	4.76
Temperature (T)	2	354.67	177.34	3.7	5.14
MT and/or error	6	291.33	48.56		
Total	11	654.67			

[a] SS materials $= \dfrac{549^2 + 542^2 + 544^2 + 545^2}{3} - \dfrac{(2180)^2}{12}$

$= \underline{8.67}$

SS temperatures $= \dfrac{696^2 + 740^2 + 744^2}{4} - \dfrac{(2180)^2}{12}$

$= \underline{354.67}$

SS total $= 185^2 + 175^2 + \ldots + 189^2 - \dfrac{(2180)^2}{12}$

$= \underline{654.67}$

SS MT and/or error $=$ SS total $-$ SS materials $-$ SS temperatures

$= 654.67 - 8.67 - 354.67$

$= \underline{291.33}$

The conclusion is that all effects are not significant at the
0.05 level. The F-tests for temperature and material effects are
based on using the MT and/or error mean square for the denominator
which is really assuming the interaction of temperatures by materials
is zero. To show this we need to derive the expected mean squares
for each source of variation so that the theoretical basis for the
F-tests can be explained. Suffice it to say here that we will write

these expected values down showing where the assumption of the
interaction being zero is made and later use an algorithm on a more
complicated model to derive the expected mean squares. A good
reference for the mathematical derivation of expected mean squares
is Anderson and Bancroft (1952, Chapters 18 and 20).

The expected mean squares (EMS) shown as a sum of σ^2 and an
unknown function of the fixed effects are given in Table 2.1.3.

TABLE 2.1.3

Expected Mean Squares for Table 2.1.2

Source	EMS
Materials (M)	$\sigma^2 + 3\phi(M)$
Temperatures (T)	$\sigma^2 + 4\phi(T)$
MT and/or error	$\sigma^2 + \phi(MT)$

In order to make a test of significance on temperature using F,
we need a denominator that has no term or component of temperature
(T). The hypothesis that there is no temperature effect may be given
as:

$$H_0 : \phi(T) = 0$$

where $\phi(T)$ is a function of the effect of temperature. To test H_0
using the F-test we need a denominator that estimates σ^2 only, but
in this experiment because we took only one observation per cell we
cannot separate the interaction functional effect $\phi(TM)$ from the
error variance, σ^2. Refer back to the model (2.1.1) and recall that
the subscripts on MT and ϵ were the same. This indicates complete
confounding.

In practice for testing temperatures in Table 2.1.2 we used

$$F_{2,6} = \frac{\text{MS temperatures}}{\text{MS MT and/or error}}$$

but to do this we must assume the interaction is zero or $\phi(MT) = 0$.
Similarly for testing for materials.

The conclusion of course, is that if the interaction, MT, may be
important one should take more than one observation per cell to
evaluate it. This concept is covered in Section 2.2. Now, however,
we wish to show various analyses to investigate a part of the
interaction separate from error where there is only one observation
per cell.

2.1.1 Both Factors Qualitative

Even if only two experimental units (pieces of material) were
used for each treatment combination, the expense of the experiment
would double if the experiment were completely randomized. Hence
sometimes it is worth examining the interaction or nonadditivity of
the model shown in Eq. (2.1.1) using a technique given by Tukey
(1949).

Table 2.1.4 shows the technique of splitting the sum of squares
291.33 for MT and/or error into one portion for MT which has one
degree of freedom and the remaining portion for error with 5 df.

TABLE 2.1.4

SS for One Degree of Freedom for Nonadditivity[a]

| Material | Temperature | | | $\bar{y}_{i\cdot}$ | $(\bar{y}_{i\cdot} - \bar{y}_{\cdot\cdot})$ |
	1	2	3		
a	185	182	182	183.000	1.333
b	175	183	184	180.667	-1.000
c	171	184	189	181.333	-0.333
d	165	191	189	181.667	0.000
$\bar{y}_{\cdot j}$	174	185	186	$\bar{y}_{\cdot\cdot} = 181.667$	
$(\bar{y}_{\cdot j} - \bar{y}_{\cdot\cdot})$	-7.667	3.333	4.333		

$$^a SS \text{ (for 1 df)} \atop \text{Nonadditivity} = \frac{[\sum\limits_{i,j} y_{ij} (\bar{y}_{i.} - \bar{y}_{..})(\bar{y}_{.j} - \bar{y}_{..})]^2}{[\sum\limits_i (\bar{y}_{i.} - \bar{y}_{..})^2][\sum\limits_j (\bar{y}_{.j} - \bar{y}_{..})^2]}$$

$$= \frac{[185(1.333)(-7.667) + \ldots + 189(0.000)(4.333)]^2}{[(1.333)^2 + \ldots + (0.000)^2][(-7.667)^2 + \ldots + (4.333)^2]}$$

$$= 73.01$$

where y_{ij} is defined as in Eq. (2.1.1); $\bar{y}_{i.}$ = the mean of the ith material; $\bar{y}_{.j}$ = mean of the jth temperature; $\bar{y}_{..}$ = overall mean.

Using the result of Table 2.1.4, the ANOVA in Table 2.1.5 is appropriate to study.

TABLE 2.1.5

ANOVA including Nonadditivity Source

Source	df	SS	MS	EMS	F	$F_{crit}^{(0.05)}$
Materials (M)	3	8.67	2.89	$\sigma^2 + 3\phi(M)$	0.07	5.41
Temperatures (T)	2	354.67	177.33	$\sigma^2 + 4\phi(T)$	4.06	5.79
MT (nonadditivity)	1	73.01	73.01	$\sigma^2 + \phi(MT)$	1.67	6.61
Error	5	218.32	43.66	σ^2		
Total	11	654.67				

It still turns out that all F-tests are not significant at the 0.05 level, including the interaction of MT, an indication that the MT term could be deleted from Eq. (2.1.1).

2.1.2 One Factor Qualitative and the Other Quantitative

There is more efficiency from a quantitative factor than from
a qualitative factor because the majority of the quantitative effect
is usually best described by a linear component with only one degree
of freedom (recall the concept of investigating trend in Section 1.4).

If one uses the concept of expanding any function into a power
series, he realizes that ordinarily the lower order terms (e.g.,
linear) contribute more than higher order ones in explaining the
true function. We use this concept here.

The levels of temperature are equally spaced quantitative
levels. Hence one can estimate the interaction of materials by
temperatures by using three degrees of freedom from materials by
linear effect of temperatures.

To obtain an estimate of nonadditivity, the interaction of
materials by linear effect of temperatures may be used. We set up
the orthogonal polynomial coefficients of temperatures for
each material as given in Table 2.1.6. For a discussion of this
procedure see Hicks (1973, p. 138).

TABLE 2.1.6

SS of Materials by Linear Temperatures[a]

| | Temperatures | | |
| | orthogonal polynomials (coefficients) | | |
Materials	-1	0	1
a	185	182	182
b	175	183	184
c	171	184	189
d	165	191	189

[a]SS of materials by linear temperatures:

$$= \frac{1}{(-1)^2 + (0)^2 + (1)^2} \left\{ [(-1)(185) + (0)(182) + (1)(182)]^2 \right.$$

$$\left. + \dots + [(-1)(165) + 0\,(191) + (1)\quad(189)]^2 \right\}$$

$$- \frac{[-185 + 182 + \dots - 165 + 189]^2}{[(-1)^2 + (0)^2 + (1)^2] + \dots + [(-1)^2 + 0^2 + (1)^2]}$$

$$= \tfrac{1}{2}[(-3)^2 + (9)^2 + (18)^2 + (24)^2] - \frac{(48)^2}{2(4)} = 207$$

The ANOVA for the data from Table 2.1.6 is given in Table 2.1.7.

TABLE 2.1.7

ANOVA Using Materials x Linear Temperatures

Source	df	SS	MS	EMS	F	$F_{crit}^{(0.05)}$
Materials (M)	3	8.67	2.89	$\sigma^2 + 3\phi(M)$	<1	9.28
Temperatures (T)	2	354.67	177.33	$\sigma^2 + 4\phi(T)$	6.31	9.55
(Materials) by (linear temp.)	3	207.00	69.00	$\sigma^2 + \phi(TM)$	2.45	9.28
Error	3	84.33	28.11	σ^2		
Total	11	654.67				

Hence there were no significant main effects or interaction in this case.

2.1.3 Both Factors Quantitative

Referring back to the example presented in Section 2.1.1 we see that the levels of temperature are equally spaced quantitative levels.

In addition, let us now state, the materials which were presented as
a qualitative factor were actually a quantitative factor, this was a
result of each type of material being different as a result of each
containing different percentages of a new alloying element. Further,
these percentages were equally spaced, hence we can again use the
concept of orthogonal polynomials introduced in Section 1.4 to
investigate nonadditivity.

Theoretically if there is one degree of freedom that represents
the nonadditivity or interaction of two quantitative factors, the
most probable one is associated with the linear by linear interaction.
The basis for this statement is that if the true function using
temperatures and materials to predict stress-rupture life could be
written and this function expanded into a power series function in
which six interaction terms of temperature and materials would appear,
the lowest order term, linear x linear, usually would be the largest
contributor.

With this concept in mind, the practitioner usually should
compute the linear by linear term to represent the nonadditive
portion if only one degree of freedom is to be used to represent
interaction. Using only one degree of freedom for nonadditivity is
a common practice because the investigator wants to retain as many
degrees of freedom for error as he can.

Utilizing the same data as was used in Sections 1.2.1 and 1.2.2
we use the orthogonal polynomials in Appendix 11 and refer to Section
1.4 as a reminder of the technique. Here, however, we must consider
the orthogonal polynomials for both variables and cross multiply them
to get the appropriate coefficients for each observation. The
procedure is shown in Table 2.1.8.

The ANOVA using this linear x linear SS estimate of
nonadditivity is given in Table 2.1.9.

TABLE 2.1.8

SS for 1 df for MT Linear x Linear[a]

Orthogonal coefficients	Temperatures		
	-1	0	1
Materials -3	185(3)	182(0)	182(-3)
-1	175(1)	183(0)	184(-1)
1	171(-1)	184(0)	189(1)
3	165(-3)	191(0)	189(3)

$$^a\text{SS for MT}(\ell \times \ell) = \frac{[185(3) + 175(1) + \ldots + (189)(3)]^2}{40}$$

$$= 202.5$$

TABLE 2.1.9

ANOVA of Data in Table 2.1.7

Source	df	SS	MS	EMS	F	$F_{crit}^{(0.05)}$
Materials (M)	3	8.67	2.89	$\sigma^2 + 3\phi(M)$	<1	5.41
Temperature (T)	2	354.67	177.33	$\sigma^2 + 4\phi(T)$	10.0	5.79
MT($\ell \times \ell$)	1	202.50	202.50	$\sigma^2 + \phi(MT)$	11.4	6.61
Error	5	88.83	17.77	σ^2		
Total	11	654.67				

Therefore both temperatures and the interaction of temperatures by materials are significant. The next step in the analysis of the data after this ANOVA would be to examine the interaction carefully. Since the linear x linear term is significant an exploration of the response surface dealing with stress rupture life must be made. Plotting the observations and making some prediction of maximum life should be planned for the next part of the experiment.

2.1.4 Summary of Section 2.1

The problem of examining nonadditivity or interaction in a two
way classification with one observation per cell is not straightforward. We have presented three methods which cover the use of both
qualitative and quantitative factors. As more information is
available on the levels of the factors (that is as one has more
quantitative information on the factors) the better will be the
estimate of the interaction contribution or nonadditivity. There is
no reason to stop at linear x linear for the orthogonal polynomial
investigation except to keep the number of degrees of freedom for
error reasonably high. One can remove the linear by quadratic,
quadratic by linear and so on to find out what each degree of freedom
provides if he is not satisfied with the variability in y accounted
for by the linear x linear term.

As for the Tukey procedure, it seems to be the best method to
estimate nonadditivity if all factors are qualitative. The linear
of one quantitative factor by the qualitative factor should be used
if there is a mixture of quantitative and qualitative factors.

Problem 2.1.1. Using the following data from a similar
experiment as given in Section 2.1:

| | Temperatures | | |
Materials	1	2	3
a	169	182	188
b	174	183	190
c	181	188	194
d	184	191	195

(a) Run all the analyses: Tukey, materials x linear on
temperatures, and linear x linear.

(b) Comment on the results from these three analyses.

2.2 TWO-FACTOR EXPERIMENTS WITH MORE THAN ONE OBSERVATION PER CELL

In this section it seems most reasonable to introduce the

concept of random as well as fixed factors and their consequences in
ANOVA. This concept broadens the usefulness of ANOVA to many
problems including multi-factor experiments to be covered in Chapter
4. To show how an experimenter should test for the various main
effects and interactions of these factors we must first derive the
expected mean squares. In this book we will merely show an algorithm
given by Bennett and Franklin (1954) and not actually derive expected
mean squares for certain combinations of random and fixed factors.
Anderson and Bancroft (1952, Chapters 18, 20, 22, and 23) provide a
procedure for those readers who wish to see the mathematical proof.

2.2.1 Expected Mean Square Algorithm

As indicated in Section 1.2, a factor is declared "fixed" if
all levels of interest to the investigator are included in the
experiment. This concept, then, is that the experiment contains the
population of levels of a fixed factor. For this fixed case the
experimenter is primarily interested in analyzing means.

If, however, fewer than the population of levels of a factor
are to be examined in an experiment and those levels are selected at
random from all possible levels of interest to the investigator, the
factor is called "random." Many times the number of all possible
levels is finite and the sampling rate, or the ratio of the number
of levels of the factor to be included in the experiment divided by
the number of possible levels of interest, is a fraction greater than
zero. This concept is the basis for part of the expected mean squares
algorithm.

Consider the case in which there are Q levels of the factor in
the population and only q < Q are selected at random to represent the
Q levels in the experiment. It follows, Cochran (1963, Chapter 2),
that a finite population correction $(1 - q/Q)$ is the coefficient to
account for the finiteness encountered in the variance estimate.

Observe that this coefficient is equal to zero in the case of
the fixed model and approaches one whenever Q is much larger than q
as is experienced in most applications of the random model.

In this book we will assume that the finite population
correction is zero if the factor is fixed and one if it is random.
The basis for this choice is a conservative one, for if the
population is truly finite where $Q < \infty$, the variance would actually
be smaller than the one used assuming the population is infinite.
In extreme cases, for example if $q/Q = 1/2$, extra care must be taken
not to be too conservative and the finite adjustment should be taken
into account. We believe this ultra finite case happens so
infrequently in practice that it is ignored in the remaining part of
this book. The Bennett and Franklin (1954) presentation of the
algorithm utilizes the finite population concept.

The discussion of the expected mean square algorithm is divided
into three subsections; fixed, random, and mixed models. Ostle (1963,
Chapter 11), provides a rather detailed discussion of these models.

Example 2.1:   Fixed Model. Consider the case where both factors
are fixed and the model is $y_{ijk} = \mu + \alpha_i + \beta_j + \alpha\beta_{ij} + \varepsilon_{(ij)k}$ $i = 1,$
..., a $j = 1, \ldots, b$ $k = 1, \ldots, n$ where

$\quad y_{ijk}$ = variable to be analyzed from the kth experimental unit
$\qquad$ associated with the ith level of factor α and j^{th} level
$\qquad$ of factor β

$\quad\quad \mu$ = overall mean

$\quad\quad \alpha_i$ = effect of the ith level of factor α (fixed)

$\quad\quad \beta_j$ = effect of the jth level of factor β (fixed)

$\quad \alpha\beta_{ij}$ = effect of the interaction of the ith level of factor
$\qquad$ α with the jth level of factor β

$\quad \varepsilon_{(ij)k}$ = experimental error of the kth experimental unit
$\qquad$ associated with the ith level of factor α and j^{th} level
$\qquad$ of factor β, within the ij^{th} cell, assume NID $(0, \sigma^2)$

Set up the following table:

| | a | b | n |
| | F | F | R |
Source	i	j	k
α_i	0		
β_j		0	
$\alpha\beta_{ij}$	0	0	
$\varepsilon_{(ij)k}$	1	1	1

where

(1)   The sources of variation include all the variable terms on the right-hand side of the model together with their associated subscripts

(2)   All subscripts are placed as column headings next to the source with F for fixed and R for random above the appropriate subscript and the number of levels; a, b, and n placed above the appropriate letter and subscripts

(3)   Ones are entered in the slots where the bracketed subscripts in the row headings match the subscripts in the column heading

(4)   Zeros are entered in the slots where the row and column subscripts match for the F type columns and ones for the R type columns.

(5)   Fill in the remaining slots with the number of levels shown in the respective column heading.

The table then becomes

| | a | b | n |
| | F | F | R |
Source	i	j	k
α_i	0	b	n
β_j	a	0	n
$\alpha\beta_{ij}$	0	0	n
$\varepsilon_{(ij)k}$	1	1	1

and we are ready to derive the EMS.

Begin at the bottom and multiply the coefficients for $\varepsilon_{(ij)k}$.

The product turns out to be 1. The EMS for this source is then $1 \cdot \sigma^2 = \sigma^2$. Next to determine the EMS for $\alpha\beta$ interaction, we only consider those rows which contain at least the same subscripts as $\alpha\beta_{ij}$, in this case only $\varepsilon_{(ij)k}$ and $\alpha\beta_{ij}$. Then cover all columns with headings the same as those contained in the $\alpha\beta$ row heading, i.e., columns i and j. For each row under consideration, determine the product of the remaining coefficients. In this case it is the number 1 for the $\varepsilon_{(ij)k}$ row and n for the $\alpha\beta_{ij}$ row. These values then become the coefficient for the corresponding component. The sum of these terms is then the EMS for $\alpha\beta$.

So far the algorithm has produced the following:

| | a | b | n | |
| | F | F | R | |
Source	i	j	k	EMS
α_i	0	b	n	
β_j	a	0	n	
$\alpha\beta_{ij}$	0	0	n	$\sigma^2 + n\phi(\alpha\beta)$
$\varepsilon_{(ij)k}$	1	1	1	σ^2

We will use the notation given by Cochran (1951) for fixed components as ϕ which is a non-negative function dependent upon the effect contained within the parentheses. Hence, following the same rules of allowing only those components that have at least the same subscripts as the source being investigated to enter the EMS and using the same column covering rule with multiplication of coefficients we finally complete the EMS as:

Source	i	j	k	EMS
α_i	0	b	n	$\sigma^2 + bn\phi(\alpha)$
β_j	a	0	n	$\sigma^2 + an\phi(\beta)$
$\alpha\beta_{ij}$	0	0	n	$\sigma^2 + n\phi(\alpha\beta)$
$\varepsilon_{(ij)k}$	1	1	1	σ^2

Example 2.2:   Random Model. For this case all factors are
random and we place R's in all column headings. Using the same rules
as before we obtain the following:

	a	b	n	
	R	R	R	
Source	i	j	k	EMS
α_i	1	b	n	$\sigma^2 + n\sigma^2_{\alpha\beta} + bn\sigma^2_\alpha$
β_j	a	1	n	$\sigma^2 + n\sigma^2_{\alpha\beta} + an\sigma^2_\beta$
$\alpha\beta_{ij}$	1	1	n	$\sigma^2 + n\sigma^2_{\alpha\beta}$
$\varepsilon_{(ij)k}$	1	1	1	σ^2

Notice that since all effects are random the σ^2 notation
(rather than any ϕ for fixed) is used for all components. This
indicates sampling variance and the experimenter's interest is in
estimation of these components of variance as well as testing the
hypotheses that the components, σ^2_α, σ^2_β and $\sigma^2_{\alpha\beta}$ are zero.

It can easily be seen that the test for $\sigma^2_{\alpha\beta} = 0$ requires the ε
or within error mean square as the denominator and both main effects
are tested using the interaction $(\alpha\beta_{ij})$ mean square. The latter is
quite different from the previous tests for fixed models (the means
are of interest) in which the within error mean square is used as
the denominator for all tests.

Example 2.3:   Mixed Model. Let us allow α to be random and β
to be fixed in the mixed model. The expected mean square algorithm
then becomes:

	a	b	n	
	R	F	R	
Source	i	j	k	EMS
α_i	1	b	n	$\sigma^2 + bn\sigma^2_\alpha$
β_j	a	0	n	$\sigma^2 + n\sigma^2_{\alpha\beta} + an\phi(\beta)$
$\alpha\beta_{ij}$	1	0	n	$\sigma^2 + n\sigma^2_{\alpha\beta}$
$\varepsilon_{(ij)k}$	1	1	1	σ^2

In this case ϕ is used only for fixed and σ^2 is used for both random and mixed ($\sigma^2_{\alpha\beta}$ is mixed here) components.

It is interesting to note that for the mixed case the fixed source, β, is tested by the interaction and the random source, α, is tested by the within error mean square. This seems contradictory because for the fixed model the fixed main effects were tested using the within error mean square and for the random model they were tested using the interaction mean square.

This can be explained by considering the appropriate inference spaces. If both α and β are fixed, the inference space is confined to only those levels a = A and b = B of the two factors and all the sampling is carried out within each ij cell.

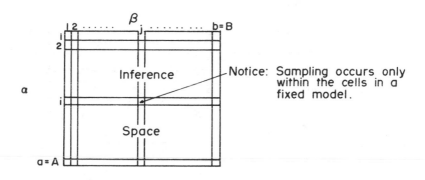

Hence all tests made within the inference space are based on the within error.

If both factors are random the inference is over all combinations of levels of both α and β. Hence the interaction is the basis for both tests since intuitively one is required to use a larger mean square in the denominator because the inference is to a larger population than in the fixed model case.

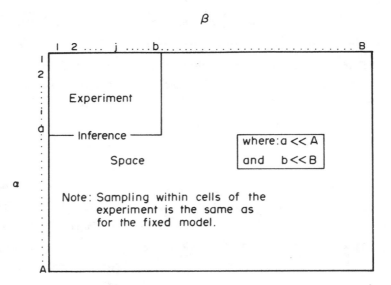

If α is random and β is fixed, the inferences for α is only over the b = B levels of β and within the cells, but the inference for β is over the A levels for α as well as the within cells. Pictorially, it follows that the test for β should be against the

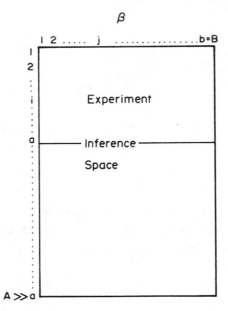

mixed interaction component since it must be valid for all the A
levels of α. But the test for α should be made using the within
error source only because the inference is only over those levels
of β which were included in the experiment.

Example 2.4: Fixed Model (One experimental unit per treatment
combination.) Referring back to Section 2.1 and Table 2.1.3, let us
derive the EMS for Eq. (2.1.1). Using the algorithm and letting
k = 1 for one observation per cell we have Table 2.2.1.

TABLE 2.2.1

ANOVA: Derived EMS for Table 2.1.3

df	Source	4 F i	3 F j	1 R k	EMS
3	M_i	0	3	1	$\sigma^2 + 3\phi(M)$
2	T_j	4	0	1	$\sigma^2 + 4\phi(T)$
6	MT_{ij}	0	0	1	$\sigma^2 + \phi(MT)$
0	$[\varepsilon_{(ij)k} = \varepsilon_{(ij)1} = \varepsilon_{(ij)}]$	1	1	1	σ^2

We could use this result for the EMS, but since there are no
degrees of freedom for the within error, $\varepsilon_{(ij)}$, many authors prefer
dropping this term in the ANOVA. From Table 2.2.1 above it can be
seen that Table 2.1.3 combines the components of both MT_{ij} and $\varepsilon_{(ij)}$
in the same source, MT_{ij}. This consolidation of effects is necessary
since there are no degrees of freedom for error. Notice, however,
in Tables 2.1.5, 2.1.7 and 2.1.9 that there are separate estimates
of MT and ε mean squares each of which has nonzero degrees of freedom.

Example 2.5: Fixed Model (More than one observation per cell).
If in the same example given in Section 2.1, there were two
experimental units per treatment combination, then the algorithm
method of obtaining the EMS would be that presented in Table 2.2.2.

Notice that the two main effects and interaction are tested
using the within error.

TABLE 2.2.2

ANOVA: Two Observations per Cell Extension of Table 2.1.1

df	Source	4 F i	3 F j	2 R k	EMS
3	M_i	0	3	2	$\sigma^2 + 6\phi(M)$
2	T_j	4	0	2	$\sigma^2 + 8\phi(T)$
6	MT_{ij}	0	0	2	$\sigma^2 + 2\phi(MT)$
12	$\varepsilon_{(ij)k}$	1	1	1	σ^2

Example 2.6: Random Model. Using the same example but assuming both factors are random we have Table 2.2.3.

TABLE 2.2.3

ANOVA: Random for Table 2.2.2

df	Source	4 R i	3 R j	2 R k	EMS
3	M_i	1	3	2	$\sigma^2 + 2\sigma^2_{MT} + 6\sigma^2_M$
2	T_j	4	1	2	$\sigma^2 + 2\sigma^2_{MT} + 8\sigma^2_T$
6	MT_{ij}	1	1	2	$\sigma^2 + 2\sigma^2_{MT}$
12	$\varepsilon_{(ij)k}$	1	1	1	σ^2

The df and source are identical to those of Example 2.5, but the remaining part of the table is different. The algorithm uses all ones in the slots for matching subscripts in this case and the results of the EMS are quite different from those of the previous two cases, namely, the interaction component σ^2_{MT} appears in both main effect EMS's. For the F-tests on the main effects, then, it follows that the interaction mean square should be used as the denominator.

Example 2.7: Mixed Model (M fixed, T random). Continuing with

the same example but letting T be random and M be fixed we get
Table 2.2.4.

TABLE 2.2.4

ANOVA: Mixed for Table 2.2.2

df	Source	4 F i	3 R j	2 R k	EMS
3	M_i	0	3	2	$\sigma^2 + 2\sigma_{MT}^2 + 6\phi(M)$
2	T_j	4	1	2	$\sigma^2 + 8\sigma_T^2$
6	MT_{ij}	0	1	2	$\sigma^2 + 2\sigma_{MT}^2$
12	$\varepsilon_{(ij)k}$	1	1	1	σ^2

In this case the df and source are identical to both Examples
2.5 and 2.6, but the other parts of the table are different from both
of the previous examples. Referring back to Example 2.3, we can see
that the inference space for temperatures (T_j) is only over the four
levels of materials (M_i) and the test for T_j requires the mean square
for $\varepsilon_{(ij)k}$ (the within error) only. On the other hand, it is easily
seen that the test for materials (M_i) uses the mean square for
interaction, indicating the inference is broader. In fact the
inference space is over the T >> 3 levels of the population of
temperatures from which the three in the experiment were chosen.

For demonstration purposes this example is satisfactory to show
the mechanism of the algorithm and indicate the various F-tests, but
for the practitioner, temperatures are almost always chosen at
definite levels and cannot possibly be random. Explanations to guide
the investigator in deciding whether a factor is random or fixed in a
given situation will be given throughout the book. The decision of
whether a factor is random or fixed is necessarily part of designing
an experiment.

The standard error of a fixed treatment mean when the model is

mixed includes the random components and is quite often impossible
to estimate from the data. In general for those cases in the
completely randomized design in which there is a direct test for the
fixed treatment in the ANOVA, the standard error of the difference of
two means is estimated using the same mean square that was used to
make the test in the ANOVA. This allows one to use the Newman-Keuls
test by using

$$\left(\frac{\text{Mean square used to test the fixed treatment}}{\text{Number of observations in each mean of the treatment}}\right)^{1/2}$$

for the standard error of the mean. Refer to Appendix 12 for an
example of derivational techniques.

From now on throughout this book we will place indices on the
symbols under Source in the ANOVA tables only if we believe the
indices are needed to obtain the EMS or if we believe a correspondence
between tables warrants them.

2.2.2 ANOVA of Two-Way Classification with More than One
Observation per Cell

Utilizing the various appropriate techniques described prior to
this section, we now show data from a completely randomized designed
experiment on weld breaks in steel manufacturing. Two factors were
investigated; time of welding (time of automatic weld cycle) and gage
bar setting (distance weld die travels during automatic weld cycle).
Detailed description of weld process is given in Problem 9.1.4. The
measured variable was the breaking strength of the weld, where two
welds were made per treatment combination and the objective was to
find the treatment combination that produced maximum strength. The
coded variable to be analyzed and levels of both factors are given
in Table 2.2.5 where the circled number is the total of the two
observations for that cell.

Since the levels of both factors were selected purposely, the
model is completely fixed.

TABLE 2.2.5

Gage Bar Setting

		1	2	3	Total
	1	10,12 ㉒	15,19 ㉞	10, 8 ⑱	74
Time	2	13,17 ㉚	14,12 ㉖	12, 9 ㉑	77
of	3	21,30 ㉛	30,38 ㉘	10, 5 ⑮	134
Welding	4	18,16 ㉞	15,11 ㉖	14,15 ㉙	89
	5	17,21 ㉚	14,12 ㉖	19,11 ㉚	94
Total		175	180	113	468

$$y_{ijk} = \mu + T_i + G_j + TG_{ij} + \varepsilon_{(ij)k} \qquad (2.2.1)$$

$i = 1, 2, .., 5 \quad j = 1, 2, 3 \quad k = 1, 2 \quad$ where

y_{ijk} = coded value of strength of the kth weld at the ith time of welding and j^{th} gage bar setting

μ = overall mean

T_i = effect of the ith time of welding (fixed)

G_j = effect of the jth gage bar setting (fixed)

TG_{ij} = effect of the interaction of the ith time with the jth gage bar setting

$\varepsilon_{(ij)k}$ = within error of the kth weld in the ith time and j^{th} gage bar setting, NID $(0, \sigma^2)$

All assumptions for the ANOVA are assumed met, and the ANOVA data are given in Table 2.2.6

TABLE 2.2.6

ANOVA of Data in Table 2.2.5[a]

df	Source	5 F i	3 F j	2 R k	EMS	MS	F
4	Time (T_i)	0	3	2	$\sigma^2 + 6\phi(T)$	96.4	8.8*
2	Gage (G_j)	5	0	2	$\sigma^2 + 10\phi(G)$	139.3	12.8*
8	T x G (TG_{ij})	0	0	2	$\sigma^2 + 2\phi(TG)$	74.6	6.8*
15	Within error ($\varepsilon_{(ij)k}$)	1	1	1	σ^2	10.9	
29	Total						

*Significant at the $\alpha = 0.05$ level

$$^a\text{MS time} = \frac{\text{SS time}}{df = 4} = \frac{1}{4} \left[\frac{74^2 + 77^2 + 134^2 + 89^2 + 94^2}{6} - \frac{468^2}{30} \right]$$

$$= 96.4$$

$$\text{MS gage} = \frac{\text{SS gage}}{df = 2} = \frac{1}{2} \left[\frac{175^2 + 180^2 + 113^2}{10} - \frac{468^2}{30} \right] = 139.3$$

$$\text{MS T x G} = \frac{\text{SS T x G}}{df = 8} = \frac{1}{8} \left[\frac{22^2 + 30^2 + \ldots + 30^2}{2} - \frac{468^2}{30} \right.$$

$$\left. - \text{SS time} - \text{SS gage} \right] = 74.6$$

$$\text{MS}^{\text{within}}_{\text{error}} = \frac{\text{SS}^{\text{within}}_{\text{error}}}{df = 15} = \frac{1}{15}\left[10^2 + 12^2 - \frac{(22)^2}{2} + \ldots + 19^2 + 11^2 - \frac{(30)^2}{2}\right]$$

$$= 10.9$$

Of course all these numbers come from Table 2.2.5.

Since there is a significant interaction of time of welding and gage bar setting, care must be taken to interpret the results. The

first step for further analysis of the results is usually to plot
the cell means. Table 2.2.7 displays the 15 means identified by the
combinations of the levels of time of welding and gage bar setting.
To obtain Fig. 2.2.1 Table 2.2.7 of means and standard errors (refer
to Section 1.2.1 for notation) is given.

TABLE 2.2.7

Means and Standard Errors[a]

| Time of Welding | Gage bar setting | | | Mean |
	1	2	3	
1	11.0	17.0	9.0	12.3
2	15.0	13.0	10.5	12.8
3	25.5	34.0	7.5	22.3
4	17.0	13.0	14.5	14.8
5	19.0	13.0	15.0	15.7
Mean	17.5	18.0	11.3	15.6

[a]Standard errors: cell (body) $= s_{\bar{y}_{ij.}} = \sqrt{\frac{10.9}{2}} \cong 2.33$

$$\text{Gage bar setting} = s_{\bar{y}_{.j.}} = \sqrt{\frac{10.9}{10}} \cong 1.00$$

$$\text{Time of welding} = s_{\bar{y}_{i..}} = \sqrt{\frac{10.9}{6}} \cong 1.34$$

$$\text{Overall} = s_{\bar{y}_{...}} = \sqrt{\frac{10.9}{30}} \cong 0.60$$

The individual comparisons of these means depends upon the
interests of the experimenter. The following subsections show a
few of the possible further analyses that could be carried out on
the means of Fig. 2.2.1 and Table 2.2.7.

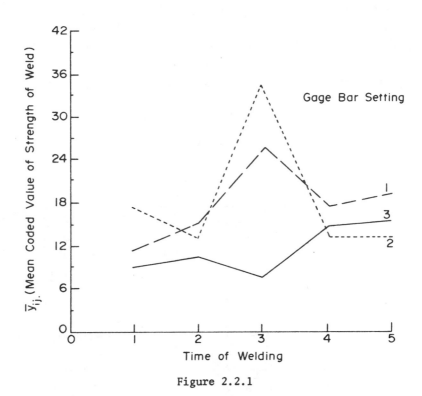

Figure 2.2.1

Example 2.8: Both Qualitative Factors. If both factors are qualitative and there is no preference from a cost viewpoint, quite frequently the investigator wants to find out the combination of levels that provides him with the best response. In this case the interaction is significant and there is crossing over in Fig. 2.2.1. Hence the experimenter usually would not be very interested in the main effects even though they are significant (Table 2.2.6). In fact under the assumption that he is only interested in comparing the cell means he has really changed his model to

$$y_{\ell m} = \mu + C_\ell + \epsilon_{\ell m} \qquad \begin{array}{l} \ell = 1, 2, \ldots, 15 \\ m = 1, 2 \end{array} \qquad (2.2.2)$$

where

$$y_{\ell m} = y_{ijk} \text{ of Eq. (2.2.1)}$$

$$\mu = \mu \text{ of Eq. (2.2.1)}$$

$$\varepsilon_{(\ell)m} = \varepsilon_{(ij)k} \text{ of Eq. (2.2.1)}$$

and

$$C_\ell = \text{effect of the } \ell^{th} \text{ combination of the } i^{th} \text{ level of}$$
$$\text{time with } j^{th} \text{ level of gage bar setting}$$

In other words the model is really a one way classification with 15 levels of the combinations of time of welding and gage bar setting and two observations per cell. Marascuilo and Levin (1970) point out that if one uses Eq. (2.2.2) as the descriptive model rather than retaining Eq. (2.2.1) (as most experimenters do), the analysis of differences between cell means includes both the interaction effects and the main effects. We believe this is precisely what many experimenters really want to do.

Using Eq. (2.2.2) we can rank the cell means and use the Newman-Keuls test as we did in Section 1.2 (See Fig. 2.2.2).

Using Appendix 6 and the $s_{\tilde{y}_{ij\cdot}} \cong 2.33$ from Table 2.2.7 the least significant ranges for this example are:

$$R_{15} = q_{0.05} \, (15,15) \, s_{\tilde{y}_{ij\cdot}} \cong (5.65)(2.33) \cong 13.2$$

$$R_{14} = q_{0.05} \, (14,15) \, s_{\tilde{y}_{ij\cdot}} \cong (5.58)(2.33) \cong 13.1$$

$$R_{13} = q_{0.05} \, (13,15) \, s_{\tilde{y}_{ij\cdot}} \cong (5.49)(2.33) \cong 12.9$$

$$R_{12} = q_{0.05} \, (12,15) \, s_{\tilde{y}_{ij\cdot}} \cong (5.40)(2.33) \cong 12.7$$

$$R_{11} = q_{0.05} \, (11,15) \, s_{\tilde{y}_{ij\cdot}} \cong (5.31)(2.33) \cong 12.4$$

$$R_{10} = q_{0.05} \, (10,15) \, s_{\tilde{y}_{ij\cdot}} \cong (5.20)(2.33) \cong 12.2$$

$$R_9 = q_{0.05} \, (9,15) \, s_{\tilde{y}_{ij\cdot}} \cong (5.08)(2.33) \cong 11.9$$

$$R_8 = q_{0.05} \, (8,15) \, s_{\tilde{y}_{ij\cdot}} \cong (4.94)(2.33) \cong 11.5$$

$$R_7 = q_{0.05} \, (7,15) \, s_{\tilde{y}_{ij\cdot}} \cong (4.78)(2.33) \cong 11.2$$

Rank Order the Combinations (i time, j gage) as follows:

	1	2	3	4	5	6	7	8	9	10	11	12	13	14	15
(i,j)	(3,2)	(3,1)	(5,1)	(1,2)	(4,1)	(2,1)	(5,3)	(4,3)	(2,2)	(4,2)	(5,2)	(1,1)	(2,3)	(1,3)	(3,3)
Mean	34	25.5	19	17	17	15	15	14.5	13	13	13	11	10.5	9	7.5

The Differences of the Means are then

NUMBER OF MEANS SPANNED

Least Significant Range		k (15)	k-1 (14)	k-2 (13)	k-3 (12)	k-4 (11)	k-5 (10)	k-6 (9)	k-7 (8)	k-8 (7)	k-9 (6)	k-10 (5)	k-11 (4)	k-12 (3)	k-13 (2)
13.2	1	•26.5	•25	•23.5	•23	•21	•21	•21	•19.5	•19	•19	•17	•17	•15	•8.5
13.1	2	•18	•16.5	•15	•14.5	•12.5	•12.5	•12.5	11	10.5	••10.5	8.5	8.5	6.5	
12.9	3	11.5	10	8.5	8	6	6	6	4.5	4	4	2	2		
12.7	4	9.5	8	6.5	6	4	4	4	2.5	2	2	0			
12.4	5	9.5	8	6.5	6	4	4	4	2.5	2	2				
12.2	6	7.5	6	4.5	4	2	2	2	0.5	0					
11.9	7	7.5	6	4.5	4	2	2	2	0.5						
11.5	8	7	5.5	4	3.5	1.5	1.5	1.5							
11.2	9	5.5	4	2.5	2	0	0								
10.7	10	5.5	4	2.5	2	0									
10.2	11	5.5	4	2.5	2										
9.5	12	3.5	2	0.5											
8.6	13	3	1.5												
7.0	14	1.5													

• = Significant at the α = .05 Level

•• Note this value 10.5 > R₅ = 10.2 , however, we do not declare it significant due to the rule in section 1.2.

Figure 2.2.2

$$R_6 = q_{0.05} \ (6,15) \ s_{\tilde{y}_{ij \cdot}} \ \tilde{=} \ (4.60)(2.33) \ \tilde{=} \ 10.7$$

$$R_5 = q_{0.05} \ (5,15) \ s_{\tilde{y}_{ij \cdot}} \ \tilde{=} \ (4.37)(2.33) \ \tilde{=} \ 10.2$$

$$R_4 = q_{0.05} \ (4,15) \ s_{\tilde{y}_{ij \cdot}} \ \tilde{=} \ (4.08)(2.33) \ \tilde{=} \ 9.5$$

$$R_3 = q_{0.05} \ (3,15) \ s_{\tilde{y}_{ij \cdot}} \ \tilde{=} \ (3.67)(2.33) \ \tilde{=} \ 8.6$$

$$R_2 = q_{0.05} \ (2,15) \ s_{\tilde{y}_{ij \cdot}} \ \tilde{=} \ (3.01)(2.33) \ \tilde{=} \ 7.0$$

Hence, from the differences of means triangular table, it is seen that the only conclusions are the following:

(1) The mean of combination (3,2) > all others

(2) The mean of combination (3,1) > all others except (4,3), (5,3), (2,1), (4,1), (1,2), and (5,1)

(3) All others are not significantly different at the $\alpha = 0.05$ level

From a practitioner's view this means that time of welding of 3 in combination with gage bar setting of 2 would be preferred if the cost of using this combination was no greater than any other.

If the cost of using time of welding of 3 in combination with gage bar setting of 1 were less than (3,2), the investigator should consider combination (3,1) for future production of this type of steel.

Another analysis that may be considered when both factors are qualitative and interaction is significant is to look at the means of the different levels of one factor for just one level of the other factor. For example, in this case, it may be of interest to the experimenter to examine the three gage bar settings at only the 5th time of welding. The best estimate of the error variance is still the within error from Table 2.2.6, namely, 10.9, because the assumption of homogeneity of variance was accepted indicating the basic error variance is the same over all these treatment combinations and this procedure provides us with 15 degrees of freedom for the error estimate. If only the three levels of gage bar setting for 5th level of time of welding were used, only 3 degrees of freedom would be available for the error estimate.

The basic model would now be only appropriate for the 5th level of time of welding:

$$y_{pq} = \mu + G_p + \epsilon_{(p)q} \quad \begin{matrix} p = 1, 2, 3 \\ q = 1, 2 \end{matrix} \qquad (2.2.3)$$

where

y_{pq} = variable to analyze the qth experimental unit in the pth gage bar setting

μ = overall mean for those three levels of gage bar setting

G_p = effect of the pth level of gage bar setting

$\epsilon_{(p)q}$ = within error of the qth experimental unit in the pth level of gage bar setting NID $(0, \sigma^2)$. The error estimate used for analysis of these three levels, however, is from all 15 treatment combinations of the original experiment

Hence run a Newman-Keuls test on these three means as follows:

(1) Refer to Table 2.2.7 to get the means:

	Gage bar setting		
	1	2	3
Means	19	13	15

(2) Rank the means as:

1	3	2
19	15	13

(3) Differences:

	Levels	
+ \ −	2	3
1	6	4
3	2	

(4) Using Appendix 6 and $s_{\bar{y}_{5p}}$ = 2.33 as before

$$R_3 = q_{0.05} \ (3,15) \ s_{\bar{y}_{5p.}} = (3.67)(2.33) = 8.6$$
$$R_2 = q_{0.05} \ (2,15) \ s_{\bar{y}_{5p.}} = (3.01)(2.33) = 7.0$$

Conclusion:

No significant differences exist at the $\alpha = 0.05$ level for time of welding of 5 among the three means for the gage bar settings.

Problem 2.2.1. Compare the means of gage bar settings for level 3 of time of welding and give a full explanation of the results. Include the model, the reasoning for the error variance used and how the investigator can use the results.

Problem 2.2.2. Compare the means of time of welding for level 2 of gage bar setting and give a full explanation of the results. Include the model and so on as in Problem 2.2.1 above.

Example 2.9: One Factor Qualitative and One Quantitative. When the interaction is significant in a two-way ANOVA and one factor is qualitative and the other quantitative, the experimenter may be interested in the curvature over each level of the qualitative factor.

Let us assume gage bar setting is qualitative and time of welding is quantitative. If the experimenter wants to perform a trend analysis for level 2 of gage bar setting he may test for significance with the error variance from the whole experiment because homogeneity of variances was accepted. He may also use orthogonal polynomials assuming the levels of time of welding are equally spaced. Using this assumption the procedure to investigate the trend is to use Appendix 11 for n = 5 and set up the following:

	Time				
	1	2	3	4	5
z_1	-2	-1	0	1	2
z_{11}	2	-1	-2	-1	2
z_{111}	-1	2	0	-2	1
y	34	26	68	26	26

Using the procedure given in Section 1.4.2 to analyze for trend we obtain the following results:

$$SS \; \hat{\alpha}_1 = \frac{[-2(34) \; -1(26) \; + \; 1(26) \; + \; 2(26)]^2}{[(-2)^2 + (-1)^2 + 1^2 + 2^2] \cdot (2)} = 12.8$$

$$SS \; \hat{\alpha}_{11} = \frac{[2(34) \; -1(26) \; -2(68) \; -1(26) \; + \; 2(26)]^2}{[2^2 + (-1)^2 + (-2)^2 + (-1)^2 + 2^2] \cdot (2)} = 165.1$$

$$SS \; \hat{\alpha}_{111} = \frac{[-1(34) \; + \; 2(26) \; - \; 2(26) \; + \; 1(26)]^2}{(10) \cdot (2)} = 3.2$$

SS time over gage level 2 only

$$= \frac{34^2 + 26^2 + 68^2 + 26^2 + 26^2}{2} - \frac{(180)^2}{10} = 664$$

TABLE 2.2.8

ANOVA for Trend in Time over Gage Level 2

Source	df	SS	MS
Linear	1	12.8	
Lack of fit	3	651.2	217.1*
Quadratic	1	165.1	
Lack of fit	2	486.1	243.0*
Cubic	1	3.2	
Lack of fit	1	482.9	482.9*
Error	15	164.0	10.9

*Significant at $\alpha = 0.05$ level

Conclusion:

The polynomial trend is higher order than cubic. Hence the experimenter is almost certainly not interested in the polynomial fit to the data. He may have a theoretical function in mind that would not be polynomial or he may just run a Newman-Keuls test to check for the better levels of time for that particular level 2 of gage.

Problem 2.2.3. Run a trend analysis on level 3 of gage bar setting and make conclusions about the results.

Example 2.10: Both Factors Quantitative. The general approach to investigating the means of the two way table after the ANOVA has shown a significant interaction when both factors are quantitative is to generalize the approach given in Section 2.1.3. This means that the experimenter usually wants to investigate the polynomial breakdowns of the interaction.

By this we mean for the example on time and gage that the 8 degrees of freedom for interaction may be broken into the following parts each with 1 df:

1. Time linear x gage linear
2. Time linear x gage quadratic
3. Time quadratic x gage linear
   and so on to
8. Time quartic x gage quadratic

From these results the experimenter must decide whether or not he is interested in this polynomial approximation for the interaction parts. If he is, he can then introduce the main effect parts also and investigate the response surface. Usually he will not want to go to an order much higher than two for the whole inference or factor space here since the model is completely fixed. He could, however, discover that a certain part of the factor space is different from the others and he may want to investigate this part only with a polynomial approach.

Utilizing the information obtained in Example 2.9 along with an inspection of Fig. 2.2.1, it is suggested that in order to obtain a low order polynomial to approximate the response surface represented by the data of Table 2.2.5, it is necessary to eliminate welding times 1 and 5. The determination of the optimum conditions will then be done utilizing the data shown in Table 2.2.9. Note that the original values of welding times are now T''s which are coded into T's for ease in using orthogonal polynomials. Bear in mind that the within error mean square of Table 2.2.6 will be used for any tests of hypothesis.   Reference should also be made to Section 1.4 for methods of calculation.

TABLE 2.2.9

Subset of Data from Table 2.2.5[a]

Time of Welding		Gage bar setting						Total
T'	T	1		2		3		
2	1	13,17	30	14,12	26	12, 9	21	77
3	2	21,30	51	30,38	68	10, 5	15	134
4	3	18,16	34	15,11	26	14,15	29	89
Total		115		120		65		300

[a]The circle indicates the total of the two observations in that cell.

The ANOVA for the data of Table 2.2.9 is given in Table 2.2.10 for reference purposes in the derivation of the response surface approximation.

TABLE 2.2.10

ANOVA of Data in Table 2.2.9

Source	df	SS	MS
Time (T_i)	2	301.0	150.5*
Gage (G_j)	2	308.3	154.1*
T x G (TG_{ij})	4	460.7	115.2*
Within error	9	110.0	12.2
Total	17	1180.0	
Within error from Table 2.2.6	15	164.0	10.9

*Significant at $\alpha = 0.05$ level.

A brief review of the MS values in Table 2.2.10 indicates that both main effects as well as the interaction are contributing significantly to the variation in the dependent variable. Consequently, to determine the response surface equation, one must determine the importance of the linear and quadratic terms for each main effect as well as the significant portions of the interaction effect.

The procedures used in Section 1.4.2 may be used to calculate the sum of squares and the estimate of the appropriate regression coefficient for each of the main effects. As a check on calculations it will be noted that the linear and quadratic sums of squares add to the sum of squares given in Table 2.2.10 for the corresponding factor. This type of check may always be applied whenever all factors are equally spaced with equal number of observations per cell as long as one computes as many sums of squares for linear, quadratic, cubic, and etc., as there are degrees of freedom for that particular factor.

The T x G interaction sum of squares may also be partitioned into four independent pieces with one degree of freedom each. These four pieces are denoted as T_{linear} x G_{linear} (T_ℓ x G_ℓ), T_{linear} x $G_{quadratic}$ (T_ℓ x G_q), $T_{quadratic}$ x G_{linear} (T_q x G_ℓ), and $T_{quadratic}$ x $G_{quadratic}$ (T_q x G_q). The calculation for the sum of squares due to T_ℓ x G_ℓ is similar to that shown in Table 2.1.8 for MT (ℓ x ℓ). The calculation for SS T_ℓ x G_q and the estimate of the coefficient of the corresponding orthogonal polynomial is given in Table 2.2.11. The label given to this polynomial is $Z_1 Z_{22}$ where the single subscript 1 indicates linear for the first factor Time of Welding (T) and the double subscript 22 indicates quadratic for the second factor gage bar setting (G). The symbol for the coefficient is α_{122}. The sums of squares and the estimate for all coefficients are given in Table 2.2.12.

TABLE 2.2.11

SS for 1 df $T_\ell \times G_q$

Orthogonal coefficients	G_q					
	1		-2		1	
T_ℓ -1	30	(-1)	26	(2)	21	(-1)
0	51	(0)	68	(0)	15	(0)
1	34	(1)	26	(-2)	29	(1)

$$\text{SS } T_\ell \times G_q = \frac{[30(-1) + 51(0) +...+ 29(1)]^2}{2[(-1)^2 + (0)^2 +...+ (1)^2]} = \frac{144}{24}$$

$$= 6.0$$

The leading 2 in the denominator of SS $T_\ell \times G_q$ represents the number of observations in each cell.

$$\hat{\alpha}_{122} = \frac{[30(-1) + 51(0) +...+ 29(1)]}{2[(-1)^2 + (0)^2 +...+ (1)^2]} = \frac{12}{24} = 0.5$$

TABLE 2.2.12

SS and Coefficient Estimates for All Possible Polynomials

Factor	Polynomial	SS	Coefficient	Coefficient estimate
T_ℓ	z_1	12.0	α_1	1.00
T_q	z_{11}	289.0	α_{11}	-2.83
		301.0		
G_ℓ	z_2	208.3	α_2	-4.17
G_q	z_{22}	100.0	α_{22}	-1.67
		308.3		
$T_\ell \times G_\ell$	$z_1 z_2$	2.0	α_{12}	0.50
$T_\ell \times G_q$	$z_1 z_{22}$	6.0	α_{122}	0.50
$T_q \times G_\ell$	$z_{11} z_2$	140.2	α_{112}	2.42
$T_q \times G_q$	$z_{11} z_{22}$	312.5	α_{1122}	2.08
		460.7		

Each of the terms in Table 2.2.12 has one degree of freedom so the SS is also equal to the mean square. Comparing these against the error mean square of 10.9 indicates the terms that should be included in the regression model. The linear term for time is included for completeness only. The resulting orthogonal model is

$$y = \hat{\alpha}_o + \hat{\alpha}_1 Z_1 + \hat{\alpha}_{11} Z_{11} + \hat{\alpha}_2 Z_2 + \hat{\alpha}_{22} Z_{22} + \hat{\alpha}_{112} Z_{11} Z_2 + \hat{\alpha}_{1122} Z_{11} Z_{22}$$

where the various Z values are as defined in Section 1.4.1. For example,

$$Z_{11} Z_2 = \lambda_2 [(T - \overline{T})^2 - \frac{9-1}{12}] \lambda_1 (G - \overline{G})$$

$$= 3 [(T - 2)^2 - \frac{2}{3}] (G - 2)$$

Substitution of all these terms and their respective coefficients into the above equation results in the following equation:

$$y = 138 - 157 T - 210 G + 39.5 T^2 + 57.5 G^2 + 271 GT - 67.75 GT^2$$
$$- 75 T G^2 + 18.75 T^2 G^2$$

The determination of the point in the factor space which yield the maximum value of y in the previous equation is rather difficult without the use of a computer. One may attempt to find the optimum operating condition by trial and error or resort to some elaborate optimization procedures. In either case one would determine that T = 2.023 and G = 1.746 is an estimate of the optimum operating conditions which produces an estimated response value of 35.17. Note that this value does not differ very much from the average value of 34 observed in the cell T = 2 and G = 2.

Problem 2.2.4. An animal feed manufacturing company was interested in investigating the influences of five different amounts of a certain feed supplement and three temperatures at which the supplements were put into the feed on the amount of vitamin A retained by the feed. The data were transformed so that the assumptions in the ANOVA were met.

From a completely randomized design with two randomly chosen feed samples per treatment combinations the following data are to be analyzed:

Temperatures

Supplements	40		80		120	
2	33,37	(70)	60,64	(124)	46,50	(96)
4	50,56	(106)	79,73	(152)	72,58	(130)
6	72,64	(136)	80,88	(168)	72,80	(152)
8	77,73	(150)	79,87	(166)	81,77	(158)
10	67,73	(140)	84,80	(164)	69,83	(152)

(a)   Run the appropriate ANOVA, using the algorithm to show the correct tests, and make the tests of significance on the main effects and interaction.

(b)   Determine the regression equation which is the best fit for these data.

2.3   REFERENCES

Anderson, R. L. and Bancroft, T. A.   Statistical Theory in Research McGraw-Hill, New York, 1952.

Bennett, C. A. and Franklin, N. L.   Statistical Analysis in Chemistry and the Chemical Industry, Wiley, New York, 1954.

Cochran, W. G.   Sampling Techniques, 2nd ed., Wiley, New York, 1963.

Cochran, W. G.   Biometrics 7:17 (1951).

Hadley, W. O., Hudson, W. R., Kennedy, T. W. and Anderson, V. L. Association of Asphalt Paving Technology 1:1 (Feb. 1969).

Hicks, C. R.   Fundamentals in the Design of Experiments, 2nd ed., Holt, Rinehart and Winston, New York, 1973.

Marascuilo, L. A. and Levin, J. R.   Am. Ed. Res. J. 7:397 (1970).

Ostle, B.   Statistics in Research, 2nd ed., Iowa State University Press, Ames, Iowa, 1963.

Tukey, J. W.   Biometrics 5:232 (1949).

Chapter 3

A SCIENTIFIC APPROACH TO EXPERIMENTATION

Experiments are performed by people in nearly all walks of life.
The basic reason for running most experiments is to find out something
that is not known. Unfortunately, there are cases in which the sole
purpose of the experiment is to "prove" what the experimenter already
"knew." This type of experiment frequently is conducted so that the
"known" result will occur no matter whether it should or not. This
type of experiment cannot be condoned by persons seeking the truth.
On the other hand, a worse condition may exist where people run
experiments fully intending to be honest but being completely unaware
of their incompetency in conducting experiments intelligently.
Frequently experiments are run so that the effect of the factor of
interest is disguised by the effect of another factor not considered.
This latter factor is then ignored or considered unimportant; yet,
in the long run, it is the real cause.

An example of this occurred a number of years ago in a large
manufacturing company. A man working in the production area set up
an experiment to test a new alloy, possibly one to replace an old
one in production. He ran one heat of metal with the new alloy and
another heat with the old one. Taking one ingot from each heat and
30 pieces of metal from each ingot, he proceeded to test each of the
60 pieces for the property in which he was interested. With the data
he made a one-way ANOVA on the alloys using the pieces within ingots
with 58 degrees of freedom as the error. The results showed that the
new alloy was "better" than the old one, and the experimenter
convinced the vice president in charge of production to change the

production procedures so that the new alloy would be used in the future. Since the experimenter had used a "designed experiment" and had tested the data "statistically," the vice president concluded there could be no doubt that the new one was better.

The change cost the company $200,000, and after two years in the field there was as much trouble with the product made from the new alloy as there had been with the old product. The vice president was disgusted and called one of the authors of this book to say he would never allow his company to use designed experiments again. After some discussion, the vice president allowed the author to talk with the experimenter to find out how the experiment was conducted.

In wanting to keep the cost of the experiment low, the experimenter did not consider the possibility that the property in which he was interested varied considerably both from heat to heat and from ingot to ingot within a heat. From an ANOVA point of view, his EMS should have been what is shown in Table 3.1.1.

TABLE 3.1.1

ANOVA of Alloy Problem

Source	df	EMS
Alloys	1	$\sigma_p^2 + 30\sigma_I^2 + 30\sigma_H^2 + 30\phi(A)$
Pieces in ingots	58	σ_p^2

Hence, rather than testing that the alloy effect was zero $[\phi(A) = 0]$, he was really testing that the total effect for ingots, heats, and alloy was zero $[30\sigma_I^2 + 30\sigma_H^2 + 30\phi(A) = 0]$. Since the long-run production of the new alloy did not produce the improvement seen in the experiment, $\phi(A)$ must equal zero. Thus, it must have been that σ_I^2 and/or σ_H^2 were/was not zero.

When the results were explained, the vice president was willing to use design of experiments again on this type of problem but insisted upon having more than one heat for each alloy.

The conclusion is that if experiments are to be run, and
assuredly they will be, a scientific approach to designing them is
needed. The following is an ordered list of requirements for
scientific experimentation. The requirements will be described in
more detail in the following sections.

Recognition that a problem exists.

Formulation of the problem.

Agreeing on factors and levels to be used in the experiment.

Specifying the variables to be measured.

Definition of the inference space for the problem.

Random selection of the experimental units.

Assignment of treatments to the experimental units.

Outline of the analysis corresponding to the design before the
data are taken.

Collection of the data.

Analysis of the data.

Conclusions.

Implementation.

3.1   RECOGNITION THAT A PROBLEM EXISTS

In many industrial plants the accountants are able to show which
projects are possibly having production trouble by arraying the costs
in the various departments (sometimes called a "pareto"). The
indication is of course, that the relative costs in these areas
either have changed recently or are large relative to some production
criterion.

In other plants the foreman may recognize trouble and bring it
to the attention of his superior. In any event, at least one area
always seems to need attention, and alert management recognizes that
a problem exists.

In research departments of universities or in industry, people
are continually encountering problems that can be attacked successfully by using scientific experimentation. Usually the recognition
phase in these cases is short.

3.2 FORMULATION OF THE PROBLEM

Committee action is the most successful method of formulating the problem after management agrees that a problem exists. The word "action" is most important because the members of a committee for attacking a problem must be ones who know the technical parts involved and be willing to state their thoughts. A person who knows the requirements for scientific experimentation and keeps the thoughts flowing is the best moderator for such a group. There must never be reticence by any participants even if the ideas are in conflict with those of a supervisor who may be present; hence, a moderator from outside the plant who stimulates thought and prevents conflicts is usually preferred to a knowledgeable in-plant participant.

Usually this discussion period lasts from 2 to 8 hours; every committee member has a chance of expressing his view. Frequently 45 or 50 possible causes of the problem will be put forth. The next step for the moderator is to get the participants to reduce these to a reasonable number of causes. Hopefully, the list will be reduced to eight or ten, or preferably four or five. Rarely is the ultimate list only two or three. Of course, the greatest difficulty in this actual formulation of the problem is to get the committee members to agree on the ultimate causes to be tried in the forthcoming experiment.

An example is an investigation related to the fabrication of men's synthetic felt hats. The manufacturer had experienced extreme difficulty in producing these hats so that the flocking appeared on the molded rubber base in a uniform fashion to simulate the real felt hats. In order to approach this problem, a committee was formed consisting of a development engineer, a manufacturing foreman, a chief operator, a sales representative and a statistician. The statistician's job was to obtain from these people all possible causes of imperfect hats. Factors which were thrown out for discussion were as follows: thickness of foam rubber base, pressure of molding, time of molding, viscosity of the latex used to glue the flocking to the molded rubber base, age of the latex, nozzle size of several different spray guns, direction of spraying, condition of the flocking, speed of drying, and the effect of location within the drying furnace.

After considerable discussion, the committee finally decided that the most serious problems were probably connected with the nozzle size and the pressure under which the latex was sprayed. In arriving at these various factors the committee essentially forced a review of the entire production process. This in itself led to a better understanding of the production process and an eventual solution to the problem.

3.3 AGREEING ON FACTORS AND LEVELS TO BE USED IN THE EXPERIMENT

In arriving at the most reasonable causes of defective hats in the previous section, the committee action essentially required a critical review of all production process factors. In this review the chief operator brought out the standard operating levels of the nozzle size as well as the pressure under which the latex was sprayed. Talk with the chief operator revealed that the latex pressure varied considerably due to the viscosity of the latex and from this information the pressure levels were eventually obtained. Additional inquiry ascertained that the manufacturing area had two different nozzle sizes that had been used interchangeably; consequently, the two sizes became the basis for these factor levels.

The authors feel that through committee action of this type one is almost always able to find realistic levels for all major factors. Occasionally considerable effort is needed to find out just how shoddy one's manufacturing operation really is, and this example shows that actual production operations will allow their process to operate at various levels as long as it works. The determination of the optimal levels is, of course, the desired end of the experimental investigation.

3.4   SPECIFYING THE VARIABLES TO BE MEASURED

As one can imagine in the manufacturing of synthetic hats the measuring of product quality is a very difficult task. Consequently, the method that was used in the experiment referred to above was to visually grade the finished hat on the following items: the hungry appearance of the flocking, the starchy appearance of the flocking, and the appearance of the brim. During the course of the investigation these responses were found to be essentially independent of each other and consequently could be treated as three separate dependent variables, which could be investigated on a one-at-a-time basis. The standards for grading each of these factors were arrived at again through committee action which eventually resulted in a visual display board and gave the inspectors a realistic means of grading each of these dependent variables.

One of the most difficult things in certain types of industrial experimentation is the specification of a dependent variable. It is usually quite obvious what the variable should be; however, the means of measurement is sometimes quite difficult. In ideal cases this would be merely the measured value obtained by some simple inspection tool while in other cases it can be a value that is almost impossible to measure and will have to be graded by one or more inspectors.

3.5   DEFINITION OF THE INFERENCE SPACE FOR THE PROBLEM

After the experimenter has defined his problem he must make decisions on the limits of the inferences to be made from the results. Those limits within which the results will apply may be called the inference space. For example, a chemist may want the results of his experimentation to apply to all laboratories in the world. If this is the case, the experiment should be conducted in such a way that a sufficiently large random sample of the possible laboratories should be included in the experiment. Great care should be placed at the initial stages of formulating the experiment to assure the wide applicability desired by the experimenter (Cox, 1958, Section 1.2, pp. 9-11).

In genetic studies using <u>Drosophila</u> (fruit flies), the variation between bottles containing the flies is usually larger than the variation within bottles and the inference space is usually over bottles. Hence for certain simple laboratory experiments the desired experimental error must contain between bottle variation as well as variation from flies within bottles in order to make the results apply to all bottles. The general ideas involving the correct error terms have direct association with the inference space for any factor or combinations of factors (or treatments).

3.6 RANDOM SELECTION OF THE EXPERIMENTAL UNITS

Throughout this book the term "experimental unit" will refer to the type of experimental material that is being used to receive the application of the various treatments and be representative of the desired inference space. For example, if one is interested in investigating the effects of various types of fertilizers on potted plants, the potted plant would be the experimental unit. In this case extreme care should be taken before extending the inference to field conditions. If one is interested in investigating the effect of a certain type of teaching device then the experimental unit would be an individual student. In agronomy experiments, many times one uses a plot of ground as an experimental unit. In this case the size of the plot is frequently dependent upon the growing habits of the particular plant under consideration because the results must apply to the desired population.

Once the experimental unit has been selected, the investigator usually has several of the units at his disposal. Consequently it becomes necessary to randomly select a sufficient number of experimental units to be utilized in the experiment. The random selection is necessary at this point in order to protect against any

bias in our experiment which could be the result of some unknown
factor having had prior influence on the experimental units in some
systematic fashion. The number of experimental units is naturally
dependent upon the size of the desired standard error of individual
cell means. In certain investigations the available experimental
units will not be of the same homogenous material. Consequently,
an investigator will suspect that he is apt to get a different
response for his treatments as a result of these experimental units
being stratified together based on certain inherent characteristics.
In agronomy experiments these inherent conditions could be soil
conditions. In Section 3.5 we discussed a genetic study using
Drosophila. If the individual fruit fly is the experimental unit,
then as discussed before these fruit flies will be stratified
according to their respective bottle. This type of stratification
will be referred to as blocking in the forth coming chapters.

3.7 ASSIGNMENT OF TREATMENTS TO THE EXPERIMENTAL UNITS

Experimenters want assurance that the probability statements
they make are as nearly correct as possible and to give them this
assurance, laws of chance must enter into the experimentation. The
research worker wants each treatment to have as good a chance as any
other treatment to show its effect on the experimental unit through
the variable to be analyzed. The idea of complete randomization is
to allow every experimental unit an equal chance of receiving each
treatment. For example, in an electroplating experiment, magnetic
material was electroplated to material with two different substrates,
copper and brass. The variable to be analyzed was thickness of
plating after a given time of current flow. The design of the
experiment or randomization was associated with the order of the
test pieces with the different substrates (manufacturing was assumed
to have been handled correctly just as other extraneous variables).
If all material with copper substrates were plated before any of the
brass substrates, there may be thicker plating just because of the

environmental conditions at that time, not because of the metal of the substrates. Hence, to insure against unknown peculiarities, the order of the plating should be randomized.

As a result of randomization, the estimates of the parameters such as average thickness of plating should be unbiased and the tests of significance of the effects of treatments on the variable to be analyzed should be valid (Fisher, 1966, pp. 62-64). In addition, the manner in which randomization occurs (for example completely at random or inside blocks of experimental units), determines the analysis to be used on data (Ostle, 1963, p. 251). Randomization is then the basis for the design of the experiment, but it never eliminates variation caused by extraneous variables. References are Yates (1964, pp. 317-321), and Finney (1960, Chapter 1).

The definition of design of experiments that will be used in this book is the method by which the treatments are placed on the experimental units. Many times due to the nature of the experiment, restrictions on the randomization are inherent (sometimes rather cleverly disguised) in the design. This was the case in the synthetic hat example introduced in Section 3.2. In this case the committee decided that it would not be economically feasible in terms of time to interchange spray gun nozzles in a completely randomized fashion. Consequently, when a certain configuration of nozzles was set up several treatment combinations were run by adjusting various gun pressures in a random fashion. Restrictions on the randomization as discussed above gives rise to the term "restriction error" which must be acknowledged in the analysis of the resulting data. This acknowledgement is accomplished by inserting an appropriate term into the model (Anderson, 1970).

3.8 OUTLINE OF THE ANALYSIS CORRESPONDING TO THE DESIGN BEFORE THE DATA ARE TAKEN

The concept of committee action was introduced in Section 3.2. We now come to another critical step in this committee action. At this point, the statistician must write down the mathematical model that has evolved as a result of the committee activity in the preceding sections. This mathematical model will give rise to the ANOVA table. The ANOVA table will now consist of the degrees of freedom and the expected mean squares for each of the specific factors selected in Section 3.3 above. These expected mean squares will indicate to the statistician and experimenter all of the various factors which will have tests available. In addition, if the test is not directly available, it will be clear as to what assumptions must be made in order to obtain some type of conservative test. Once the statistician has reviewed this ANOVA table he must then present his findings to the committee to see if these assumptions and conclusions are realistic from a practical point of view. Should the committee decide that some of the necessary assumptions (which must be made in order to test the effects of various factors) are not realistic, the design will have to be changed before proceeding further in the experimentation process.

As one investigates various designs by comparing expected mean squares from the respective analysis of variance tables, it will become obvious to the statistician as well as to the experimenter when they have to compromise (for economical reasons) the availability of a clear-cut test on the effect of a certain factor. It is possible that by a certain type of restriction on the randomization, which would lead to a conservative-type test, that one would be able to run an experiment in a much more economical fashion. One must recall, however, that there is economics involved in the cost of the experiment as well as in the cost of making an incorrect decision as a result of running an experiment which does not give an explicit test for the effect of an important factor. These costs, of course, would arise when an incorrect manufacturing process is implemented on a production basis.

3.9 COLLECTION OF THE DATA

To the inexperienced investigator this particular topic seems like a rather simple and straightforward procedure. However, extreme care must be taken at this particular point. In some cases, carrying out of the experiment may require an organized team with many observers. Team planning meetings may become quite extensive and detailed instructions may have to be prepared for operators, supervisors, and technical observers. The whole success of this scientific investigation depends upon the validity of all data obtained. Even if the experiment is not too complicated, the authors have found that one of the most successful means of collecting data is for the statistician to make a specific data form which would be filled out by the inspector and further verified by the person in charge of running the experiment, possibly the manufacturing foreman. The data form itself must be easy to use, relatively understandable, and in general foolproof with respect to any misinterpretations. For a complicated experiment it is sometimes wise to pretest the experimental plan and the data collection form. By this we mean to actually process some experimental units that are designed specifically for pretesting the inspection and data recording procedures.

Considerable thought should be given to the collection of the data. For example, if the data are going to be keypunched for use in a particular computer program it is well known that keypunch errors can be reduced by having an efficient data form. In case one is collecting extreme amounts of data, possibly the data can be obtained on either magnetic tapes, paper tape, punch cards, or even possibly some type of mark sense cards. In general, the size and the importance of this experiment will dictate the amount of care that should be taken in the collection of the data.

3.10   ANALYSIS OF THE DATA

The initial phases of the analysis of data will be primarily the statistician's responsibility. Naturally the analysis of the data is highly dependent upon the design of the experiment. Modern computing facilities have many so-called canned programs which will analyze almost every type of design that is available. The primary input to these canned programs is the experimental model that we have discussed previously. Once the model has been written down and the number of levels decided upon, it is a relatively simple procedure to process the data through the computer. If a computer program is not available then one must resort to the use of a desk calculator or possibly even a slide rule and an adding machine. Formulas for these calculations may be readily found in such references as Hicks (1973), Ostle (1963), and Steel and Torrie (1960).

Once the basic calculations have been performed the statistician should review the data and possibly make graphs of the individual cell means in order to display the results of this experiment. The authors have found that graphical presentation of the data is the easiest way to show management the important aspects of the experiment.

3.11   CONCLUSIONS

Once the statistician has completed the analysis of data it is again time for various members of the committee to meet and formalize their conclusions. In almost every case the experimenter will be able to physically interpret the statistical findings that have been brought to light by this experiment. Then based on the magnitude of these findings and the cost of implementing any new procedures which might be required, the committee must make recommendations to management. These recommendations are usually in some form of written report and/or oral presentation to the interested members of management.

3.12   IMPLEMENTATION

The implementation of the recommendations developed by the committee is of course a management decision. The purpose of the scientific investigation was to develop a realistic type of experiment, assure proper running of the experiment and to find dependable results. These findings of course have to be presented in such a fashion that management can make an intelligent decision upon the implementation of these findings.

Once management decides to make a process change which has been recommended by the committee, the committee has another tremendous task facing it. This task is to assure management that the experimental procedure will be implemented as well as possible into the production process. In many cases it is not a simple thing to implement a procedure developed in a pilot study into a full-fledged production process. In no case should a new procedure be put into production without further evaluation by some type of statistical study. It is the responsibility of the committee to assure management that the revisions to the process have indeed improved the production operation. Many times a prototype of the actual production set up is devised and tried out on a small-scale production basis. This gives an indication of how the results of the experiment may work in production without using actual production equipment as a testing device.

If results from the prototype are successful then the results from the experiment may be put into production; however, many times changes are indicated by the prototype before implementation.

3.13   SUMMARY

It should be pointed out that the actual design of experiments part of the scientific approach to experimentation is covered in Sections 3.5, 3.6, and 3.7. The sections prior to these three prepare for the design, and Section 3.8 allows the experimenter to modify the design before putting it into operation.

Since the following chapters will be concerned mainly with the design of experiments we will summarize the topics covered in Sections 3.5, 3.6, and 3.7.

1. Inference Space

Before the experiment is really designed the investigator must decide how widely he wants his results to apply. The experimenter must compare the cost of the experiment with the usefulness of the results he intended to obtain. This demands that he thoroughly understand what inferences he can make before he actually undertakes the experiment. This stage, in practice, is the most neglected because many people do not take the time to think the problem through before setting up the design.

2. Randomization

This stage is necessary in order for the experimenter to construct probability statements about the results. Randomization allows for valid tests and unbiased estimates after the inference space has been decided upon.

3. Replication

Before the design is finalized, the experimenter must decide on the amount of replication to use in the experiment. Present-day design practices utilize existing information to the fullest and replication is sometimes not needed at all. In fact fractional replication (which will be covered later) actually means there are less than a complete representation of all the treatment combinations. In industry, usually there will be pressure from management (in universities, this pressure comes from the major professor or investigator) to run the experiment as quickly as possible with a minimum amount of expense. To protect the validity of the experiment, the statistician must make it clear to the investigator what minimum replication is demanded to obtain the estimates of the effects the experimenter desires with an adequate error. This decision quite frequently is based on the number of degrees of freedom that will be available for the error to test the various effects of interest.

The remainder of this book really utilizes these concepts in different arrangements of the experimental units and systematizes these layouts into the so called "designs." In all instances the expected mean square derivation (the analysis) is immediately available after the design has been made and shows the experimenter what he can cover in his inferences. The results of the design will not be realized until after the data are analyzed.

3.14 REFERENCES

Anderson, V. L. Biometrics 26:255 (1970).

Cox, D. R. Planning of Experiments. Wiley, New York, 1958, Chapters 1, 4, 5, and 8.

Finney, D. J. An Introduction to the Theory of Experimental Designs, University of Chicago Press, Chicago, Ill., 1960, Chapters 1, 8 and 9.

Fisher, R. A. The Design of Experiments. 8th ed., Hafner, New York, 1966, Chapters I-IV.

Hicks, C. R. Fundamental Concepts in the Design of Experiments, 2nd ed., Holt, Rinehart and Winston, New York, 1973, Chapter 1.

Ostle, B. Statistics In Research, 2nd ed., Iowa State University Press, Ames, Iowa, 1963, Chapter 10.

Steel, R. G. D. and Torrie, J. H. Principles and Procedures of Statistics, McGraw-Hill, New York, 1960, Chapter 6.

Yates, F. Biometrics 20:307 (1964).

Chapter 4

COMPLETELY RANDOMIZED DESIGN (CRD)

Determination of the proper design of the experiment to be used
by a research worker depends upon the inference he desires and the
randomization procedures available to him. If the experimenter wants
to investigate the effects of a factor in a narrow area, say within
his laboratory, the design may require only simple randomization and
the completely randomized design is appropriate. If, on the other
hand, he wishes the results to be applicable over an extremely wide
range, say over the entire United States, the design may need to be
quite elaborate and the completely randomized design may not be
appropriate. References on these concepts include Cox (1958, Chapter
7), Cochran and Cox (1957, Chapter 4), and John (1971, Chapter 3).

4.1 ONE FACTOR

This is the simplest design and is characterized by the
experimental units being drawn as completely at random as possible.
One procedure for constructing a completely randomized design is to
allot numbers to the experimental units and draw the numbers at
random assigning the levels of the factor or treatments (which may
be random or fixed) in order of the draw. For example, in an
experiment on firing time of an explosive switch, let us assume that
packing pressure of the primary initiator is the only factor involved.
If there are three pressures (12,000 psi; 20,000 psi; 28,000 psi) to
be investigated, it would be desirable to show the design and
analysis of this experiment if there are 21 explosive switches
available and no restrictions on randomization in the experiment.

Let us look at the inference space. The experimenter would
like to imply that if the effects of packing pressures show up in
this experiment, he can be reasonably sure that the same effects
will occur on all explosive switches from which these 21 were a
sample. This requires that these 21 switches be selected from the
population of interest, at random, even before the experiment is
begun and then numbered 1 through 21 (Hader, 1973).

The next step is to be sure these 21 switches are assigned at
random to the packing pressures in such a manner that there is an
equal number (in this case 7) of switches for each packing pressure.
To do this we can use a set of random numbers (Appendix 1) and draw
a number between 1 and 21, say 10. Then the number 10 switch is
assigned to packing pressure 12,000 psi and an explosive switch is
constructed with 12,000 psi. This procedure is continued so that
the next explosive switch drawn at random will be constructed with
20,000 psi, the following one with 28,000 psi, the one after that
with 12,000 psi and so on. This will finally give a design with
seven switches with 12,000 psi, seven with 20,000 psi and seven with
28,000 psi. The only restriction on randomization (which is minor
and ignored in the analysis) is that if the same number, say 10,
were drawn the second time it would be disregarded. In other words,
there is only one switch for each number. The sampling is really
without replacement but the correction for this in the analysis is
ignored. One reason for ignoring the sampling without replacement
is that the inference is over all switches from which these 21 are
drawn.

In some experiments only one randomization takes place, but
here one must consider the firing order of the constructed explosive
switches. If, by chance, there was some reason for the switches
fired early to have a longer firing time due to atmospheric or
environmental conditions and all the 12,000 psi were fired first,
it would not be possible to distinguish between the effect of packing
pressures and the effect of the environment. This demonstrates that
the effect of packing pressure would then be completely confounded

with the environmental effect. Hence, to assure the experimenter
that the results are as near correct as he can make them, he must
randomize the firing order. In this case, he can renumber the 21
explosive switches as

Packing Pressure

12,000 psi	20,000 psi	28,000 psi
1	8	15
2	9	16
3	10	17
4	11	18
5	12	19
6	13	20
7	14	21

and draw a random number between 1 and 21 again. Say the number is
12, then.the number 12 explosive switch (20,000 psi pressure) is
fired.first. Continue in this manner until all 21 switches have an
order number to be fired. All of this randomization must be
completed before the firing is begun so the experiment can be carried
off smoothly. One possible firing order is

Firing order	Switch number	Pressure
1	10	20,000
2	2	12,000
3	11	20,000
4	20	28,000
5	15	28,000
6	3	12,000
7	9	20,000
8	17	28,000
.	.	.
.	.	.
.	.	.
21	8	20,000

Notice the mixup in the order of the column headed "Pressure." Of course order could be a classified variable and controlled like a factor but then the experiment would not be run in a completely randomized design (Prairie and Zimmer, 1964).

The model for the analysis of variance of this experiment is

$$y_{ij} = \mu + P_i + \varepsilon_{(i)j} : [i = 1, 2, 3; j = 1, 2, \ldots, 7] \quad (4.1.1)$$

where

y_{ij} = observed firing time of the jth explosive switch with the ith packing pressure (assumed normally distributed and used as the variable to be analyzed)

μ = overall mean

P_i = effect of the ith packing pressure

$\varepsilon_{(i)j}$ = effect of the jth switch within the ith pressure assumed to be NID $(0, \sigma^2)$

The corresponding ANOVA for Eq. 4.1.1 is given in Table 4.1.1.

TABLE 4.1.1

ANOVA for Pressures

Source	df	EMS
Pressures	2	$\sigma^2 + 7\phi(P)$
Residual	18	σ^2
Total	20	

The estimated standard error of each pressure mean is

$$s_{\bar{y}_{i\cdot}} = (\frac{\hat{\sigma}^2}{7})^{\frac{1}{2}}$$

Problem 4.1.1. Show how you would examine the trend of firing time over packing pressures.

4.2 MORE THAN ONE FACTOR

When more than one factor is used in an experiment and all
levels of each factor are represented in combination with all levels
of every other factor, the experiment is called a "complete
factorial." This property (complete factorial) allows the factorial
experiment to have inherent or hidden replication because any level
of one factor occurs as often as the product of the number of levels
in the other factors. Since, in reality, a given combination of
levels of all the factors acts as a treatment on an experimental
unit, these combinations are frequently called "treatment
combinations."

In this chapter on completely randomized designs it will be
assumed that all treatment combinations are placed completely at
random with the experimental units (regardless of the number of
factors), In addition, if there is more than one experimental unit
for each treatment combination, the designs of the experiments for
combinations of fixed and random factors are easily extended from
the one factor completely randomized design given in Section 4.1.
In this section designs with one experimental unit for each treatment
combination will be emphasized because this condition requires more
design techniques (to get the design efficient) than is required for
more than one experimental unit per treatment combination.

It should be understood that the analysis of a complete factorial
with two factors whose combination of levels have been completely
randomized is sometimes called a two-way classification, and with
q-factors is a q-way classification.

Problem 4.2.1. Show a completely randomized design and analysis
for a two factor factorial experiment when the model is mixed
if there is more than one experimental unit per treatment
combination.

4.2.1 All Fixed Factors

If all factors in a completely randomized design are fixed and

there is only one experimental unit per treatment combination, there
may be no good estimate of the experimental error for investigation
of the main effects and interactions. If, however, higher factor
interactions are negligible these sources of variation may actually
be excellent estimates of the experimental error.

Consider a three-factor factorial experiment example on explosive
switches in which the three factors are fixed. The experimenter is
interested in two metals for pistons (2 s al and Teflon). This factor
is called qualitative. The other two factors are amount of primary
initiator with levels 5, 10, and 15 mg and packing pressures 12,000,
20,000, and 28,000 psi which are quantitative. The purpose of the
experiment is to investigate the effects of the factors on the
dependent variable, firing time.

An example of hidden replication can be shown in this explosive
switch experiment. There are nine observations of the firing time
for each metal and six observations of firing time for each amount
of primary initiator and for each packing pressure.

The design of the experiment or the randomization procedure is
to take the 18 treatment combinations, that is the combinations of
metals, initiators, and pressures, in an order such as that given in
the following tabulation.

	Metals	Initiator	Pressure
1.	2 s al	5	12,000
2.	Teflon	5	12,000
3.	2 s al	5	20,000
4.	Teflon	5	20,000
.	.	.	.
.	.	.	.
.	.	.	.
.	.	.	.
18.	Teflon	15	28,000

Draw a random number between 1 and 18, say 3. Then the first
combination or explosive switch to be fired will be number 3

(2 s al, 5 mg of primary initiator at 20,000 psi packing pressure).
Continue the random procedure until all 18 are placed in firing
order, at random.

The model considering all 18 treatment combinations as separate
treatments could be written as:

$$y_i = \mu + T_i + \varepsilon_{(i)1} \; ; \; i = 1, 2, \ldots, 18 \qquad (4.2.1)$$

where

y_i = firing time of the switch with the ith treatment
(combination) considered the variable to be analyzed

μ = overall mean

T_i = effect of the ith treatment (combination)

$\varepsilon_{(i)1}$ = effect of the other extraneous variables (consider
random because of the randomization) on that experimental
unit with the ith treatment, assumed NID $(0, \sigma^2)$

TABLE 4.2.1

ANOVA using Equation (4.2.1)

Source	df	EMS
Treatments	17	$\sigma^2 + \phi(T)$
Experimental Error	0	σ^2

The analysis in Table 4.2.1 obviously allows no tests on treatments
because there are zero df for experimental error (to be called
error from now on).

If, however, the factorial structure is utilized in the
analysis, the model for the 18 treatment combinations with one
experimental unit per treatment combination is:

$$y_{ijk} = \mu + M_i + I_j + MI_{ij} + P_k + MP_{ik} + IP_{jk}$$
$$+ MIP_{ijk} + \varepsilon_{(ijk)}$$
$$i = 0, 1 \quad j = 0, 1, 2 \quad k = 0, 1, 2 \qquad (4.2.2)$$

where y_{ijk} = the firing time required for the switch made up of the ith metal, j^{th} initiator and k^{th} pressure, as indicated in the treatment combination table, and all components on the right-hand side of the equation are the main effects and interactions of metals, initiators, and pressures plus the mean and error.

In Eq. (4.2.1) the term T_i is merely extended into more meaningful components known in the experiment as effects of initiators, pressures and so on, in Eq. (4.2.2). Since all factors are fixed the analysis is as shown in Table 4.2.2.

TABLE 4.2.2

ANOVA using Equation (4.2.2)

Source	df	EMS
Metals (M)	1	$\sigma^2 + 9\phi(M)$
Initiators (I)	2	$\sigma^2 + 6\phi(I)$
M x I	2	$\sigma^2 + 3\phi(MI)$
Pressures (P)	2	$\sigma^2 + 6\phi(P)$
M x P	2	$\sigma^2 + 3\phi(MP)$
I x P	4	$\sigma^2 + 2\phi(IP)$
M x I x P	4	$\sigma^2 + \phi(MIP)$
Error	0	σ^2
Total	17	

In order to describe the inference space, the experimenter must recognize that no true replication exists in this experiment and if there is no prior estimate of σ^2, no exact test of main effects or interactions is possible. If the experimenter is willing to assume that the three-factor interaction is negligible, $\phi(MIP) = 0$, then the mean square for the source, MIP, may be used for the error estimate. An additional possibility exists even if $\phi(MIP)$ cannot be assumed negligible. That is to use the mean square for MIP as the denominator for testing all other main effects and interactions. Since the true experimental error should be less than or equal to the MIP mean square the tests are conservative and those effects and interactions that show significance should be significant.

If the experimenter had no prior knowledge of σ^2 and could not assume $\phi(MIP) = 0$, he may consider repeating treatment combinations on randomly chosen experimental units or repeating a few of the treatment combinations in the experiment. The usual practice in analyzing the experiment with only a few of the treatment combinations repeated on other experimental units is to use the additional observations to estimate the experimental error, but not to use these additional observations in estimating the effects of the treatments. If a program for a digital computer were available to use the appropriate weights, it may be desirable to utilize all the information; but if the investigator does not have the program and is in a hurry for the results, he usually does not use these additional observations in the analysis of effects.

Consider the example on explosive switches in which six of the treatment combinations are repeated. The model is

$$y_{ijk\ell} = \mu + M_i + I_j + MI_{ij} + P_k + MP_{ik}$$
$$+ IP_{jk} + MIP_{ijk} + \varepsilon_{(ijk)\ell} \qquad (4.2.3)$$

where the components are the same as in Eq. (4.2.2) except that $\ell = 1, 2$ for only six of the treatment combinations, and $\ell = 1$ for the remaining twelve treatment combinations. The analysis of variance is the same as that given in Table 4.2.2 except that we use only the first observation for the factorial analysis, but use the two observations per treatment combination to calculate the error sums of squares (SS). For each treatment combination with two observations, calculate the SS with 1 df by squaring the difference between the observations and dividing by 2 [see Example 7.2 for proof]. Then merely add the six SS to obtain the SS for error. The ANOVA is given in Table 4.2.3.

TABLE 4.2.3

Error Analysis Using Equation (4.2.3)

Source	df	EMS
Error	6	σ^2

If the levels of the factors are equally spaced and quantitative, the experimenter may use a more efficient design. If he can assume a second-order model then he can utilize the df from the higher order terms to estimate the error and no repeats of treatment combinations will be necessary. A model to describe the effects and interactions of metals, initiators, and pressures on firing time follows:

$$y = \alpha_0 + \alpha_1 Z_1 + \alpha_2 Z_2 + \alpha_{22} Z_{22} + \alpha_{12} Z_1 Z_2 + \alpha_3 Z_3 + \alpha_{33} Z_{33}$$

$$+ \ \alpha_{13} Z_1 Z_3 + \alpha_{23} Z_2 Z_3 + \epsilon \qquad\qquad (4.2.4)$$

where

    y = firing time (assumed normally distributed)

    α_0 = overall constant

    α_1 = partial regression of firing time on metals

    α_2 = partial regression of firing time on linear part of initiators

    α_{22} = partial regression of firing time on quadratic part of initiators

    α_{12} = partial regression of firing time on the interaction of metals by the linear part of initiators and so on to

    α_{23} = partial regression of firing time on the interaction of the linear part of initiators and the linear part of pressures

    $Z_1, \ldots, Z_2 Z_3$ = the corresponding orthogonal independent variables

ε = residual from the higher order terms, assumed NID $(0, \sigma^2)$

This model is obtained using orthogonal polynomials since the spacings on the quantitative factors are equal.

To use this equation as a prediction equation the correct transformed variables must be used for the X's as indicated in Chapters 1 and 2. Davies (1971, p. 519 and 520) explains the analysis through a corresponding model and Edwards (1962, Chapter 14), writes on trend analysis. All of this is useful in designing an experiment which will have as few observations as possible, yet obtain the information desired by the investigator.

Table 4.2.4 summarizes the 1 to 1 correspondence between the equation using independent variables, Z, and the analysis of variance.

TABLE 4.2.4

ANOVA using Equation (4.2.4)

Regression coefficient	Source	df	EMS
α_1	Metals (M)	1	$\sigma^2 + \phi(M)$
α_2	Linear initiators (I_ℓ)	1	$\sigma^2 + \phi(I_\ell)$
α_{22}	Quadratic initiators (I_q)	1	$\sigma^2 + \phi(I_q)$
α_{12}	M x I_ℓ	1	$\sigma^2 + \phi(MI_\ell)$
α_3	Linear pressure (P_ℓ)	1	$\sigma^2 + \phi(P_\ell)$
α_{33}	Quadratic pressure (P_q)	1	$\sigma^2 + \phi(P_q)$
α_{13}	M x P_ℓ	1	$\sigma^2 + \phi(MP_\ell)$
α_{23}	I_ℓ x P_ℓ	1	$\sigma^2 + \phi(I_\ell P_\ell)$
ε	Residual (higher order terms from the interactions)	$\dfrac{9}{17}$	$\left. \begin{array}{cc} & \text{df} \\ M \times I_q & 1 \\ M \times P_q & 1 \\ P_\ell \times I_q & 1 \\ P_q \times I_\ell & 1 \\ P_q \times I_q & 1 \\ MIP & 4 \end{array} \right\} \sigma^2$

The estimate of the standard error of the ith metal mean is

$$s_{\bar{y}_{i..}} = \left(\frac{\text{residual mean square}}{\text{number of observations for that mean = 9}}\right)^{\frac{1}{2}}$$

and similarly for other means.

Hence, with the higher order terms in the model being zero, there is an estimate of the error variance from the data with nine degrees of freedom. If the assumptions are correct, the design of this experiment is very "efficient," that is, few observations have provided the estimates desired for the model; therefore there is a reasonably good error estimate.

If, on the other hand, the experimenter is uncertain that the higher order terms are zero and feels that he must be assured that the residual source with the 9 degrees of freedom is a good estimate of the experimental error, a few treatment combinations, say six, could be repeated (the same concept as given before). After this better estimate of experimental error is obtained, say with the 6 degrees of freedom, the residual from regression (or the lack of fit) could be tested using the estimate of the experimental error from the six repeated treatment combinations as the error. It would be advisable to use the probability of a Type I error of 0.25 before pooling the residual error with the experimental error because under these experimental conditions and reasonable alternative hypotheses the probability of the Type II error, β, is usually quite large for small values of α. Pooling is accomplished by adding together the SS of each source and the corresponding degrees of freedom.

References for pooling in analysis of variance are Bozivich, et al. (1956) and Winer (1971, Section 5.16).

A less desirable approach is to obtain one degree of freedom for nonadditivity given in Chapter 1 to break certain interaction sources into nonadditive and error components.

Example 4.1: Calculation of the ANOVA table and investigation of the relationship between an orthogonal and nonorthogonal regression model for a completely randomized design, and the data for this two factor design with coded levels of temperature and calcium content are given below. The measured response, y, is time-measured in hours (cell totals are circled).

Model $y_{ijk} = \mu + C_i + T_j + CT_{ij} + \varepsilon_{(ij)k}$

          $i = 1, 2, 3 \quad j = 1, 2, 3, 4 \quad k = 1, 2$

Temperature

		1		2		3		4	Total
Calcium Content	1	6 8	(14)	7 9	(16)	15 19	(34)	26 27	(53)	$T_{1..} = 117$
	2	7 9	(16)	8 8	(16)	15 17	(32)	25 24	(49)	$T_{2..} = 113$
	3	5 8	(13)	10 9	(19)	21 23	(44)	32 35	(67)	$T_{3..} = 143$
Total		$T_{.1.} = 43$		$T_{.2.} = 51$		$T_{.3.} = 110$		$T_{.4.} = 169$	$T_{...} = 373$

The calculation of the sums of squares and the ANOVA table follows:

$$\sum_i \sum_j \sum_k y^2_{ijk} = 6^2 + 8^2 + \ldots + 35^2 = 7683$$

$$CT = \frac{T^2_{...}}{2\times3\times4} = \frac{(373)^2}{24} = 5797.04$$

$$SS\ Calcium = \frac{1}{8} (117^2 + 113^2 + 143^2) - CT = 66.33$$

$$SS\ Temperature = \frac{1}{6} (43^2 + 51^2 + 110^2 + 169^2) - CT = 1721.46$$

$$SS\ Cells = \frac{1}{2} (14^2 + 16^2 + \ldots + 67^2) - CT = 1857.46$$

SS C x T = SS Cells - SS Calcium - SS Temperature = 69.67

SS Total = $6^2 + 8^2 + 7^2 + \ldots + 35^2$ - CT = 1885.96

SS Error = SS Total - SS Cells = 28.50

Source	df	SS	MS	EMS	F
Calcium (C)	2	66.33	33.17	$\sigma^2 + 8\phi(C)$	13.96*
Temperature (T)	3	1721.46	573.82	$\sigma^2 + 6\phi(T)$	241.61*
C x T	6	69.67	11.61	$\sigma^2 + 2\phi(CT)$	4.89*
Error	12	28.50	2.38	σ^2	
Total	23	1885.96			

*Significant at the α = 0.05 level.

The 11 degrees of freedom for the above treatments will be broken down into single degrees of freedom with the method of orthogonal polynomials which was demonstrated in Chapter 2. This will then be followed by various nonorthogonal regression models in order to relate the correspondence between the two methods. The method of calculation for the orthogonal polynomials was given in Chapter 2 and will not be reproduced here. The nonorthogonal regression equations were generated with a canned computer program and so no method of calculation will be shown here. The sums of squares and coefficient estimates for all possible orthogonal polynomials are given in the following tabulation.

Factor	Polynomial	SS	Coefficient	Coefficient estimate
C_ℓ	Z_1	42.25	α_1	1.62
C_q	Z_{11}	$\dfrac{24.08}{66.33}$	α_{11}	0.71
T_ℓ	Z_2	1591.41	α_2	3.64
T_q	Z_{22}	108.38	α_{22}	2.12
T_c	Z_{222}	$\dfrac{21.67}{1721.46}$	α_{222}	-0.42
$C_\ell \times T_\ell$	$Z_1 Z_2$	33.80	α_{12}	0.65
$C_\ell \times T_q$	$Z_1 Z_{22}$	0.00	α_{122}	0.00
$C_\ell \times T_c$	$Z_1 Z_{222}$	0.45	α_{1222}	-0.08
$C_q \times T_\ell$	$Z_{11} Z_2$	35.27	α_{112}	0.38
$C_q \times T_q$	$Z_{11} Z_{22}$	0.00	α_{1122}	0.00
$C_q \times T_c$	$Z_{11} Z_{222}$	$\dfrac{0.15}{69.67}$	α_{11222}	-0.02

Fitting a regression equation to the calcium content factor produces the following equation and ANOVA table:

$$y = 19.38 - 6.88C + 2.12C^2$$

Source	df	SS	MS
Regression	2	66.33	33.16
Residual	21	1819.63	86.65
Total	23	1885.96	

Note that the sum of squares due to regression on calcium content is the same as that obtained for the orthogonal fit and that obtained for calcium content in the analysis of variance. A similar result is obtained for the temperature effect as shown in the following equation and ANOVA table:

$$y = 23.072 - 27.376\ T + 12.917\ T^2 - 1.439\ T^3$$

Source	df	SS	MS
Regression	3	1721.46	573.82
Residual	20	164.50	8.22
Total	23	1885.96	

One might think that a regression equation with the cross product terms might produce the sums of squares due to regression equal to the CT interaction sum of squares (69.67). The equation for all the cross products terms and the ANOVA table follow:

$$y = 19.830 - 21.902\ CT + 9.495\ CT^2 - 0.953\ CT^3$$
$$+ 4.700\ C^2T - 1.828\ C^2T^2 + 0.168\ C^2T^3$$

Source	df	SS	MS
Regression	6	1773.00	295.50
Residual	17	112.96	6.64
Total	23	1885.96	

As seen in the above ANOVA table, the sum of square due to regression is 1773, not 69.67. The reason for this result is that the cross product sum of squares includes some of the main effects of both calcium content and temperature as well as their interaction effect for this nonorthogonal regression case.

Fitting a regression model containing all 11 terms produces the following:

$$y = 11.914 + 12.35\ C - 2.91\ C^2 - 18.67\ T + 11.31\ T^2$$
$$- 1.23\ T^3 - 6.99\ CT - 0.44\ CT^2 + 0.06\ CT^3 + 1.15\ C^2T$$
$$+ 0.53\ C^2T^2 - 0.07\ C^2T^3$$

Source	df	SS	MS
Regression	11	1857.46	168.86
Residual	12	28.50	2.37
Total	23	1885.96	

The sum of squares due to regression is now the same as the sum of
squares due to the calcium content, temperature main effects and
the interaction combined.

In summary it should be pointed out that Marascuilo and Levin
(1970) show that one cannot examine interaction effects individually
without removing the influence of the main effects first. The
problem encountered here is of this general nature. Given that the
SS due to total regression for calcium content + temperature
+ CT interaction is 1857.46 and knowing the SS due to regression
for calcium content and temperature is 1787.79, we find that the SS
due to regression for CT interaction is the difference, 69.67. This
CT interaction SS cannot be obtained directly using nonorthogonal
regression.

Problem 4.2.2. (a) Utilizing the data of Problem 2.2.4,
determine by means of nonorthogonal regression procedures the
sums of squares due to individual effects of temperatures,
supplements, TS interaction, and TS cross-product terms.
     (b) Interpret your results.
     (c) If you answered Problem 2.2.4, explain what additional
information, if any, you obtained in (a) above.

Problem 4.2.3. If an experimenter ran an analysis of variance
on data from his experiment and all main effects and interactions
turned out to be significant, describe the procedure you would
recommend before the results are written up for publication.

Problem 4.2.4. An agronomy experimenter was interested in the
amylose content of rice. Three varieties of rice each given
five different levels (equal spaced) of fertilizer were to be
grown on an agronomy experimental farm. The experimenter had
the option of using from 15 to 30 plots for the experiment and
sampling the plants as he desired from each plot before analyzing
the rice for amylose in the laboratory.
     The laboratory analysis procedure involves gradually heating
starch granules under a microscope until they lose their
reflecting ability (birefringence) and recording the temperature

at the time of birefringence. The variable to be analyzed is
a function of the recorded temperature which in turn is a
function of the amylose content.
    (a) Discuss how you would design this experiment
describing the various places randomization should occur.
    (b) Show the model and outline the analysis for your
experiment including inferences you want to make.
    (c) If the experimenter wanted a second-degree regression
model, show the procedure you would use to handle the
qualitative factor and write out the model.
    (d) Consider the role of locations and explain how the
experimenter can expand his inference space if more locations
were used.

Problem 4.2.5. Consider a two factor experiment in which factor
A has five equally spaced levels and factor B has four
qualitative levels. If the third and higher degree terms are
considered negligible, design a good experiment and show the
analysis for it, including appropriate means and estimated
standard errors.

Problem 4.2.6. Fabricate data for a three factor experiment
which has a significant three factor interaction but the sums
of squares for all main effects and two factor interactions are
equal to zero. For simplicity use only two observations per
cell and two levels per factor.

4.2.2 Some Factors Fixed and Some Random

    For the explosive switch example consider the influence on the
design of the experiment if the factor, metals, is random. The
randomization procedure for the factorial experiment is the same as
that for the fixed model once the two metals are picked from a large
population of possible metals, but the inference space has changed.
Let us assume that there are at least one hundred possible metals so
that, for practical purposes, the model is infinite. Again we have
only one observation per treatment combination; the analysis is
given in Table 4.2.5.

    For mixed models see Anderson and Bancroft (1952, Chapter 23),
Hicks (1973, Chapter 10), Winer (1971, Chapter 7), Edwards (1962,
Chapter 17) and John (1971, Section 4.7).

TABLE 4.2.5

ANOVA

Source	df	EMS
Metals (M)	1	$\sigma^2 + 9\,\sigma_M^2$
Initiators (I)	2	$\sigma^2 + 3\,\sigma_{MI}^2 + 6\phi(I)$
M x I	2	$\sigma^2 + 3\,\sigma_{MI}^2$
Pressures (P)	2	$\sigma^2 + 3\,\sigma_{MP}^2 + 6\phi(P)$
M x P	2	$\sigma^2 + 3\,\sigma_{MP}^2$
I x P	4	$\sigma^2 + \sigma_{MIP}^2 + 2\phi(IP)$
M x I x P	4	$\sigma^2 + \sigma_{MIP}^2$
Error	0	σ^2

To test initiators one should use the interaction of metals x initiators, to test pressures the correct test is against metals x pressures, and to test the interaction of initiators by pressures the correct error is the mean square of metals x initiators x pressure. In general, for any factorial experiment with one factor random and the remaining factors fixed, all the main effects and interactions of fixed factors are tested by the corresponding interaction of the fixed part and the random one. The random main effect and the interactions of the random factor by the fixed parts are tested against the error.

If experimenters are not interested in the random components other than to provide an inference base for the fixed factors, there usually is little need to have more than one experimental unit per treatment combination. If however, there is interest in estimating the variance component of the random main effect (here σ_M^2), then an estimate of σ^2 must be provided. In this latter case a regression model dealing only with the fixed factors may be desired and a residual could be obtained to estimate σ^2; but it would be preferable to take a partial replication to obtain an estimate of σ^2 directly (not assuming the higher order terms are zero for the residual).

Here, as in the fixed model, the design procedures demonstrated are easily extended to any complete factorial; however, in the mixed factorial, attention must be given to the various combinations of random and fixed factors in the analysis (the expected mean squares) before the assumptions on higher order terms are made and pooling procedures tried.

Example 4.2: Calculation of the ANOVA table for the three factor experiment on explosive switches, and the coded data for firing time are given in the following table. The only difference between this example and that shown in Table 4.2.5 is that in this example there are two switches per treatment combination (cell totals are circled) instead of one.

Metals (random)

		1				2		
		Initiators (fixed)				Initiators		
		1	2	3		1	2	3
	1	44 39 (83)	27 20 (47)	35 30 (65)		12 7 (19)	15 10 (25)	22 15 (37)
Pressures (fixed)	2	48 40 (88)	25 21 (46)	29 34 (63)		6 11 (17)	12 17 (29)	27 22 (49)
	3	43 41 (84)	28 22 (50)	31 38 (69)		7 12 (19)	11 13 (24)	21 19 (40)

Model: $Y_{ijk\ell} = \mu + M_i + I_j + MI_{ij} + P_k + MP_{ik} + IP_{jk} + MIP_{ijk} + \varepsilon_{(ijk)\ell}$

$i = 1,2 \qquad j = 1, 2, 3 \qquad k = 1, 2, 3 \qquad \ell = 1, 2$

Totals:

Metals	Initiators	Pressure	Grand
$T_{1...} = 595$	$T_{.1..} = 310$	$T_{..1.} = 276$	$T_{....} = 854$
$T_{2...} = 259$	$T_{.2..} = 221$	$T_{..2.} = 292$	$n = 36$
	$T_{.3..} = 323$	$T_{..3.} = 286$	

Metals by Initiators   Metal by Pressure

$T_{11..} = 255$ $T_{21..} = 55$ $T_{1.1.} = 195$ $T_{2.1.} = 81$

$T_{12..} = 143$ $T_{22..} = 78$ $T_{1.2.} = 197$ $T_{2.2.} = 95$

$T_{13..} = 197$ $T_{23..} = 126$ $T_{1.3.} = 203$ $T_{2.3.} = 83$

Initiators by Pressure

$T_{.11.} = 102$ $T_{.21.} = 72$ $T_{.31.} = 102$

$T_{.12.} = 105$ $T_{.22.} = 75$ $T_{.32.} = 112$

$T_{.13.} = 103$ $T_{.23.} = 74$ $T_{.33.} = 109$

$$\sum_i \sum_j \sum_k \sum_\ell y^2_{ijk\ell} = 25196$$

$$CT = \frac{T^2_{....}}{n} = 20258.78$$

Sums of squares:

$$SS\ Total = 25196 - CT = 4937.22$$

$$SS\ Metals = \frac{1}{18}(595^2 + 259^2) - CT = 3136.00$$

$$SS\ Initiators = \frac{1}{12}(310^2 + 221^2 + 323^2) - CT = 513.72$$

$$SS\ Pressure = \frac{1}{12}(276^2 + 292^2 + 286^2) - CT = 10.89$$

$$SS \ M \ x \ I = \frac{1}{6} \sum_i \sum_j T^2_{ij..} - CT - SS \ Metals - SS \ Initiators$$

$$= \frac{1}{6} (255^2 + 143^2 + 197^2 + 55^2 + 78^2 + 126^2)$$

$$- CT - SS \ M - SS \ I$$

$$= 24878.00 - 20258.78 - 3126.00 - 513.72$$

$$= 969.50$$

$$SS \ M \ x \ P = \frac{1}{6} (195^2 + 197^2 + 203^2 + 81^2 + 95^2 + 83^2)$$

$$- CT - SS \ Metals - SS \ Pressure$$

$$= 23419.67 - 20258.78 - 3136.00 - 10.89$$

$$= 14.00$$

$$SS \ I \ x \ P = \frac{1}{4} (102^2 + 105^2 + \ldots + 109^2)$$

$$- CT - SS \ Initiators - SS \ Pressures$$

$$= 20788 - CT - SS \ I - SS \ P$$

$$= 4.61$$

$$SS \ M \ x \ I \ x \ P = \frac{1}{2} \sum_i \sum_j \sum_k T^2_{ijk.} - CT - SS \ M - SS \ I - SS \ P - SS \ M \ x \ I$$

$$- SS \ M \ x \ P - SS \ I \ x \ P$$

$$= \frac{1}{2} (83^2 + 88^2 + 84^2 + 47^2 + \ldots + 40^2) - 24907.50$$

$$= 24946 - 24907.50$$

$$= 38.50$$

$$SS \ Error = SS \ Total - SS \ M - SS \ I - SS \ P - SS \ M \ x \ I$$

$$- SS \ M \ x \ P - SS \ I \ x \ P - SS \ M \ x \ I \ x \ P$$

$$= 4937.22 - 4687.22$$

$$= 250.00$$

The resulting ANOVA table is as follows:

Source	df	SS	MS	EMS	F
Metals	1	3136.00	3136.00	$\sigma^2 + 18\,\sigma_M^2$	225.8*
Initiators	2	513.72	256.86	$\sigma^2 + 6\,\sigma_{MI}^2 + 12\phi(I)$	<1
M x I	2	969.50	484.75	$\sigma^2 + 6\,\sigma_{MI}^2$	34.9*
Pressure	2	10.89	5.44	$\sigma^2 + 6\,\sigma_{MP}^2 + 12\phi(P)$	<1
M x P	2	14.00	7.00	$\sigma^2 + 6\,\sigma_{MP}^2$	<1
I x P	4	4.61	1.15	$\sigma^2 + 2\,\sigma_{MIP}^2 + 4\phi(IP)$	<1
M x I x P	4	38.50	9.62	$\sigma^2 + 2\,\sigma_{MIP}^2$	<1
Error	18	250.00	13.89	σ^2	
Total	35	4937.22			

*Significant at the 0.05 level.

Problem 4.2.7.   Interpret the results of Example 4.2.

Problem 4.2.8.   Let metals and initiators be random and pressures be fixed.
     (a) Show the model and analysis with one experimental unit for each treatment combination.
     (b) Using (a) above, design an efficient experiment using more than one experimental unit for each treatment combination if it is warranted, and show the appropriate analysis.

4.2.3 All Random Factors

Using the same explosive switch example as given previously, and assuming an infinite model (the levels of the quantitative factors will not be evenly spaced if selected at random), we have the analysis shown in Table 4.2.6.

TABLE 4.2.6

ANOVA

Source	df	EMS
Metals (M)	1	$\sigma^2 + \sigma_{MIP}^2 + 3\,\sigma_{MI}^2 + 3\,\sigma_{MP}^2 + 9\,\sigma_M^2$
Initiators (I)	2	$\sigma^2 + \sigma_{MIP}^2 + 2\,\sigma_{IP}^2 + 3\,\sigma_{MI}^2 + 6\,\sigma_I^2$
M x I	2	$\sigma^2 + \sigma_{MIP}^2 + 3\,\sigma_{MI}^2$
Pressures (P)	2	$\sigma^2 + \sigma_{MIP}^2 + 2\,\sigma_{IP}^2 + 3\,\sigma_{MP}^2 + 6\,\sigma_P^2$
M x P	2	$\sigma^2 + \sigma_{MIP}^2 + 3\,\sigma_{MP}^2$
I x P	4	$\sigma^2 + \sigma_{MIP}^2 + 2\,\sigma_{IP}^2$
M x I x P	4	$\sigma^2 + \sigma_{MIP}^2$
Error	0	σ^2
Total	17	

The two-factor interactions are all tested against the three factor
interaction, and all main effects require approximate F tests which
are sometimes called F' tests, Cochran (1951), Anderson and Bancroft
(1952, p. 350), Ostle (1963, p. 302).   For example, to test
H: $\sigma_M^2 = 0$ use

$$F' = \frac{MS(\text{metals}) + MS(M \times I \times P)}{MS(M \times I) + MS(M \times P)}$$

since the EMS(metals) + EMS(M x I x P) contains only the term $9\,\sigma_M^2$
in addition to the terms in EMS(M x I) + EMS(M x P).

To obtain the degrees of freedom for either the numerator or
the denominator, one may use the Satterthwaite (1946) approximation.

$$df = \frac{M^2}{[\dfrac{(a_1 M_1)^2}{f_1} + \dfrac{(a_2 M_2)^2}{f_2} + \ldots + \dfrac{(a_k M_k)^2}{f_k}]}$$

where

$$M = \sum_{i=1}^{k} a_i M_i$$

M_i = i^{th} mean square in the numerator if the degrees of freedom
are being calculated for the numerator or similarly if the
denominator degrees of freedom are being calculated

a_i = coefficient of the ith mean square

f_i = degrees of freedom for the ith mean square

Hence all main effects and two-factor interactions can be tested
using only one experimental unit per treatment combination. This is
a very efficient design because no assumptions on residuals are
necessary. Of course, if the three factor interaction is of interest
there must be partial replication to provide an estimate of σ^2.

Of all the models fixed, mixed, and random; the random model
allows a design with the fewest assumptions after the factors are
declared random. Thus it is absolutely necessary for the experimenter
to know whether each factor will be random or fixed before an
appropriate design is prescribed for factorial experiments.

If all factors are random there should be no regression trend
analysis. Usually the estimation of the variance components is more
important than the test of hypotheses for a random model. It is
clear that the above design provides estimates of the variance
components except for σ^2_{MIP} and σ^2, separately, with one experimental
unit per treatment combination. A good reference for variance
component analysis is Anderson and Bancroft (1952).

As in the fixed model, the design methods demonstrated for the
random three factor-factorial are extended in a straight forward
manner to any sized complete factorial. Remember the design is
completely randomized.

Problem 4.2.9. The random model example on explosive switches had the variable to be analyzed as a transformation of firing time. The data were analyzed with the following results:

ANOVA

Source	df	MS
Metals	1	530
Initiators	2	225
M x I	2	46
Pressures	2	35
M x P	2	28
I x P	4	10
M x I x P	4	5

(a) Test the various hypotheses, $\sigma_M^2 = \sigma_I^2 = \sigma_{MI}^2 = \sigma_P^2 = \sigma_{MP}^2 = \sigma_{IP}^2 = 0$, separately, using the F' test and pooling techniques where appropriate.
(b) Explain the results.
(c) Comment on a possible new model.

4.2.4 Spacing Factor Levels

For random factors there is no problem on spacing because the levels are not purposely spaced and there is no interest in means. For fixed qualitative factors it is obvious that there is no problem on spacing, but there usually is interest in the means. Finally, for fixed quantitative factors there is a problem on spacing, and there usually is interest in the means.

Most experimenters think they should arithmetically equally space the levels of fixed quantitative factors. For example, the four levels of the factor or independent variable (X) for fertilizer treatments may be the percent of nitrogen 0, 20, 40, 60. The main reason for using equally spaced factor (X) levels is to use orthogonal polynomials when the trend is not known in hopes that the trend may be predicted from the data. If the trend is known to be linear, and

the research worker is only interested in the slope of the line,
placing half the observations at either extreme of the factor space
(e.g., at zero and 60 for the nitrogen example) would give him
maximum information because $\sigma_{b_1}^2 = \dfrac{\sigma^2}{\Sigma(X_i-\overline{X})^2}$ and $\Sigma(X_i-\overline{X})^2$ is
maximized when the X's are at the extremes. As more information is
available to indicate the function is other than polynomial, the
experimenter should depart from the usual arithmetic equal-spacing
idea.

If the trend is exponential in nature the polynomial may be
quite a poor predictor and arithmethic equal spacing is not the best.
Usually it is better to follow a geometric spacing rather than an
arithmetic one if an exponential relationship is suspected. In
general, there should be a method for testing departure from the
exponential. After transforming the independent variable (or factor
levels) to logarithms the usual orthogonal polynomial trend analysis
may be used, but the analysis is on the transformed independent
variable, Li (1964, pp. 205-209), and Finney (1960, pp. 170-173).

Consider the example in which the relationship between y, the
surviving offspring, and X, the dosage of X-rays to the parents, is
thought to be of the general form $e^y = aX^b$.

The experimenter may select doses of 100, 200, 400, and 800
roentgens for a fixed time and examine the trend by testing a
departure from the exponential. Of course it can be seen that if
the logarithm of dosages are taken and the departure from a straight
line is then investigated, the experimenter has really examined the
departure from the exponential. If the trend is not found to depart
significantly from the exponential, the coefficients a and b can be
estimated using the transformed X variable.

Problem 4.2.10. In an experiment using flour beetles
(Tribolium castaneum), the effect of X-rays on progeny numbers
was to be investigated. The relationship between progeny
number (y) and dosage of X-ray (X) was thought to be

$$e^y = aX^b.$$

The lower limit of X-rays was set at 100 roentgens for a fixed time and the upper limit was set at 1000 roentgens for the same fixed time. If there were going to be five levels of X-ray doses;

(a)   What should the other three levels of roentgens be in order to allow an easy test for departure from linearity on the transformed scale?

(b)   Using the levels of (a) above, the experimenter obtained the following data:

Dosage level	Average progeny number per family
1	25.8
2	22.5
3	21.2
4	19.0
5	17.5

Assuming there were 10 families for each dosage level in a CR design and that the error mean square was 13.0 show the analysis of the data including the prediction model for an exponential investigation. That is, check for linearity on the transformed scale. If you accept linearity estimate a and b and then write down the appropriate exponential prediction model.

4.3   REFERENCES

Anderson, R. L. and Bancroft, T. A.   Statistical Theory in Research, McGraw-Hill, New York, 1952.

Bozivich, H., Bancroft, T. A. and Hartley, H. O.   Ann. Math. Stat. 27:1017 (1956).

Cochran, W. G.   Biometrics 7:17 (1951).

Cochran, W. G. and Cox, G. M.   Experimental Designs, 2nd ed., Wiley, New York, 1957

Cox, D. R.   Planning of Experiments, Wiley, New York, 1958.

Davies, O. L. (Ed.).   Design and Analysis of Industrial Experiments, 2nd ed., Oliver and Boyd, London, or Hafner, New York, 1971.

Edwards, A. L.   Experimental Design in Psychological Research, Rinehart, New York, 1962.

Finney, D. J.   An Introduction to the Theory of Experimental Designs, University of Chicago Press, Chicago, Ill., 1960.

Hader, R. J.   The Am. Statistician 27, 82 (1973).

Hicks, C. R. Fundamental Concepts in the Design of Experiments, 2nd
ed., Holt, Rinehart and Winston, New York, 1973.

John, P. W. M. Statistical Design and Analysis of Experiments,
Macmillan, New York, 1971.

Kempthorne, O. Design and Analysis of Experiments, Wiley, New York,
1952. Distributed by Krieger Pub. Co., Huntington, New York.

Li, Jerome C. R. Statistical Inference II. Edwards Brothers, Inc.
Ann Arbor, Michigan, 1964.

Mandel, J. Statistical Analysis of Experimental Data, Wiley, New
York, 1964.

Marascuilo, L. A. and Levin, J. R. Am. Ed. Res. J. 7:397 (1970).

Ostle, Bernard Statistics in Research, 2nd ed., Iowa State University
Press, Ames, Iowa, 1963.

Prairie, R. R. and Zimmer, W. J. JASA 59:1205 (1964).

Satterthwaite, F. E. Biometrics 2:110 (1946).

Winer, B. J. Statistical Principles in Experimental Design, 2nd ed.,
McGraw-Hill, New York, 1971.

Chapter 5

RANDOMIZED COMPLETE BLOCK DESIGN (RCBD)

Basically, the randomized complete block design is a group of completely randomized designs (usually, however, each CRD has only one experimental unit per treatment combination). Ordinarily each member of the group contains a sufficient number of homogeneous experimental units to accommodate a complete set of treatment combinations. This collection of experimental units is referred to as a block. One major reason for blocking is that the experimenter does not have a sufficient number of homogeneous experimental units available to run a completely randomized design with several observations per treatment combination.

A realistic situation where blocking would be required would be in an industrial type experiment where there is only sufficient time to run one set of treatment combinations per day. In such an experiment one would be blocking on time and the number of days that the experiment is carried out would be the number of blocks. This type of experiment points out another reason for blocking which is the technique that is utilized to expand the inference space. Even if it was possible to run a large number of each treatment combination on one day the experimenter might still choose to block over time so that he could make inferences over time rather than to just one particular day.

If an experimenter were to run a RCBD and analyze it as a CRD, then any effects which should have been attributed to blocks would end up in the error term of the model. Thus another major reason for blocking is to remove a source of variation from the error. It should be observed that if the experimental unit variation among

123

blocks is not larger than the variation within blocks, there is no
reason to run a RCBD rather than a CRD.

The group of blocks, which may be random or fixed, make up the
design called the randomized complete block design (RCBD). These
blocks are called "complete" because all treatments or treatment
combinations appear in each block. As a result of the construction
of this design, there is one restriction on the randomization in that
the randomization of treatments onto the experimental units is carried
out within each block separately, not over all the experimental units
at one time as with a CRD.

The single most important concept introduced in this text is the
usefulness of the random component called a "restriction error"
(Anderson, 1970). We place this component (where appropriate) into
the mathematical model in order to present a realistic picture of the
experimental situation. The component is common to a group of
experimental units isolated by the experimental procedure. It is
random in the sense that if the experimenter were to recreate the
same group of units, this component would be different.

By the very nature of the confinement of the experimental units
to a given group, any randomization of the treatments onto the
experimental units must be restricted to only those units that are
present in that grouping. This situation occurs in all RCBD, and
the restriction error component is completely confounded with blocks.

In many other experimental situations treatments cannot be
randomized onto the experimental units within the blocks or groupings.
In this nonrandom case the restriction error is still present because
the grouping of the experimental units still exists and the effect
of blocks is still completely confounded with the restriction error.

The restriction error is not estimable from the data but it is
placed in the model (appropriately indexed) and is allowed to appear
in the corresponding analysis of variance as a source of variation.
There are no degrees of freedom (df) and no sum of squares (SS) for
the error; however, since the restriction error appears in the

theoretical linear model, the variance component for it does appear in the expected mean squares (EMS). This variance component in the EMS forces the experimenter to recognize the restriction on randomization and account for it in the F-tests. The real power of the restriction error is that it makes the experimenter think about the restriction on randomization he has imposed on his design (usually to save time and/or money) and to understand its effect on the overall results of the experiment. Almost always that effect is to decrease the number of degrees of freedom for the appropriate error to test the most important factor.

It is interesting to note that the restriction error does not disturb the algorithm presented in Chapter 2, and referenced in Bennett and Franklin (1954) and more recently by Hicks (1973), to derive the expected mean squares given the linear model. The EMS are still obtained in the same straightforward manner and are correct when restriction errors are in the model.

To explain the influences that the restriction error has on analysis of data from a RCBD, treatments will be considered fixed and the two cases, blocks fixed and random, will be covered in detail in the following sections. It is left to the reader to use the material from Chapter 2 to handle the cases in which treatments are random.

5.1 BLOCKS FIXED

Suppose that b levels of one fixed set of treatments and t levels of another fixed set are run in all possible bt combinations on bt experimental units completely at random. The design of such an experiment is called a completely randomized design (CRD). The linear model for the analysis of the data from such an experiment, if the interaction of blocks and treatments is zero, is usually expressed as:

$$y_{ij} = \mu + B_i + T_j + \varepsilon_{(ij)} \qquad \begin{matrix} i = 1, 2, \ldots, b \\ j = 1, 2, \ldots, t \end{matrix} \qquad (5.1.1)$$

where

y_{ij} = the response from the experimental unit treated with the ith level of treatment B and the jth level of treatment T

μ = overall mean

B_i = the effect of the ith level of (fixed) treatment B

T_j = the effect of the jth level of (fixed) treatment T

$\varepsilon_{(ij)}$ = the (random) within error

An estimate of the error mean square may come from the interaction source. Assume all the analysis of variance assumptions hold, so that the $\varepsilon_{(ij)}$ is NID $(0, \sigma^2)$, i.e., normally and independently distributed with mean zero and variance σ^2. For this case the appropriate analysis of variance (ANOVA) is given in Table 5.1.1.

To be able to write Eq. (5.1.1) one must assume there are no restrictions on randomizations in the design of the experiment. This may be called a two-way factorial completely randomized design.

Suppose, next, there are b blocks in an experiment, and t fixed treatments are completely randomized onto t experimental units in each of the b blocks such that there is a different randomization in each block. The equation for analyzing the data from such an experiment has been given by most authors and implied by many others to be the same as Eq. (5.1.1). This design is certainly not completely randomized because there is a different randomization of the treatments in each block, not just one randomization over the whole experiment as would be demanded for a CRD. In general, statisticians agree that this is a randomized complete block design (RCBD); however, they do not always agree on the model to analyze the data. Wilk (1955), Addelman (1969), and others have suggested using a model for a generalized randomized block design which includes the usual CRD and RCBD. We wish to stress restriction errors and in this section we deal with an equation that accounts for errors between blocks in contrast to Eq. (5.1.1) that does not account for the errors.

TABLE 5.1.1

ANOVA for the Data from the CRD Using Equation (5.1.1)

Source	df	EMS
Treatments (B_i)	$b - 1$	$\sigma^2 + t\phi(B)$
Treatments (T_j)	$t - 1$	$\sigma^2 + b\phi(T)$
Within error $(\varepsilon_{(ij)})$	$(b - 1)(t - 1)$	σ^2
(assuming interaction BT is zero)		

Since there is a different randomization of treatments within each block, we recognize this peculiarity by making sure that the error to test the hypothesis of equality among the treatment means comes from within the blocks. On the other hand, if there is interest in testing the hypothesis that block means are equal, we insist that there must be an error among the blocks. The former test (for treatment means) would be satisfied using the error designated in Eq. (5.1.1), but the latter test (for block means) is not satisfied using the same error.

One linear model that expresses all the sources of variation mentioned above with the assumption of no interaction of blocks by treatments and also provide intuitively correct tests is:

$$y_{ijk\ell} = \mu + B_i + \delta_{(i)k} + T_j + \varepsilon_{(ijk)\ell}$$

$$i = 1,\ldots, b \quad j = 1,\ldots, t \quad k = 1 \quad \ell = 1$$

(5.1.2)

where

$y_{ijk\ell}$ = the response from the ℓth experimental unit in block i, the kth randomization, and given the jth treatment

μ = overall mean

B_i = effect of the ith block. In this case B_i may be random or fixed, but we will assume it is fixed for the ANOVA

$\delta_{(i)k}$ = the kth restriction error within the ith block, NID $(0, \sigma_\delta^2)$, and completely confounded with B_i since k must always equal 1. In general one does not include a subscript when the range is 1; however, it is included here to indicate that there are many errors possible but only one is realized. This term will have zero degrees of freedom and no sum of squares, but must appear in the model and expected mean squares in order to distinguish this design from the CRD related to Eq. (5.1.1). This term is the result of the restriction on the randomization of the treatments onto the ith block's experimental units,

T_j = the effect of the jth treatment (fixed),

$\varepsilon_{(ijk)\ell}$ = effect of the ℓth random error associated with the ij^{th} experimental unit and subjected to the kth restriction on the ith block. Again, ℓ must always equal 1, and hence is usually not included in the model. It is assumed that $\varepsilon_{(ijk)\ell}$ is NID $(0, \sigma^2)$. The variance σ^2 is estimated in the analysis of variance from the usual source since we assumed that the block x treatment interaction is zero in this experiment.

The analysis of variance for this model, along with the EMS derivation is given in Table 5.1.2.

Ordinarily one does not include the subscripts k and ℓ in a model of this type since they have a range of only 1. Note, however, that it is necessary to include these subscripts in order to make the EMS algorithm function properly. For all models of this type from now on we will not include the subscripts that have the range of one except for cases that are more clearly described using them, but it will be understood that the procedure used in Table 5.1.2 will be inherent in the derivation of the EMS.

The corresponding model for Eq. (5.1.2) not using the subscripts k and ℓ is:

TABLE 5.1.2

ANOVA for Data from RCBD Using Equation (5.1.2)

df	Source	b F i	1 R k	t F j	1 R ℓ	EMS
b - 1	B_i	0	1	t	1	$\sigma^2 + t\sigma_\delta^2 + t\phi(B)$
0	$\delta_{(i)k}$	1	1	t	1	$\sigma^2 + t\sigma_\delta^2$
t - 1	T_j	b	1	0	1	$\sigma^2 + b\phi(T)$
(b - 1)(t - 1)	$\varepsilon_{(ijk)\ell}$	1	1	1	1	σ^2

$$y_{ij} = \mu + B_i + \delta_{(i)} + T_j + \varepsilon_{(ij)} \qquad \begin{array}{l} i = 1, 2, \ldots, b \\ j = 1, 2, \ldots, t \end{array} \qquad (5.1.3)$$

where the terms are defined as they were for Eq. (5.1.2) except that k and ℓ are deleted.

To emphasize the effect of the restriction error on the various F-tests in the ANOVA, we introduce a line immediately below each source of variation called "restriction error."

It can be seen from Table 5.1.3 that the correct error for blocks is the restriction error with zero df not the within error with (b - 1)(t - 1) df as suggested in Table 5.1.1.

TABLE 5.1.3

ANOVA for Data from RCBD Using Equation 5.1.3

Source	df	EMS
Blocks (B_i)	b - 1	$\sigma^2 + t\sigma_\delta^2 + t\phi(B)$
Restriction error $(\delta_{(i)})$	0	$\sigma^2 + t\sigma_\delta^2$
Treatments (T_j)	t - 1	$\sigma^2 + b\phi(T)$
Within error $(\varepsilon_{(ij)})$	(b - 1)(t - 1)	σ^2

Note: The restriction error is represented in the theoretical model but cannot be estimated from the data in this experiment. All of this is recognized in Table 5.1.3 by writing down the source (restriction error), showing zero (0) df to depict the lack of data and expressing the theory in the EMS; also, there is no sum of squares or mean square. Of course, the other three sources have sums of squares that can be computed from the data.

Since there are zero df and no sum of squares for the restriction error, there is no test for blocks in this experiment. Many authors[*] have indicated that there would be no test for blocks for various reasons, but none has been specific to indicate it from the model as we attempt to do here. It should be pointed out that many times the experimenter is interested only in whether or not blocks have been effective in reducing the estimated error to test treatments. Under these circumstances the investigator may test the combination of block effects and the block error (our restriction error) to see if it is zero. That is, the hypothesis could be stated from Table 5.1.2 as H_o : $\sigma_\delta^2 + \phi(B) = 0$. It is obvious, then, that the within error is appropriate for the test and it would make no difference in the test whether Eq. (5.1.1) or (5.1.2) is considered.

It should be emphasized that the inference from such an experiment is only for this set of treatments and blocks, and that the estimated standard error of the jth treatment mean is

$$s_{\bar{y}_{.j}} = \left(\frac{\hat{\sigma}_\delta^2 + \hat{\sigma}^2}{b} \right)^{\frac{1}{2}}$$

which is impossible to estimate from this RCBD. It is interesting to notice, however, that the estimated standard error of the difference of two treatment means (which is more important for individual or multiple comparisons once treatments show significance in the analysis of variance) is:

---

[*]For example, Ostle (1963, p. 368).

$$s_{(\bar{y}_{\bullet p} - \bar{y}_{\bullet q})} = \left(\frac{2\hat{\sigma}^2}{b}\right)^{\frac{1}{2}}$$

and is estimable.

The real danger in using Eq. (5.1.1) in analyzing data from RCBD occurs when the experimenter is interested in the block means in addition to the treatment means.   To be more explicit, frequently experimenters want to use treatments to represent the blocks in a RCBD and restrict randomization of another set of treatments inside the ones used as blocks.   Recognition of this restriction on randomization demands an equation other than Eq. (5.1.1).   Our suggestion is that Eq. (5.1.2) provides the basis for a more nearly correct analysis of the data from such a design.

5.1.1   A Medical-Engineering Example

An example to demonstrate this danger occurred in a medical-engineering problem (Beeson, 1965).   An engineer constructed a mechanical apparatus to simulate the circulatory system of human beings.   A storage tank was used to control the pressure of the liquid, simulating the blood, in the system.   A pulse pump squeezed flexible rubber tubing in the system to create the pulsing action needed to simulate the heart actions.   The motor on the pump could vary the pulse rate between 0 and 220 beats per minute.   In the experiment 4 prosthetic cardiac valve types and 6 different pulse rates were to be used.   If a completely randomized design were run so that each treatment combination would occur twice, the experimenter would have needed 48 valves (12 of each type).

The purpose of the experiment was to select the best valve type out of the four for all pulse rates and/or best valve type for particular pulse rates.   To a statistician this means that the main effect of valve types and the interaction of valve types by pulse rates should be examined carefully.

One of the variables to be measured and analyzed was maximum
flow gradient (mmHg) and it was found to have reasonably good
statistical properties for analysis.

The experimenter thought it was too difficult to obtain as many
as 12 valves per valve type; and, too, the experiment would take too
long to run. After all, the CRD would require that the mechanism
holding the valve type would have to be dismantled and reassembled
47 times. An example of such a CRD is the following:

Run	Valve type	Pulse rate
1.	2	4
2.	4	3
3.	3	5
4.	2	1
5.	2	4
6.	1	2
.	.	.
.	.	.
.	.	.
48.	1	6

Note: In this design a new valve and reassembly operation or
set up of the machine would be necessary for each run. Even for run
5 it would be necessary to remove the valve for type 2 and replace
it with another valve of type 2.

5.1.2 RCBD More Than One Observation Per Cell (Example)

The investigator wanted to run the experiment by drawing, at
random, one of the valve types and running, in a random order, all
12 pulse rates (6 actual pulse rates each repeated once) on only one
valve of this valve type. Next he wanted to select one of the
remaining three valve types at random and with a new random order,
run the 12 pulse rates as before on only one valve of this new valve
type. He wanted to continue this procedure for the other two valve
types and complete the experiment using this RCBD with two

observations per treatment combination of valve type by pulse rate, assuming that there is no correlation of the within treatment combination errors. Since he has used six different treatments (pulse rates) twice on each experimental unit (valve), this assumption may be violated. Detailed discussion of this topic is given in Section 6.2.1. Suffice it to say here that this assumption of no correlated errors will be used for the remainder of this chapter.

Pictorially an example of this RCBD is shown in Table 5.1.4.

TABLE 5.1.4

Layout for RCBD Cardiac Valve Example

		Pulse rates											
	2	1	3	2	1	4	2	3	5	4	6	6	5
		Run (1)	(2)	(3)	(4)	(5)	(6)	(7)	(8)	(9)	(10)	(11)	(12)
	4	5	2	3	4	2	6	1	4	5	3	6	1
		Run (13)	(14)	(15)	(16)	(17)	(18)	(19)	(20)	(21)	(22)	(23)	(24)
Valve types	1	4	2	1	6	4	3	5	2	1	6	5	3
		Run (25)	(26)	(27)	(28)	(29)	(30)	(31)	(32)	(33)	(34)	(35)	(36)
	3	6	1	5	4	1	2	3	3	5	6	2	4
		Run (37)	(38)	(39)	(40)	(41)	(42)	(43)	(44)	(45)	(46)	(47)	(48)

The model for this example of a RCBD in Table 5.1.4, if there are no correlations among errors, is:

$$y_{ijk} = \mu + V_i + \delta_{(i)} + P_j + VP_{ij} + \varepsilon_{(ij)k} \qquad (5.1.4)$$

$$i = 1, 2, 3, 4 \quad j = 1, 2, \ldots, 6 \quad k = 1, 2$$

where

y_{ijk} = k^{th} maximum flow gradient from the jth pulse rate of the ith valve type

μ = overall mean

V_i = effect of the ith valve type (fixed)

$\delta_{(i)}$ = restriction error (random) caused by all 12 pulse rates being run on the ith valve type before running any other valve type without reseating the valve type (completely confounded with the effect of the ith valve type) NID $(0, \sigma_\delta^2)$

P_j = effect of the jth pulse rate (fixed)

VP_{ij} = interaction effect of the ith valve type with the jth pulse rate

$\epsilon_{(ij)k}$ = random error caused by the kth observation of the jth pulse rate in the ith valve type, NID $(0, \sigma^2)$

The corresponding analysis of variance for this experiment is set out in Table 5.1.5. It is seen that the mean square for the restriction error (which is really a between block error and intuitively correct) would be the denominator to test valve types. In this case, however, there are zero df and no sum of squares for the restriction error; consequently there is no mean square and there is no test for one of the most important factors, valve types, when this design is used even if the correlation assumption is correct. Hence if the experimenter uses Eq. (5.1.4) in his preliminary outline of the analysis from this proposed design he will know he should change the design of his experiment before he has run any part of it.

TABLE 5.1.5

ANOVA Using Equation (5.1.4)

Source	df	EMS
Valve types (V_i)	3	$\sigma^2 + 12\sigma_\delta^2 + 12\phi(V)$
Restriction error $(\delta_{(i)})$	0	$\sigma^2 + 12\sigma_\delta^2$
Pulse rates (P_j)	5	$\sigma^2 + 8\phi(P)$
Interaction (VP_{ij})	15	$\sigma^2 + 2\phi(VP)$
Within error $(\varepsilon_{(ij)k})$	24	σ^2
Total	47	

Example 5.1: Calculation of the ANOVA table for the experiment described by Eq. (5.1.4). Even though the observations of maximum flow gradient for each pulse rate were taken randomly for each valve type (block), the coded values will be shown in the following ordered array.

Valve type (fixed blocks)	Pulse rate (treatments)						Totals
	1	2	3	4	5	6	
1	$^{12}_{\ 7}$ ⑲	$^8_5$ ⑬	$^4_7$ ⑪	$^1_5$ ⑥ $^{\ 8}_{13}$ ㉑		$^{14}_{20}$ ㉞	104
2	$^{20}_{14}$ ㉞	$^{15}_{12}$ ㉗	$^{10}_{\ 7}$ ⑰	$^8_6$ ⑭	$^{14}_{18}$ ㉜	$^{25}_{21}$ ㊻	170
3	$^{21}_{13}$ ㉞	$^{13}_{14}$ ㉗	$^8_7$ ⑮	$^5_9$ ⑭	$^{15}_{19}$ ㉞	$^{27}_{23}$ ㊿	174
4	$^{15}_{12}$ ㉗	$^{10}_{14}$ ㉔	$^8_5$ ⑬	$^6_9$ ⑮	$^{10}_{14}$ ㉔	$^{21}_{17}$ ㊳	141
Totals	114	91	56	49	111	168	589

The ANOVA table for these data follows.

Source	df	SS	MS	EMS	F
Valve types (V)	3	261.90	87.30	$\sigma^2 + 12\sigma_\delta^2 + 12\phi(V)$	None
Restriction error	0	None		$\sigma^2 + 12\sigma_\delta^2$	
Pulse rate (P)	5	1192.35	238.47	$\sigma^2 + 8\phi(P)$	28.7*
Interaction (VP)	15	55.73	3.72	$\sigma^2 + 2\phi(VP)$	< 1
Within error	24	199.50	8.31	σ^2	
Total	47	1709.48			

*Significant at the 0.05 level.

Thus one would conclude that there is a significant difference among pulse rates but no valve-pulse rate interaction. A multiple comparison test may be carried out to find out which pulse rates are different from one another. The best estimate of σ^2 would be a pooled mean square obtained as follows:

$$\hat{\sigma}^2 = \frac{199.50 + 55.73}{24 + 15} = \frac{255.23}{39}$$

$$= 6.54$$

Thus the standard error to use for the multiple range test is

$$\sqrt{\frac{6.54}{8}} = \sqrt{0.82} \doteq 0.91$$

The calculation of the sums of squares are as follows:

$$\sum_i \sum_j \sum_k y_{ijk}^2 = 12^2 + 7^2 + 20^2 + \dots + 17^2 = 8937$$

$$T_{...} = 589 \qquad CT = \frac{589^2}{48} = 7227.52$$

$$\text{SS total} = \Sigma\ \Sigma\ \Sigma y^2_{ijk} - CT = 1709.48$$

$$\text{SS valves} = \frac{1}{12} \Sigma\ T^2_{i\cdot\cdot} - CT = \frac{1}{12}\ (104^2 + 170^2 + 174^2 + 141^2)$$

$$- CT$$

$$= 7489.42 - CT = 261.90$$

$$\text{SS pulse rates} = \frac{1}{8} \Sigma\ T^2_{\cdot j\cdot} - CT = \frac{1}{8}\ (114^2 + \ldots + 168^2) - CT$$

$$= 8419.87 - CT = 1192.35$$

$$\text{SS VP} = \frac{1}{2} \Sigma\ T^2_{ij\cdot} - CT - SS\ V - SS\ P$$

$$= \frac{1}{2}\ (19^2 + 13^2 + 11^2 + \ldots + 38^2) - CT$$

$$- SS\ V - SSP$$

$$= 8737.50 - 7227.52 - 261.90 - 1192.35$$

$$= 55.73$$

$$\text{SS error} = \text{SS total} - SS\ V - SS\ P - SS\ VP$$

$$= 199.50$$

5.1.3   RCBD When Interaction Is Present and One Observation Per Cell

   In some experiments there is only one experimental unit per
treatment and only one observation on that experimental unit in a
block. If, in addition, the interaction between blocks and treatments
cannot be assumed zero the model will be

$$y_{ij} = \mu + B_i + \delta_{(i)} + T_j + BT_{ij} + \varepsilon_{(ij)} \qquad (5.1.5)$$

$$i = 1,\ 2,\ \ldots,\ b \qquad j = 1,\ 2,\ \ldots,\ t$$

where the usual definitions of the components hold. The corresponding
ANOVA is given in Table 5.1.6.

TABLE 5.1.6

ANOVA Using Equation (5.1.5)

Source	df	EMS
Blocks (B_i)	$(b - 1)$	$\sigma^2 + t\sigma_\delta^2 + t\phi(B)$
Restriction error (δ_i)	0	$\sigma^2 + t\sigma_\delta^2$
Treatments (T_j)	$(t - 1)$	$\sigma^2 + b\phi(T)$
Interaction (BT_{ij})	$(b - 1)(t - 1)$	$\sigma^2 + \phi(BT)$
Within error $(\varepsilon_{(ij)})$	0	σ^2
Total	$(bt - 1)$	

If treatments are tested by the interaction mean square, the test is conservative. By conservative we mean that the denominator of the F test may be somewhat larger than it should be. As a result, the actual α level of a conservative test may be smaller than the selected α level. If treatments turn out to be not significant, it would be desirable to use a nonadditivity test or to partially replicate in the original design.

5.2 BLOCKS RANDOM

Many experimenters would like to have the blocks in the experiment represent all possible blocks or be random. To do this they must be careful in the selection of the blocks so that what they want in their inference actually exists in the experiment. For example, if a chemist wants the results from an experiment run in his laboratory to apply to all chemistry laboratories in the world, he must verify these results under a large enough random sample of laboratories over the world. The chemist must be the one to determine how many and which laboratories are to represent all laboratories because he must be concerned about his inference space before the scientific investigation is complete.

In this case (blocks random) even if interaction between blocks and treatments exists there is a direct test for treatments when there is only one experimental unit for each treatment in a block. The general model for the analysis of b blocks and t treatments is

$$y_{ij} = \mu + B_i + \delta_{(i)} + T_j + BT_{ij} + \varepsilon_{(ij)} \qquad (5.2.1)$$

where the symbols are defined as before except that B_i is random and BT_{ij} is mixed.

TABLE 5.2.1

ANOVA Using Equation (5.2.1)

Source	df	EMS
Blocks (B)	b - 1	$\sigma^2 + t\sigma_\delta^2 + t\sigma_B^2$
Restriction error	0	$\sigma^2 + t\sigma_\delta^2$
Treatment (T)	(t - 1)	$\sigma^2 + \sigma_{BT}^2 + b\phi(T)$
B x T	(b - 1)(t - 1)	$\sigma^2 + \sigma_{BT}^2$
Within error	0	σ^2
Total	(bt - 1)	

where: σ_{BT}^2 is the "mixed" interaction component and cannot be estimated with only one experimental unit per treatment, unless one is willing to use a test for nonadditivity.

The estimate of the standard error of the treatment mean coming from this mixed model is:

$$s_{\bar{y}_{\cdot j}} = \left(\frac{\hat{\sigma}^2 + \hat{\sigma}_{BT}^2 + \hat{\sigma}_\delta^2 + \hat{\sigma}_B^2}{b} \right)^{\frac{1}{2}}$$

which is not estimable. The estimate of the standard error of the difference, however, is estimable and is:

$$s_{(\bar{y}_{\cdot p} - \bar{y}_{\cdot q})} = \left[\frac{2(\hat{\sigma}^2 + \hat{\sigma}_{BT}^2)}{b}\right]^{\frac{1}{2}}$$

which is the usual basis for making individual or multiple comparisons
after treatments are declared significant in the analysis of variance.
A derivation for this type of standard error is given in Appendix 12.

Let us now extend the mixed model case discussed in Example 2.7
to factorials. Consider the model

$$y_{ijkl} = \mu + B_i + \delta_{(i)} + M_j + BM_{ij} + I_k + BI_{ik}$$

$$+ MI_{jk} + BMI_{ijk} + P_l + BP_{il} + MP_{jl} + BMP_{ijl} \qquad (5.2.2)$$

$$+ IP_{kl} + BIP_{ikl} + MIP_{jkl} + BMIP_{ijkl} + \varepsilon_{(ijkl)}$$

where $i = 1, 2, 3$; $j = 1, 2$; $k = 1, 2, 3$, and $l = 1, 2, 3$

y_{ijkl} = firing time of the switch in the ith block or replicate
made up of the jth metal, k^{th} initiator, and l^{th}
pressure and all other symbols are defined as before,
except that blocks are random and all interactions
with blocks are similar to BT_{ij} in Eq. (5.1.5).

Allowing blocks to be random and metals, initiators and
pressures to be fixed, the appropriate denominator for the F-tests
for all fixed effects and their interactions will be the
corresponding interaction with blocks mean square. If only three
replicates are run there will be only 2 degrees of freedom for the
error to test metals. Hence the investigator must be careful to
outline the analysis before running the experiment. In this case
he almost surely would want more than two degrees of freedom for
error.

Some experimenters know before the experiment is run that all
the interactions with blocks or replicates will be about the same
size. If this is the case here, there would be 34 degrees of
freedom for a common pooled error. The degrees of freedom for the
corresponding pieces of the error are:

Source: Error = BM + BI + BMI + BP + BMP + BIP + BMIP

df: 34 = 2 + 4 + 4 + 4 + 4 + 8 + 8

If not as much information is available to the experimenter, an optional procedure is to use the sometimes pooling procedure with $\alpha = 0.25$ (Bozivich et al., 1956), after the data are obtained hoping enough mean squares will be poolable to provide sufficient degrees of freedom for the various tests to give reasonable power.

Another example in which blocks are random occurred in a quantitative genetics study. Drosophila were used to investigate chromosomal effects on body weight. The entire experiment had four genome samples drawn at random which formed the blocks or replications. Within each genome sample or block, eight heterozygous states of chromosomes, acting as treatments, were examined. The model is:

$$y_{ij} = \mu + G_i + \delta_{(i)} + H_j + GH_{ij} + \varepsilon_{(ij)} \qquad (5.2.3)$$

$$i = 1, 2, 3, 4 \qquad j = 1, 2, \ldots, 8$$

with the usual definitions of the terms in the equations.

The analysis is given in Table 5.2.2.

TABLE 5.2.2

ANOVA Using Equation (5.2.3)

Source	df	EMS
Blocks (genome sample)	3	$\sigma^2 + 8\sigma_\delta^2 + 8\sigma_G^2$
Restriction error	0	$\sigma^2 + 8\sigma_\delta^2$
Treatments (heterozygous states)	7	$\sigma^2 + \sigma_{GH}^2 + 4\phi(H)$
Interaction (genome sample by heterozygous states)	21	$\sigma^2 + \sigma_{GH}^2$
Within error	0	σ^2
Total	31	

The test on heterozygous states is against the interaction source mean square with or without $\sigma_{GH}^2 = 0$.

Example 5.2: Consider the case of Eq. (5.2.3) when blocks are genome samples and treatments are heterozygous states. The coded value of Drosophila body weight is as follows:

Genome sample	Heterozygous states (treatments)								
(random block)	1	2	3	4	5	6	7	8	Totals
1	16	13	8	7	15	3	14	7	83
2	15	15	10	6	13	6	15	10	90
3	20	18	14	10	22	9	20	14	127
4	12	14	9	5	13	1	12	5	71
Totals	63	60	41	28	63	19	61	36	371

The ANOVA table for these data follows and is based on the analysis shown in Table 5.2.2.

Source	df	SS	MS	EMS	F
Genome sample	3	218.59	72.86	$\sigma^2 + 8\sigma_\delta^2 + 8\sigma_G^2$	None
Restriction error	0	None		$\sigma^2 + 8\sigma_\delta^2$	
Heterozygous states	7	543.97	77.71	$\sigma^2 + \sigma_{GH}^2 + 4\phi(H)$	46.4*
Interaction	21	35.16	1.67	$\sigma^2 + \sigma_{GH}^2$	None
Within error	0	None		σ^2	
Total	31	797.72			

*Significant at the 0.05 level.

Thus one would conclude that there is a significant difference among States but as indicated in the test there is no exact test for Genome Samples.

Multiple comparison tests among heterozygous states means may be carried out using

$$s_{\bar{y}} = \sqrt{1.67/4} = \sqrt{0.42} \ .$$

The sums of squares calculations are as follows:

$$\sum_i \sum_j y_{ij}^2 = 16^2 + 15^2 + \ldots + 5^2 = 5099$$

$$\sum_i \sum_j y_{ij} = 371 \qquad CT = \frac{371^2}{32} = 4301.28$$

$$SS\ total = \sum_i \sum_j y_{ij}^2 - CT = 5099 - 4301.28 = 797.72$$

$$SS\ blocks = \frac{1}{8} \sum_{i=1}^{4} T_{i\cdot}^2 - CT = \frac{1}{8} (83^2 + 90^2 + 127^2 + 71^2)$$

$$- CT = 218.59$$

$$SS\ treatments = \frac{1}{4} \sum_{j=1}^{8} T_{\cdot j}^2 - CT = \frac{1}{4} (63^2 + \ldots + 36^2)$$

$$- CT = 543.97$$

$$SS\ interaction = SS\ total - SS\ blocks - SS\ treatments = 35.16$$

5.3 ALLOCATION OF EXPERIMENTAL EFFORT IN RCBD

For all RCBD there may be more than one observation for each treatment in each block if there is enough time (and/or money) and a reason from an efficient design point of view to handle the extra observations. An extra observation may just be another measurement of the same experimental unit, possibly by another technician. This type of observation may not have much value for the inference because the variation measured may be only an operator and measurement error variation. Under some circumstances it is of value to have this information, but to allow the experimenter to make wider inferences, over all possible experimental units, the experimenter must repeat the treatments on other experimental units for his additional observations.

This concept of the experimenter knowing his inference space before he decides the method of obtaining repeated observations is basic for designing any experiment. An example of this idea occurred in an agronomy experiment in which the cellulose content of alfalfa was measured from plants in each pot (the experimental unit). The blocks (locations in the greenhouse) were fixed and the treatments were fixed so the inference would be made for those blocks and treatments only. The repetitions within each treatment-block combination could be just another reading for the same experimental unit (that is dividing the material from the same experimental unit and getting two observations) or randomly assigning the same treatment to another pot in the same block and getting another observation. The variation within (B - T) combinations for the former design should be smaller than, or at most equal to that variation for the latter design. The inference for the former should be more limited than for the latter. The model for the former design is

$$y_{ijk\ell} = \mu + B_i + \delta_{(i)} + T_j + BT_{ij} + \rho_{(ij)k} + \eta_{(ijk)\ell} \qquad (5.3.1)$$

where

$i = 1, 2, \ldots, b \quad j = 1, 2, \ldots, t \quad k = 1 \quad \ell = 1, 2$

$y_{ijk\ell}$ = cellulose content of the ℓth observation of the plants in the kth pot from the jth treatment in the ith location in the greenhouse

B_i = effect of the ith location

$\delta_{(i)}$ = restriction error

T_j = effect of jth treatment

BT_{ij} = effect of the interaction of the ith location with the jth treatment

$\rho_{(ij)k}$ = random error due to the kth pot in the ith block and j^{th} treatment

$\eta_{(ijk)\ell}$ = random error due to the ℓth observation in the kth pot in the jth treatment in the ith location.

The analysis of the data from the former design (observations from the same experimental unit) is given in Table 5.3.1.

TABLE 5.3.1

ANOVA When Pots Are Not Repeated or Using Equation (5.3.1)[a]

Source	df	EMS
Blocks (B)	b - 1	$\sigma_\eta^2 + 2\sigma_\rho^2 + 2t\sigma_\delta^2 + 2t\phi(B)$
Restriction error	0	$\sigma_\eta^2 + 2\sigma_\rho^2 + 2t\sigma_\delta^2$
Treatments (T)	(t - 1)	$\sigma_\eta^2 + 2\sigma_\rho^2 + 2b\phi(T)$
Interaction (B x T)	(b - 1)(t - 1)	$\sigma_\eta^2 + 2\sigma_\rho^2 + 2\phi(BT)$
Between pots within (B - T) combinations	0	$\sigma_\eta^2 + 2\sigma_\rho^2$
Between observations within pots	bt	σ_η^2
Total	2bt - 1	

[a] σ_η^2 = variance component between samples within experimental units (pots) (a type of sampling error)

σ_ρ^2 = variance component between experimental units (pots) treated alike

σ_δ^2 = variance component due to extraneous variables on the blocks

Hence there is no estimate of the error to test treatment effects and the interaction. Treatments can be tested against the interaction mean square if $\phi(BT)$ is zero or a conservative test may be run if $\phi(BT)$ is not negligible. The model for the latter design is:

$$y_{ijk\ell} = \mu + B_i + \delta_{(i)} + T_j + BT_{ij} + \rho_{(ij)k} + \eta_{(ijk)\ell} \qquad (5.3.2)$$

$$i = 1, 2, \ldots, b \qquad j = 1, 2, \ldots, t \qquad k = 1, 2 \qquad \ell = 1$$

and the components are defined the same as for Eq. (5.3.1) except that $k = 1, 2$ here and $\ell = 1$ only here.

The analysis for the latter design (two experimental units randomly assigned to each treatment within each block) is given in Table 5.3.2.

TABLE 5.3.2

ANOVA When Pots Are Repeated or Using Equation (5.3.2)

Source	df	EMS
Blocks (B)	$(b - 1)$	$\sigma_\eta^2 + \sigma_\rho^2 + 2t\sigma_\delta^2 + 2t\phi(B)$
Restriction error	0	$\sigma_\eta^2 + \sigma_\rho^2 + 2t\sigma_\delta^2$
Treatments (T)	$(t - 1)$	$\sigma_\eta^2 + \sigma_\rho^2 + 2b\phi(T)$
Interaction (B x T)	$(t - 1)(b - 1)$	$\sigma_\eta^2 + \sigma_\rho^2 + 2\phi(BT)$
Between pots within (B - T) combinations	(bt)	$\sigma_\eta^2 + \sigma_\rho^2$
Total	$2bt - 1$	

Hence the tests on treatments and the interaction of blocks by treatments are possible since the appropriate error is estimated from between experimental units within the block-treatment combinations. If an estimate of σ_η^2 alone were wanted, more than one sample per treatment combination must be taken on at least a few experimental units.

This general approach of evaluating the experiment through the outline of the analysis before the experiment is run to guide the experimenter in designing his experiments should always be carried out, keeping the inference space clearly in mind.

An example of a RCBD in chemistry is that if chemists wish to make inferences about five methods over all laboratories in the United States, a random sample of the admissible laboratories may be obtained.

It should be made clear to the reader at this point that this is an example in which the experimenter is not interested in the blocks, i.e., the laboratories per se, and emphasis is made on the tests of treatments (here methods) only.

Let us say that twenty laboratories are selected over the United States at random and the same CRD is used at each laboratory. The effect of laboratories has location effects, the environmental conditions around the experiment, personnel, and so on confounded with it. It is impossible to enumerate all the causes of variation from laboratory to laboratory, but it is assumed that the factor, laboratories, has all these in it.

Within each laboratory the same five fixed methods to develop a product are used. Each method is to be run ten times in each laboratory. Assuming the 50 setups of equipment to obtain information on the five methods each run ten times are completely randomized, there is really only one place in the design that randomization occurs once the twenty laboratories are known. Of course that place is within each laboratory. The random laboratories merely allows a broad inference space for the fixed methods.

The model for this experiment is:

$$y_{ijk} = \mu + L_i + \delta_{(i)} + M_j + LM_{ij} + \varepsilon_{(ij)k} \qquad (5.3.3)$$

$$i = 1, 2, \ldots, 20 \qquad j = 1, 2, \ldots, 5 \qquad k = 1, 2, \ldots, 10$$

where

y_{ijk} = the variable to be analyzed from the kth setup of the jth method in the ith laboratory

μ = overall mean

L_i = the effect of the ith laboratory (random)

$\delta_{(i)}$ = the ith laboratory or restriction error due to the restriction on randomization of the methods in the laboratories

M_j = the effect of the jth method (fixed)

LM_{ij} = the interaction effect of the ith laboratory with the jth method

$\varepsilon_{(ij)k}$ = within error, the random effect of the error due to the kth setup within the ith laboratory and the jth method

TABLE 5.3.3

ANOVA Using Equation (5.3.3)

Source	df	EMS
Laboratories (L)	19	$\sigma^2 + 50\sigma_\delta^2 + 50\sigma_L^2$
Restriction error	0	$\sigma^2 + 50\sigma_\delta^2$
Methods (M)	4	$\sigma^2 + 10\sigma_{LM}^2 + 200\phi(M)$
L x M	76	$\sigma^2 + 10\sigma_{LM}^2$
Within error	900	σ^2
Total	999	

As indicated in Chapter 2 on mixed models the test on the fixed main effect uses the interaction mean square since in this case the inference is made to all the laboratories from which the 20 were

drawn. An inference that is broad must contain in its error a
reflection of the sampled material. In our previous example for
only one laboratory the within methods source of variation was used
as the experimental error reflecting a very narrow inference, namely
for that one laboratory, or if it is known there are no differences
between laboratories in their effect on the systems, the inferences
would be over all laboratories.

Another feature of this design is the allocation of time and
observations in experimentation. Is it necessary to have ten
observations per laboratory-method combination? Since it is so
expensive to investigate laboratories, is it permissible to reduce
the number of laboratories?

The answers to these questions must come from the interests of
the experimenter. If the experimenter is interested in estimating
σ_L^2, he must have an estimate of σ^2 and σ_δ^2. To estimate σ^2 with 900
degrees of freedom seems like a waste of observations unless it
costs almost nothing to make the experimental setups (which is most
unlikely). An estimate of σ^2 with about 100 degrees of freedom is
usually more than sufficient for most practical work. This means
that two observations per cell would be sufficient here. Of course
this assumes the variances are homogeneous because this estimate
comes from the one hundred combinations of laboratories and methods.
For estimating σ_δ^2, there must be replication of the overall
experiment at each laboratory.

As far as the number of laboratories is concerned, there should
be an ample number to make the inferences on the methods for all
the laboratories in the population. If there is very little
variation from laboratory to laboratory and the interaction of
laboratory by methods is small, there is no need for a huge sample
of laboratories. In devising a sampling plan for selecting
laboratories, the experimenter should use as much information as he
can on differences between laboratories to select the sample size.
In this case, sequential experimentation seems like a very good

procedure. That is, take a relatively small random sample of
laboratories, say seven, and estimate the variance components. If
the methods mean square seems large relative to the interaction mean
square but the experimenter can detect no statistical difference
among methods, the experimenter should take more laboratories. A
guide to the number of laboratories would be to find the number of
degrees of freedom needed in the interaction to show the F-ratio
obtained in the first experiment to be significant and calculate the
number of additional laboratories required to get a significant
effect.

Consider the following numerical example: Let us use five
laboratories, three methods and four experimental setups per
laboratory-method combination for the first part of the experiment.
The analysis is given in Table 5.3.4.

TABLE 5.3.4

ANOVA For Example Above

Source	df	MS	EMS
Laboratories (L)	4	70	$\sigma^2 + 12\sigma_\delta^2 + 12\sigma_L^2$
Restriction error	0		$\sigma^2 + 12\sigma_\delta^2$
Methods (M)	2	130	$\sigma^2 + 4\sigma_{LM}^2 + 20\phi(M)$
L x M	8	40	$\sigma^2 + 4\sigma_{LM}^2$
Within error	45	25	σ^2
Total	59		

The F-test for methods is:

$$F_{2,8} = \frac{130}{40} = 3.25$$

Using $\alpha = 0.05$, theoretical $F_{2,8} = 4.46$. The ratio in the experiment looks large, but it is difficult to draw conclusions. How many more laboratories would be suggested? A theoretical $F_{2,40} = 3.23$, so 16 more laboratories should be drawn at random which would make a total of 21 laboratories in the experiment. If similar results were obtained for the mean squares (there is no absolute assurance that this will happen), the combined analysis would allow a conclusion that the methods were different.

If the results from part one of the experiment using five laboratories were drastically different from the second part using a different group of 16 laboratories, the experimenter must thoroughly investigate the two sets of laboratories to explain the reason for the large difference.

Another possibility from the sample of five laboratories exists, and that is there is already a significant difference in methods obtained with the five laboratories. If the research worker has good reason for making the inferences to all laboratories, there may be no reason to go further.

The final possibility for this randomized complete block design with more than one experimental unit per combination is that the mean square for methods is no larger than the interaction mean square and the experimenter has good reason to believe that all assumptions for the model are met. Under these circumstances the experimentation would usually stop with the conclusion that the methods are not different.

5.4 RELATIVE EFFICIENCY OF DESIGNS

After understanding the two basic designs discussed so far it is important to compare the efficiency of one design relative to the other. In this case let us compare the efficiency of the randomized complete block relative to the completely randomized design (Fisher, 1966; and Cochran and Cox, 1957, pp. 31-34). There are various ways of doing this but we will use the ratio,

$$\frac{(n_1 + 1)(n_2 + 3) \; s_2^2}{(n_2 + 1)(n_1 + 3) \; s_1^2}$$

where for our problem

s_1^2 is the estimate of the error mean square (from the interaction mean square) to test treatments from the randomized complete block design

s_2^2 is the estimate of the error mean square to test treatments from the completely randomized design

n_1 is the number of degrees of freedom for the error in the randomized complete block design

n_2 is the number of degrees of freedom for the error in the completely randomized design

Problem 5.4.1.   A company wanted to increase light intensity of its photoflash cartridge. The variable to be analyzed was a function of light intensity.
    The four treatments:
        (1) 1/8 inch thick wall with the ignition point at the end of the cartridge
        (2) 1/16 inch thick wall with the ignition point at the end of the cartridge
        (3) 1/8 inch thick wall with the ignition point at the center of the cartridge
        (4) 1/16 inch thick wall with the ignition point at the center of the cartridge
were used in the experiment.
    Five random batches of the basic formulation used in the cartridges were made up. From each batch three cartridges with the same treatment were constructed so in all there were 12 cartridges made from each batch.
    (a) Describe the various places in the experiment that randomization should occur.
    (b) Show the layout of the experiment.
    (c) Carefully describe the inference space.
    (d) Show the appropriate model and analysis resulting from the randomizations and inference space. Describe each symbol.
    (e) Compare the inference space of this design with a design in which only one batch with fifteen cartridges for each treatment was used.
    (f) Under what conditions would the one batch design be preferred.

Problem 5.4.2.   An agronomist was interested in the effect of four different sprays on control of weeds in corn.   The variable to be analyzed was a function of the remaining weeds 1 week after application of the sprays.

Five locations on the research farm were used to run the experiment and three randomly chosen plots in each location were used for each spray.

(a)   Describe the randomization procedures and inference space for the experiment.

(b)   Show the appropriate model and analysis, describing the symbols.

(c)   Comment on the design and discuss improvements, if possible, using another inference space and model.

5.5   REFERENCES

Addelmen, S.   Amer. Statistician 23:35 (1969).

Anderson, V. L.   Biometrics 26:255 (1970).

Beeson, J.   A simulator for evaluating prosthetic cardiac valves. Unpublished M.S. thesis, Purdue University Library.   (1965).

Bennett, C. A. and Franklin, N. L.   Statistical Analysis in Chemistry and the Chemical Industry.   Wiley, New York, 1954.

Bozivich, H., Bancroft, T. A. and Hartley, H. O.   Ann. Math. Statist. 27:1017 (1956).

Cochran, W. G. and Cox, G. M.   Experimental Design, 2nd ed., Wiley, New York, 1957.

Fisher, R. A.   The Design of Experiments, 8th ed., Hafner, New York, 1966.

Hicks, Charles R.   Fundamental Concepts in the Design of Experiments, 2nd ed., Holt, Rinehart and Winston, New York, 1973.

Ostle, Bernard   Statistics in Research, 2nd ed., Iowa State University Press, Ames, Iowa, 1963.

Wilk, M. B.   Biometrika 42:70 (1955).

Chapter 6

NESTED (HIERARCHICAL) AND NESTED FACTORIAL DESIGNS

The designs presented in Chapters 4 and 5 are more conventional, a bit more easily recognized and used more frequently than the ones to be presented in this chapter. For the designs in this chapter the factorial structure is not complete and to recognize the features and how to set up reasonable linear models for the analysis of the data from these designs is not always obvious. Also, in this chapter, we use Latin letters rather than Greek letters to represent the well-defined random components in the mathematical models.

6.1 NESTED (HIERARCHICAL)

In some experiments the levels for a given factor are all different across the levels of the other factors. Frequently the levels are chosen at random for each factor, but this is not a necessary condition for the so-called "nested design." In fact, there can be random, mixed, or fixed models in the hierarchical arrangement. It should be emphasized that this is really another type of the CR design; or if there is a restriction, the associated error is understood to be part of the appropriate component.

6.1.1 All Factors Random

An example of a nested design (often called multistage sampling) involves human beings in the United States. States are sampled at random, counties are sampled at random in the states, towns and cities are sampled at random in the counties, and finally households may be sampled at random in the towns or cities.

Actually, the sampling procedures are quite a bit more complicated than this, but a linear model to analyze data from the above design could be the following:

$$y_{ijk\ell} = \mu + S_i + C_{(i)j} + T_{(ij)k} + H_{(ijk)\ell} \qquad (6.1.1)$$

$i = 1, 2, \ldots, s \quad j = 1, 2, \ldots, c \quad k = 1, 2, \ldots, t \quad \ell = 1, 2, \ldots, h$

where

μ = the overall mean

S_i = the effect of the ith state

$C_{(i)j}$ = the effect of the jth county in the ith state

$T_{(ij)k}$ = the effect of the kth town in the jth county in the ith state

$H_{(ijk)\ell}$ = the effect of the ℓth household in the kth town in the jth county in the ith state

Notice that the example shows the impossibility of the level of any factor such as counties being the same for the levels of the factor, state. Since there are never the same counties within the various states in this investigation, it is impossible to obtain an interaction between counties and states. This concept of never being able to obtain an interaction is always present in a nested design.

Table 6.1.1 shows the ANOVA resulting from Eq. (6.1.1).

TABLE 6.1.1

ANOVA Using Equation (6.1.1)

Source	df	EMS
States	$s - 1$	$\sigma_H^2 + h\sigma_T^2 + hto\sigma_C^2 + htco\sigma_S^2$
Counties in states	$s(c - 1)$	$\sigma_H^2 + h\sigma_T^2 + hto\sigma_C^2$
Towns in counties in states	$sc(t - 1)$	$\sigma_H^2 + h\sigma_T^2$
Households in ... states	$sct(h - 1)$	σ_H^2
Total	$scth - 1$	

In this table the number of levels are constant in the subclasses.
In most actual sampling and design of experiments problems the
number of levels may be quite unequal. In this case the calculations
of sums of squares is still straightforward, but the calculation of
the coefficients on the variance components may be quite involved
(Anderson and Bancroft, 1952, p. 327).

Example 6.1: Determination of the components of variance for
the model

$$y_{ijk} = \mu + S_i + T_{(i)j} + H_{(ij)k}$$

$$i = 1, 2, 3 \quad j = 1, 2 \quad k = 1, 2, 3, 4$$

where the terms in the model are as defined in Eq. (6.1.1). Two
methods of calculation are presented: one gives the sum of squares
for nested factors directly, whereas the other utilizes the method
of calculation for a factorial design and then combines appropriate
sums of squares to obtain the correct values. The data for this
example are given in the following tabulation.

	State					
	1		2		3	
	Town		Town		Town	
	1	2	1	2	1	2
Household						
1	10	7	6	6	15	12
2	13	12	5	12	18	15
3	16	11	9	7	20	18
4	12	9	3	10	19	16
Totals	51	39	23	35	72	61
	$T_{1..} = 90$		$T_{2..} = 58$		$T_{3..} = 133$	

$$T_{...} = 281$$

The ANOVA table is:

Source	df	SS	MS	EMS
States	2	354.08	177.04	$\sigma_H^2 + 4\sigma_T^2 + 8\sigma_S^2$
Towns/states	3	51.13	17.04	$\sigma_H^2 + 4\sigma_T^2$
Households/towns/states	<u>18</u>	<u>107.75</u>	5.99	σ_H^2
Total	23	512.96		

where

$$\text{SS states} = \frac{1}{2 \times 4} \sum_{i=1}^{3} T_{i..}^2 - \frac{T_{...}^2}{2 \times 4 \times 3}$$

$$= \frac{1}{8} (90^2 + 58^2 + 133^2) - \frac{281^2}{24} = 354.08$$

$$\text{SS towns/states} = \frac{1}{4} \sum_{i=1}^{3} \sum_{j=1}^{2} T_{ij.}^2 - \frac{1}{2 \times 4} \sum_{i} T_{i..}^2$$

$$= \sum_{i=1}^{3} (\frac{1}{4} \sum_{j=1}^{2} T_{ij.}^2 - \frac{1}{8} T_{i..}^2)$$

= sum of squares for towns within state summed
over states

$$= \frac{1}{4} (51^2 + 39^2 + \ldots + 61^2) - \frac{1}{8} (90^2 + 58^2$$

$$+ 133^2)$$

$$= 51.13$$

$$\text{SS households/towns/states} = \sum_i \sum_j \sum_k y_{ijk}^2 - \frac{1}{4} \sum_i \sum_j T_{ij.}^2$$

$$= 10^2 + 13^2 + \ldots + 18^2 + 16^2$$

$$- \frac{1}{4} (51^2 + 39^2 + \ldots + 61^2)$$

$$= 3803 - 3695.25$$

$$= 107.75$$

$$SS\ total = \sum_i \sum_j \sum_k y^2_{ijk} - \frac{T^2_{...}}{24}$$

$$= SS\ states + SS\ towns/states + SS\ households/$$
$$towns/states$$

$$= 3803 - 3290.04$$

$$= 512.96$$

In order to determine the method of calculating the above analysis of variance table using a factorial computing procedure we combine sums of squares in the same manner that one has to combine degrees of freedom for the model

$$y_{ijk} = \mu + S_i + T_j + ST_{ij} + \varepsilon_{(ij)k}$$

to obtain the correct degrees of freedom for the model

$$y_{ijk} = \mu + S_i + T_{(i)j} + H_{(ij)k}$$

In the nested model it is impossible to have a state by town interaction and so it seems reasonable to combine the degrees of freedom for the fictitious "towns" with 1 df with the "town by state interaction" with 2 df to obtain the 3 df for towns within states. Checking this reasoning gives 1 df + 2 df = 3 df from the factorial model which in turn is the correct degrees of freedom for towns within states. Consequently we combine the same sums of squares. One may prove this relationship algebraically, but this will not be shown here. In order to demonstrate this we calculate the two sums of squares for the factorial model.

$$SS\ "towns" = \frac{1}{4\ x\ 3} [(51 + 23 + 72)^2 + (39 + 35 + 61)^2] - \frac{T^2_{...}}{24}$$

$$= 3295.08 - 3290.04 = 5.04$$

$$SS\ "T\ x\ S" = \frac{1}{4} (51^2 + 23^2 + ... + 61^2) - CT - SS\ towns - SS\ states$$

$$= 3695.25 - 3290.04 - 5.04 - 354.08$$

$$= 46.09$$

Note that $46.09 + 5.04 = 51.13$, which is the same as that obtained for the towns within states sum of squares. Calculation of sums of squares for any nested type of sums of squares in the nested factorial models may be done in a similar fashion.

The estimates for the individual components of variance are as follows (refer to the above ANOVA table)

$$\hat{\sigma}^2_H = 5.99$$

$$\hat{\sigma}^2_T = \frac{1}{4}(17.04 - 5.99) = 2.76$$

$$\hat{\sigma}^2_S = \frac{1}{8}(177.04 - 17.04) = 20.00$$

The estimate of the total variation of a single observation becomes the total of these three variance components (Eisenhart, 1947). Hence $\hat{\sigma}^2_{Total} = 28.75$. It follows that $\frac{20.00}{28.75} = 0.70$ of the total variance is estimated to be due to variation among states. This indicates that if further information is desired one should concentrate more on states and less on towns in states.

Example 6.2: In a genetics study on fruit flies, the variance components of body weight due to males, females, bottles in which the flies are kept and offspring was investigated. The mating system involved selecting four males and twelve females at random from each of four genome samples, mating three females to each male in single-pair matings to retain identity of the female and dividing the eggs laid into two bottles for each mating. After the progeny became adult flies, five picked at random from each bottle were weighed individually.

Hence the design was completely hierarchical and a reasonable linear model is

$$y_{ijk\ell m} = \mu + G_i + M_{(i)j} + F_{(ij)k} + B_{(ijk)\ell} + I_{(ijk\ell)m} \qquad (6.1.2)$$

$i = 1, 2, 3, 4 \quad j = 1, 2, 3, 4 \quad k = 1, 2, 3 \quad \ell = 1, 2 \quad m = 1, \ldots, 5$

where each symbol is easily identified from the explanation above.

Problem 6.1.1. Considering all factors random, show the variance component estimates for σ_G^2, σ_M^2, σ_F^2, σ_B^2 and σ_I^2 in the preceding example on fruit flies.

Problem 6.1.2. In the actual experiment there were only 3 males for genome 4; two females mated to the second male for genome 1, first male from genome 2, fourth male from genome 3; and only one female mated to the first male from genome 1. In addition only three progeny existed for the mating of male 1 with female 1 in genome 1 and male 2, with female 3 in genome 3; and only one progeny existed for mating of male 1 with female 3 in genome 4.
    Write out the adjusted coefficients of the variance components for the actual experiment (Anderson and Bancroft, 1952, p. 327).

6.1.2 Fixed and Random Factors

An example in which some factors are random and one is fixed occurred in an aluminum alloy development problem. An aluminum company had four (fixed) alloys in which the chemistry was different. All the properties except strength had been evaluated and an experiment to examine this variable was wanted.

One proposal was to make four ingots, one with each chemistry, and compare the strengths of the finished product. If this were done the inference could be made to this one ingot for each chemistry made from only one heat (batch of molten aluminum). If each ingot could possibly represent an adequate random sample of ingots for each heat, the inference could be made to all ingots from that one heat with that chemistry. To infer indefinitely over heats one must replicate the experiment by having at least one more heat run per chemistry or assume the variation between heats was no greater than the variation between ingots within heats. This latter assumption, however, is usually very poor for problems of this type.

The cost of heats was quite large and the research worker decided to take three ingots per chemistry chosen at random from each heat-chemistry combination. Further he took four random samples of the material in each ingot to measure the variation between pieces within ingots. The appropriate model and ANOVA for this experiment is:

$$Y_{ijk\ell} = \mu + C_i + H_{(i)j} + I_{(ij)k} + \varepsilon_{(ijk)\ell} \qquad (6.1.3)$$

$i = 1, 2, 3, 4 \quad j = 1 \quad k = 1, 2, 3 \quad \ell = 1, 2, 3, 4$

where

$Y_{ijk\ell}$ = strength of the ℓth piece of metal in the kth ingot from the jth heat with the ith chemistry

μ = overall mean

C_i = effect of the ith chemistry (fixed)

$H_{(i)j}$ = effect of the jth heat using the ith chemistry. This is completely confounded with chemistry [random, NID $(0, \sigma_H^2)$]

$I_{(ij)k}$ = effect of the kth ingot in the jth heat in the ith chemistry [random, NID $(0, \sigma_I^2)$]

$\varepsilon_{(ijk)\ell}$ = effect of the ℓth piece of metal in the kth ingot in the jth heat in the ith chemistry [random, NID $(0, \sigma_\varepsilon^2)$]

TABLE 6.1.2

ANOVA Using Equation (6.1.3)

Source	df	EMS
Chemistries (C_i)	3	$\sigma_\varepsilon^2 + 4\sigma_I^2 + 12\sigma_H^2 + 12\phi(C)$
Heats in C ($H_{(i)j}$)	0	$\sigma_\varepsilon^2 + 4\sigma_I^2 + 12\sigma_H^2$
Ingots in heats ($I_{(ij)k}$)	8	$\sigma_\varepsilon^2 + 4\sigma_I^2$
Pieces in ingots ($\varepsilon_{(ijk)\ell}$)	36	σ_ε^2

Problem 6.1.3. Using the EMS results from Table 6.1.2 and the information in the following tabulation

Source	df	MS
Chemistries	3	160
Ingots	8	20
Pieces	36	10

(a) Give a detailed report of the results to the research worker.

(b) · If it costs $1000 to run a heat, $300 to use an ingot and $50 to examine a piece of metal, recommend a design for the future and explain to the research worker in the aluminum plant why you prefer your design over his. How much more would your experiment cost than the one that was run?

Problem 6.1.4. An agronomist was interested in finding out the amount of potassium in corn leaves, not including the stems, after four different fertilizers had been used in a field of corn.

(a) If 5 plots of corn were used per fertilizer in a completely randomized manner, show the layout of an experiment.

· (b) If in the 5 plots per fertilizer, 2 stalks of corn per plot were to be sampled from the experimental area, show the ANOVA and state the assumptions necessary to test the effects of fertilizer.

(c) Comment on this design and analysis.

(d) Show the standard error of the fertilizer means.
Refer to Section 2.2.1, Example 2.7, and Appendix 12.

6.2 NESTED FACTORIAL

The concept developed in Section 6.1 that some experiments are designed such that the levels of a factor cannot be the same across the levels of another is extended in this section. In addition, however, another factor or factors may have the same levels across other factors and be "factorial" to these factors in the experiment. This mixture of nesting and factorial structure is sometimes called nested factorial (see Hicks, 1973, Section 11.4).

6.2.1 Cardiac Valve Experiment

To appreciate the concept of this design let us return to the
example on prosthetic cardiac valves given in Chapter 5. A more
reasonable design, provided that the within cell errors are not
correlated, is to repeat the tests on the valve types at random and
to run each of the six pulse rates only once on each valve
(occurrence) for a given valve type. To further clarify this design,
consider the four valve types as treatments and completely randomize
the order that the valve types or treatments are run on the machine
so that each valve type occurs twice, each time with a different
valve of the same type. Notice that each occurrence (valve) is
still a block as defined in Chapter 5, but the blocks are now nested
within valve types. At this stage, we have a completely randomized
design and one possible randomization is:

Valve types

1		2		3		4	
4	5	2	7	6	8	1	3

Order of run on machine

Using the above schematic, after a valve of type 4 is chosen for
the first run, all six pulse rates are run at random on the valve of
valve type 4. Next a valve of type 2 is inserted in the machine
because the #2 run in the schematic indicates this. Of course all
six pulse rates are then run in a new random order using a valve of
type 2. The experiment continues in this manner until the second
valve of the third type is used (#8 in the order of run on the
machine in the schematic), and the experiment is completed.

After the experiment is completed the order of running the
experiment can be ignored and the data may be put into a table such
as Table 6.2.1.

TABLE 6.2.1

Arrangement for Cardiac Valve Data
from a Nested Factorial Experiment

	Valve types							
	1		2		3		4	
	1	2	3	4	5	6	7	8
Pulse rates								
1								
2								
3								
4								
5								
6								

The model for the analysis of the data from this experiment,
assuming the errors are not correlated, is:

$$y_{ijk} = \mu + V_i + O_{(i)j} + \delta_{(ij)} + P_k + VP_{ik} \hspace{1cm} (6.2.1)$$
$$+ OP_{(i)jk} + \varepsilon_{(ijk)}$$

$$i = 1, 2, 3, 4 \hspace{1cm} j = 1, 2 \hspace{1cm} k = 1, 2, \ldots, 6$$

where

y_{ijk} = maximum flow gradient (mmHg) obtained from the jth
occurrence of the ith valve type with the kth pulse
rate

μ = overall mean

V_i = effect of the ith valve type (fixed)

$O_{(i)j}$ = effect of the jth occurrence (random valve) in the
i^{th} valve type NID $(0, \sigma_0^2)$

$\delta_{(ij)}$ = restriction error (Anderson, 1970) caused by the 6 pulse rates being run in the jth occurrence of the ith valve type. This has zero degrees of freedom and no sums of squares. It is assumed that $\delta_{(ij)}$ is NID $(0, \sigma_{\delta}^2)$

P_k = effect of the kth pulse rate (fixed)

VP_{ik} = effect of the interaction of the ith valve type with the kth pulse rate

$OP_{(i)jk}$ = effect of the interaction of the jth occurrence in the ith valve type by the kth pulse rate, NID $(0, \sigma_{Op}^2)$

$\varepsilon_{(ijk)}$ = within error, NID $(0, \sigma^2)$, in this case since there is only one observation within the jth occurrence of the ith valve type and k^{th} pulse rate, there are zero degrees of freedom

In this type experiment, where the same experimental unit is used repeatedly for the six pulse rates, a multivariate analysis may be preferred (Morrison, 1967, Chapter 5) if the errors, $\varepsilon_{(ijk)}$, are correlated. There is a conservative test that could be made for pulse rates and the interaction of valve types by pulse rates for the situation of correlated errors. In any event the test for valve types is identical. Assuming that the effect of the correlated errors is relatively trivial in this particular experiment, the analysis is as shown in Table 6.2.2.

In Table 6.2.2 it can be seen that there is no test for occurrences in valve types since there are zero degrees of freedom for the appropriate source to test it, namely, restriction error. This is consistent with the RCBD result.

TABLE 6.2.2

ANOVA Using Equation (6.2.1)

Source	df	EMS
Valve types (V_i)	3	$\sigma^2 + 6\sigma_\delta^2 + 6\sigma_0^2 + 12\phi(V)$
Occurrences in valve types ($0_{(i)j}$)	4	$\sigma^2 + 6\sigma_\delta^2 + 6\sigma_0^2$
Restriction error ($\delta_{(ij)}$)	0	$\sigma^2 + 6\sigma_\delta^2$
Pulse rates (P_k)	5	$\sigma^2 + \sigma_{OP}^2 + 8\phi(P)$
Interaction (VP_{ik})	15	$\sigma^2 + \sigma_{OP}^2 + 2\phi(VP)$
Interaction ($OP_{(i)jk}$)	20	$\sigma^2 + \sigma_{OP}^2$
Within error ($\varepsilon_{(ijk)}$)	0	σ^2

To provide information to the reader who will not use
multivariate analysis for correlated error data, one can obtain
conservative tests. The conservative tests require that the degrees
of freedom for the main effect given repeatedly to the same
experimental units [pulse rates (P_k) with 5 df] and for the
interactions associated with that factor [(VP_{ik}) with 15 df and
($OP_{i(jk)}$) with 20 df] all be divided by the degrees of freedom for
that main effect (5 df). The mean squares are not altered. Hence
the conservative tests for Table 6.2.2 are:

(1) Pulse rates using $F_{1,4} = \dfrac{MS(P_k)}{MS(OP_{(i)jk})}$

and   (2) (VP_{ik}) using $F_{3,4} = \dfrac{MS(VP_{ik})}{MS(OP_{(i)jk})}$

(see Morrison, 1967, pp. 194-196; Winer, 1971, pp. 522-524; and
Greenhouse and Geisser, 1959).

The present tendency, however, is for experimenters to use the tests indicated by Table 6.2.2 and not use the conservative tests for those designs (including split plot type given in Chapter 7) in which different treatments are used on the same experimental unit. Even for moderate correlation among the errors, Monte Carlo studies have indicated this conclusion (Winer, 1971, p. 524).

For those cases in which the experimenter is greatly concerned about correlated errors because he has used more than one treatment on each experimental unit, we recommend using both the conservative and direct or usual ANOVA test indicated by the EMS assuming no correlation (e.g., Table 6.2.2) in the following manner:

(1) Use the results of the conservative tests if they show significance

(2) Use the results of the direct ANOVA tests if they show no significance

(3) Use theory on the size of the correlation for those results lying between (1) and (2) with a tendency to use the direct ANOVA tests unless the correlation is known to be very high.

It has been our experience that whenever a process or experimental unit is disturbed while the experimenter applies another treatment, that the correlation of errors becomes small. Hence we believe that the direct test from the ANOVA is usually appropriate.

Example 6.3: Given that Table 6.2.1 is filled in with the following data, determine the corresponding ANOVA table.

	Valve types							
	1		2		3		4	
	Occurrence		Occurrence		Occurrence		Occurrence	
Pulse rates	1	2	3	4	5	6	7	8	Totals
1	2	3	4	2	6	5	7	5	34
2	4	4	4	4	5	5	5	4	35
3	5	7	4	3	5	6	6	5	41
4	3	5	5	3	8	10	9	10	53
5	7	7	8	5	9	9	10	11	66
6	6	6	6	7	7	8	8	9	57
Totals	27	32	31	24	40	43	45	44	286
	59		55		83		89	

$$y_{ijk} = \mu + V_i + O_{(i)j} + \delta_{(ij)} + P_k + VP_{ik} + OP_{(i)jk} + \varepsilon_{(ijk)}$$

$$i = 1, 2, 3, 4 \qquad j = 1, 2 \qquad k = 1, 2, \ldots, 6$$

ANOVA

Source	df	SS	MS	EMS	F	$F_{crit}^{(0.05)}$
V_i	3	72.25	24.08	$\sigma^2 + 6\sigma_\delta^2 + 6\sigma_0^2 + 12\phi(V)$	13.76*	6.59
$O_{(i)j}$	4	7.00	1.75	$\sigma^2 + 6\sigma_\delta^2 + 6\sigma_0^2$	None	
$\delta_{(ij)}$	0	None		$\sigma^2 + 6\sigma_\delta^2$		
P_k	5	105.42	21.08	$\sigma^2 + \sigma_{OP}^2 + 8\phi(P)$	28.11*	2.71
VP_{ik}	15	38.25	2.55	$\sigma^2 + \sigma_{OP}^2 + 2\phi(VP)$	3.40*	2.20
$OP_{(i)jk}$	20	15.00	0.75	$\sigma^2 + \sigma_{OP}^2$	None	
$\varepsilon_{(ijk)}$	0	None		σ^2		
Total	47	237.92				

where

$$SS\ V_i = \frac{59^2 + 55^2 + 83^2 + 89^2}{12} - \frac{286^2}{48}$$

$$= 1776.33 - 1704.08$$

$$= 72.25$$

$$SS\ P_k = \frac{34^2 + 35^2 + \ldots + 57^2}{8} - \frac{286^2}{48}$$

$$= 1809.50 - 1704.08$$

$$= 105.42$$

$$SS\ O_{(i)j} = \frac{27^2 + 32^2 + \ldots + 44^2}{6} - \frac{59^2 + 55^2 + 83^2 + 89^2}{12}$$

$$= 1783.33 - 1776.33$$

$$= 7.00$$

$$SS\ VP_{ik} = \frac{5^2 + 6^2 + 11^2 + \ldots + 17^2}{2} - \frac{286^2}{48} - SS\ V_i - SS\ P_k$$

$$= 1920.00 - 1704.08 - 72.25 - 105.42$$

$$= 38.25$$

$$SS\ OP_{(i)jk} = \sum_i (SS\ OP(V_i))$$

where

$$SS\ OP(V_1) = 2^2 + 3^2 + \ldots + 6^2 - \frac{59^2}{12}$$

$$- \left[\frac{5^2 + 8^2 + \ldots + 12^2}{2} - \frac{59^2}{12} \right] - \left[\frac{27^2 + 32^2}{6} - \frac{59^2}{12} \right]$$

$$= 2.42$$

$$SS\ OP(V_2) = 4^2 + 2^2 + \ldots + 7^2 - \frac{55^2}{12}$$

$$- \left[\frac{6^2 + 8^2 + \ldots + 13^2}{2} - \frac{55^2}{12} \right] - \left[\frac{31^2 + 24^2}{6} - \frac{55^2}{12} \right]$$

$$= 5.42$$

$$SS\ OP(V_3) = 6^2 + 5^2 + \ldots + 8^2 - \frac{83^2}{12}$$

$$-\left[\frac{11^2 + 10^2 + \ldots + 15^2}{2} - \frac{83^2}{12}\right] - \left[\frac{40^2 + 43^2}{6} - \frac{83^2}{12}\right]$$

$$= 2.75$$

$$SS\ OP(V_4) = 7^2 + 5^2 + \ldots + 9^2 - \frac{89^2}{12}$$

$$-\left[\frac{12^2 + 9^2 + \ldots + 17^2}{2} - \frac{89^2}{12}\right] - \left[\frac{45^2 + 44^2}{6} - \frac{89^2}{12}\right]$$

$$= 4.41$$

Therefore

$$SS\ OP_{(i)jk} = \sum_{i=1}^{4} SS\ OP(V_i) = 15.00$$

$$SS\ Total = 323 + 285 + 611 + 723 - 1704.08 \text{ (from our work}$$

above)

$$= 237.92$$

Problem 6.2.1. Using the results from Example 6.3, investigate the interaction thoroughly and interpret the overall results of the experiment carefully. Be sure to consider the consequences of installing an incorrect valve into a patient who had a particular range of pulse rates. In this case assume that the data used in this analysis have the characteristic of a small number being desirable.

Problem 6.2.2. Suppose that an experimenter sets up a prosthetic cardiac valve designed experiment and the correct model for the analysis is

$$y_{ijk\ell} = \mu + V_i + O_{(i)j} + P_k + VP_{ik} + OP_{(i)jk} + \varepsilon_{(ijk)\ell}$$

with the ranges on the subscripts being

$i = 1, 2, 3, 4 \quad j = 1, 2 \quad k = 1, 2, 3, 4, 5, 6 \quad \ell = 1, 2$

(a) How many valves (occurrences) were used in this experiment?

(b) Illustrate the layout of this design and describe how

the randomization must be carried out in order to make this model valid (consider correlations among errors and where they could occur).

(c) Show the analysis for this design and compare it to the analysis shown in Table 6.2.2.

6.2.2 Social Science Application

Nested factorial experiments were recognized early in the social sciences where human beings are the experimental units. In these designs frequently subjects (experimental units) are grouped because of a common characteristic and then quite a few subjects are used in each group. Since the same individuals cannot appear in different groups, they are nested within groups. Also, many experiments then have the subjects do certain tasks (treatments) and there is a factorial arrangement of groups and tasks. The same problem, possible correlated errors, exists here as was discussed in the cardiac valve experiment, Section 6.2.1. In this case the same subjects do all four tasks.

An example is that three randomly chosen groups of 10 students each were given a different method of training. After the training period four tasks were given the students at random (this gives rise to the restriction error, $\delta_{(ij)}$). The dependent variable, y, is the time to perform the task correctly. One object of the experiment was to find out whether or not one training method was superior to the other two methods for all tasks. Another object was to find out whether or not certain methods of training were better in adapting students for certain tasks (interaction).

A model that recognizes the restriction error and assumes all errors are not correlated is

$$y_{ijk} = \mu + M_i + S_{(i)j} + \delta_{(ij)} + T_k + MT_{ik} + ST_{(i)jk} + \varepsilon_{(ijk)} \quad (6.2.2)$$

$$i = 1, 2, 3 \quad j = 1, 2, \ldots, 10 \quad k = 1, 2, 3, 4$$

where

y_{ijk} = variable to be analyzed, time to perform the kth task by the jth student in the ith training method

μ = overall mean

M_i = the effect of the ith training method (fixed)

$S_{(i)j}$ = the effect of the jth student in the ith method (random)

$\delta_{(ij)}$ = restriction error caused by the four tasks being done by the jth student in the ith method

T_k = the effect of the kth task (fixed)

MT_{ik} = the effect of the interaction of the ith method with the kth task

$ST_{(i)jk}$ = the effect of the interaction of the jth student in the ith method with the kth task

$\varepsilon_{(ijk)}$ = within error

Note that the model and analysis for this example is the same as that given in Section 6.2.1 (also see Winer, 1971).

6.2.3 Nutrition Application

Recently the nested factorial experiments have been used by research workers in other areas. An example of such an experiment occurred in a nutrition study where three (fixed) rations were used. There were eight (random) hens assigned to each ration and the gain in weight from week to week was measured on each hen at the end of each of 5 (fixed) weeks. In addition, the hens were placed in three cages, one cage for each ration and the number of eggs laid by each group of eight was kept for the 5 weeks also.

The model, assuming the correlations among all errors are zero, (Mandel, 1957), for the gain variable to be analyzed follows:

$$y_{ijk} = \mu + R_i + H_{(i)j} + \delta_{(ij)} + W_k + RW_{ik} + HW_{(i)jk} + \varepsilon_{(ijk)} \qquad (6.2.3)$$

$$i = 1, 2, 3 \qquad j = 1, 2, \ldots, 8 \qquad k = 1, 2, \ldots, 5$$

where

y_{ijk} = gain in weight of the jth hen on the ith ration during the kth week

μ = overall mean

R_i = the effect of the ith ration (confounded with cages)

$H_{(i)j}$ = the effect of the jth hen on the ith ration

$\delta_{(ij)}$ = restriction error

W_k = the effect of the kth week

RW_{ik} = the effect of the interaction of the ith ration and k^{th} week

$HW_{(i)jk}$ = the effect of the interaction of the jth hen on the ith ration and the kth week

$\varepsilon_{(ijk)}$ = within error

The analysis of variance is given in Table 6.2.3.

TABLE 6.2.3

ANOVA Using Equation (6.2.3)

Source	df	EMS
Rations (R)	2	$\sigma^2 + 5\sigma_\delta^2 + 5\sigma_H^2 + 40\phi(R)$
Hens in Rations (H)	21	$\sigma^2 + 5\sigma_\delta^2 + 5\sigma_H^2$
Restriction error	0	$\sigma^2 + 5\sigma_\delta^2$
Weeks (W)	4	$\sigma^2 + \sigma_{HW}^2 + 24\phi(W)$
R x W	8	$\sigma^2 + \sigma_{HW}^2 + 8\phi(RW)$
H x W	84	$\sigma^2 + \sigma_{HW}^2$
Within error	0	σ^2

The estimated standard error of a ration mean is derived in Part 4 of Appendix 12 and is shown to be

$$s_{\bar{y}_{i..}} = \left(\frac{\hat{\sigma}_H^2}{8} + \frac{\hat{\sigma}_\delta^2}{8} + \frac{\hat{\sigma}^2}{40} \right)^{\frac{1}{2}}$$

which can be written as $[\frac{1}{40} \text{ (mean square hens)}]^{\frac{1}{2}}$

Problem 6.2.3.   Find the estimated standard error of the difference between ration means using the method described in Appendix 12.

The model for the eggs laid variable follows:

$$y_{ik} = \mu + R_i + \delta_{(i)} + W_k + RW_{ik} + \varepsilon_{(ik)} \qquad (6.2.4)$$

$$i = 1, 2, 3 \qquad k = 1, 2, \ldots, 5$$

y_{ik} = the number of eggs laid by all hens on the ith ration in the ith cage for the kth week

μ = overall mean

R_i = the effect of the ith ration and/or i^{th} cage and/or i^{th} group of hens (Note: ration, cages and group of hens are completely confounded)

$\delta_{(i)}$ = restriction error

W_k = the effect of the kth week

RW_{ik} = the effect of the interactions of the ith ration and/or cage and/or group of hens with the kth week

$\varepsilon_{(ik)}$ = within error

The analysis of variance for the eggs laid variable is given in Table 6.2.4.

TABLE 6.2.4

ANOVA Using Equation (6.2.4)

Source	df	EMS
Rations and/or cages and/or groups of hens (R)	2	$\sigma^2 + 5\sigma_\delta^2 + 5\phi(R)$
Restriction error	0	$\sigma^2 + 5\sigma_\delta^2$
Weeks (W)	4	$\sigma^2 + 3\phi(W)$
R x W	8	$\sigma^2 + \phi(RW)$
Within error	0	σ^2

This is a very poorly designed experiment because the analysis shows that rations cannot be tested even if it were not confounded with cages and groups of hens. This example is a clear case of the experimenter not outlining his analysis before the data were taken.

The research workers for this problem should have been aware of the difficulty in the analysis of "eggs laid" and have designed the experiment differently. One major contribution at the design stage to a good analysis would have been to "trap nest" the hens and record individual hen's performance. The analysis, then, would have been similar to the analysis for the weight varaible, if the assumption of no correlated errors holds.

6.2.4 An Application in Ammunition Manufacturing

In an ammunition depot an igniter was operating ineffectively and an experiment was designed in an attempt to locate the trouble. One variable of interest had to do with whether or not the igniter functioned properly and another was related to the length of time it functioned given that the igniter was satisfactory. After thorough discussion a group of engineers agreed that the trouble should be found if the following factors were examined together in an experiment:

1. Thickness of encloser discs; as thickness increases the pressure increases and the disc may blow out causing a malfunction. On the other hand, if the disc is too thin, the pressure is too low and could also cause a malfunction.

2. Powder lots from manufacturers.

3. Containers within powder lots.

4. Moisture content of ingredients in igniter.

In each of two powder lots (1 and 2) two containers were used, but there were different containers for lot 1 (A and B containers) and for lot 2 (C and D containers). The same two thicknesses of discs were used for each igniter made from the powder. The moisture content was controlled at 0.2%, 0.4%, and 0.8% for the experiment. The usual level was 0.4%, but this may be a source of trouble.

The engineers presented this picture: (continued on next page)

Discs (thickness)

1

Powder lots											
1						2					
Container						Container					
A			B			C			D		
Moisture content (%)											
0.2	0.4	0.8	0.2	0.4	0.8	0.2	0.4	0.8	0.2	0.4	0.8
1	6										
2	7										
3	8										
4	9										
5	10										

igniters in the cells

Discs (thickness)

2

Powder lots											
1						2					
Container						Container					
A			B			C			D		
Moisture content (%)											
0.2	0.4	0.8	0.2	0.4	0.8	0.2	0.4	0.8	0.2	0.4	0.8

116
.
.
119
120

but the nesting of containers repeated for discs causes trouble in understanding the following model, because one cannot easily understand where the restriction on randomization occurs:

$$y_{ijk\ell m} = \mu + D_i + P_j + DP_{ij} + C_{(j)k} + DC_{i(j)k} + M_\ell + DM_{i\ell}$$
$$+ PM_{j\ell} + DPM_{ij\ell} + CM_{\ell(j)k} + DCM_{i(j)k\ell} + \varepsilon_{(ijk\ell)m} \qquad (6.2.5)$$

$i = 1, 2 \qquad j = 1, 2 \qquad k = 1, 2 \qquad \ell = 1, 2, 3 \qquad m = 1, 2, \ldots, 5$

where

$y_{ijk\ell m}$ = variable to be analyzed if each igniter performed properly and the length of time is used

μ = overall mean

$D_i = i^{th}$ thickness of disc effect

$P_j = j^{th}$ powder lot effect

DP_{ij} = effect of the interaction of the ith thickness with the jth powder lot

$C_{(j)k}$ = effect of the kth container in the jth powder lot

$DC_{i(j)k}$ = effect of the interaction of the ith thickness of disc with the kth container in the jth powder lot

M_ℓ = effect of the ℓth moisture content

.

.

.

and so on

$\varepsilon_{(ijk\ell)m}$ = effect of the mth igniter in the $ijk\ell^{th}$ cell assumed NID $(0, \sigma^2)$

In order to see the restriction on randomization within the containers, the following arrangement seems to be better:

Powder lots

				Container		Container	
				A	B	C	D
	1	Moisture content	0.2	1, 2, 3, 4, 5	...		
			0.4				
			0.8				
Discs							
	2		0.2				
			0.4				
			0.8				116, ..., 120

and the model is easily seen to be

$$y_{ijk\ell m} = \mu + P_i + C_{(i)j} + \delta_{(ij)} + D_k + PD_{ik} + CD_{(i)jk} + M_\ell$$

$$+ PM_{i\ell} + CM_{(i)j\ell} + DM_{k\ell} + PDM_{ik\ell} + CDM_{(i)jk\ell} \qquad (6.2.6)$$

$$+ \varepsilon_{(ijk\ell)m}$$

where all the symbols are defined as before except that $\delta_{(ij)}$ = restriction error.

Of course if percentage failures are analyzed, the variable is $y_{ijk\ell}$ instead of $y_{ijk\ell m}$ because there is only one percentage computed for the 5 observations of success or failure.

Problem 6.2.4. Referring to Eq. (6.2.6), if powder lots, containers and igniters are random and discs and moisture content are fixed, show the analysis of the experiment (including source, degrees of freedom and expected mean squares) if the variable to be analyzed is to be a function of the percent failures per treatment combination. Comment on the design of this experiment in which there are five experimental units per treatment combination but only the percent of failures of the five responses can be analyzed. Is there a better design? If so, explain your reasoning and show your design.

Problem 6.2.5. An advertising firm wanted to test the effectiveness of three fixed sizes (S) and four fixed color arrangements (C) on attractiveness of billboards in five different fixed locations (L) of the country. The variable to be analyzed was a score from each of five different random judges (J) used in each of the 20 location-color arrangement combinations. Hence there were 100 judges in this experiment. Each judge looked at and scored the three sizes of billboards for his particular location-color arrangement combination.
    Recognizing that there were 300 scores in the experiment:
    (a) Show the layout of the experiment.
    (b) Show an appropriate model and analysis of the experiment.
    (c) Indicate the tests of the various main effects and interactions.
    (d) Show the standard errors for the size and the color combination.
    (e) What would be the appropriate standard error for a Newman-Keuls test on color arrangements?
    (f) How would you change the experiment so you could get a good test on location?
    (g) Suggest changes in other parts of the experiment to make the inference space broader.

COMMENT

    It has been our experience that to understand the restriction on randomization (if one exists) in nested factorial experiments, the experimenter should place the factor that has the nesting at the top of the design layout (horizontally). This allows one to see the repeated measures (if they exist) over the other factor, down the layout (vertically). As a result of all the above one can write

down the model in a manner that provides an ANOVA (with the EMS)
that is easy to interpret.

There is an example of the nested factorial experiment with no
restriction on randomization in Hicks (1973), p. 199, Problem 11.5.

6.3 REFERENCES

Anderson, V. L. Biometrics 26:255 (1970).

Anderson, R. L. and Bancroft, T. A. Statistical Theory in Research,
  McGraw-Hill, New York, 1952.

Edwards, A. L. Experimental Design in Psychological Research, Holt,
  Rinehart and Winston, New York, 1962.

Eisenhart, C. Biometrics 3:1 (1947).

Hicks, C. R. Fundamental Concepts of Design of Experiments, 2nd ed.,
  Holt, Rinehart and Winston, New York, 1973.

Greenhouse, S. W. and Geisser, S. Psychometrika 24:95-112 (1959).

Mandel, J. JASA 52:552 (1957).

Morrison, D. F. Multivariate Statistical Methods, McGraw-Hill,
  New York, 1967.

Winer, B. J. Statistical Principles in Experimental Design, 2nd ed.,
  McGraw-Hill, New York, 1971, Chapter 7.

Chapter 7

SPLIT PLOT TYPE DESIGNS

The concept of restriction on randomization for the blocks introduced in Chapter 5 on the RCBD is extended in the use of split plot designs. Actually there can be many restrictions on randomization at various levels in experimentation that will cause many "splits" in the design. These designs are sometimes called split-split plot, split-split-split plot and so on. The general split plot type designs with their associated restrictions on the randomization are considered in this chapter. For all cases covered in this chapter we assume that the effects of any correlation among errors on the analyses given is negligible.

7.1 SPLIT PLOT DESIGNS

The major difficulty that investigators have in using the split plot design is recognition of the restrictions on randomization in their experimental setup. After they understand the restrictions they must next understand the difference between the model for the factorial experiment CRD and that for the split plot design caused by the restrictions on randomization. In Chapter 4 on the CRD, the factorial experiment was introduced where no restriction on randomization existed between the treatment combinations and the experimental units. This is not the case for a split plot type experiment. Kempthorne (1952, Chapter 19, especially p. 375) gives an excellent discussion of split plot models when interactions of treatments with blocks are assumed zero.

If one has a factorial experiment with three fixed factors, say A with 3 levels, B with 4 levels and C with 5 levels, the experiment for one experimental unit per treatment combination would demand 60 experimental units. The complete randomization procedure may take place by arraying the treatment combinations as:

	A	B	C
1.	1	1	1
2.	1	1	2
3.	1	1	3
.			
.			
.			
60.	3	4	5

and choosing a random number between 1 and 60, say 2. Then assigning the second treatment combination (A at level 1, B at level 1 and C at level 2) to the first experimental unit. Continue in this manner as expressed in Chapter 4 for a completely randomized design.

A reasonable linear model for the analysis of this experiment is:

$$y_{ijk} = \mu + A_i + B_j + AB_{ij} + C_k + AC_{ik}$$
$$+ BC_{jk} + ABC_{ijk} + \epsilon_{(ijk)} \tag{7.1.1}$$

$$i = 1, 2, 3 \quad j = 1, 2, 3, 4 \quad k = 1, 2, 3, 4, 5$$

where all effects and interactions are defined as usual. The analysis for this model is given in Table 7.1.1.

TABLE 7.1.1

ANOVA Using Equation (7.1.1)

Source	df	EMS
A	2	$\sigma^2 + 20\phi(A)$
B	3	$\sigma^2 + 15\phi(B)$
AB	6	$\sigma^2 + 5\phi(AB)$
C	4	$\sigma^2 + 12\phi(C)$
AC	8	$\sigma^2 + 4\phi(AC)$
BC	12	$\sigma^2 + 3\phi(BC)$
ABC	24	$\sigma^2 + \phi(ABC)$
Error	0	σ^2
Total	59	

In the usual split plot design, the three levels of A act as blocks and the four levels of B are called whole plot treatments. Then the five levels of C are randomized within each of the twelve combinations of A and B. A linear model for the analysis of this experiment is:

$$Y_{ijk} = \mu + A_i + \delta_{(i)} + B_j + AB_{ij} + \omega_{(ij)} + C_k + AC_{ik}$$
$$+ BC_{jk} + ABC_{ijk} + \varepsilon_{(ijk)}$$

(7.1.2)

where

Y_{ijk} = variable to be analyzed from the ith level of A, j^{th} level of B and k^{th} level of C

μ = overall mean

A_i = effect of the ith level of A

$\delta_{(i)}$ = first restriction error

B_j = effect of the jth level of B

AB_{ij} = interaction effect of ith level of A with the jth level of B

$\omega_{(ij)}$ = whole plot error or second restriction error
(Anderson, 1970), and so on to

$\varepsilon_{(ijk)}$ = split plot error or error due to repeating the
combination of the ith level of A with j^{th} level of
B and k^{th} level of C

This model may be interpreted as a split-split plot model if
one calls

$\delta_{(i)}$ a whole plot error

$\omega_{(ij)}$ a split plot error

$\varepsilon_{(ijk)}$ a split-split plot error

This model approach essentially refers to a randomized complete
block design as a special case of a split plot design. In most
books, however, this design is called a split plot design because
the first restriction error is not recognized.

The analysis of variance for data from this design is given in
Table 7.1.2.

The tests for all main effects and interactions are not
straightforward in Table 7.1.2. Some assumptions on interactions
are needed or conservative tests must be made when this fixed type
design is used. If, however, it is known by the experimenter that
all interactions with A in this experiment are zero, the tests are
all obvious.

To demonstrate how the split plot design may be used in a given
experimental situation the prosthetic cardiac experiment discussed
in Chapters 5 and 6 is continued here.

The actual design of the experiment used by Beeson (1965) was
a split plot design. He set up 2 random blocks in which he forced
all 4 valve types to appear once before the second valve of any
type was used. Of course, he randomized the order of using the
valve types in the machine and randomized the 6 pulse rates for each
valve type.

TABLE 7.1.2

ANOVA Using Equation (7.1.2)

Source	df	EMS
A	2	$\sigma^2 + 5\sigma_\omega^2 + 20\sigma_\delta^2 + 20\phi(A)$
First restriction error	0	$\sigma^2 + 5\sigma_\omega^2 + 20\sigma_\delta^2$
B	3	$\sigma^2 + 5\sigma_\omega^2 + 15\phi(B)$
AB	6	$\sigma^2 + 5\sigma_\omega^2 + 5\phi(AB)$
Second restriction error	0	$\sigma^2 + 5\sigma_\omega^2$
C	4	$\sigma^2 + 12\phi(C)$
AC	8	$\sigma^2 + 4\phi(AC)$
BC	12	$\sigma^2 + 3\phi(BC)$
ABC	24	$\sigma^2 + \phi(ABC)$
Within error	0	σ^2
Total	59	

Some authors[*] believe the analysis for such an experiment should be as in Table 7.1.3, where the arrows on the mean squares indicate that the main effect of valve types would be tested by the whole plot error and that the main effects of pulse rates and the interaction of valve types by pulse rates would be tested using the split plot error assuming the correlation of errors caused by all six pulse rates occurring on the same valve for each type is negligible. In order to calculate the whole plot error sum of squares one would use the blocks by valve types interaction sum of squares and to calculate the split plot error sum of squares one would combine the blocks by pulse rates and blocks by valve types by pulse rates sums of squares.

[*] Examples are Yates (1967, p. 785) and some of his references and Federer (1955, p. 274).

TABLE 7.1.3

ANOVA of Usual Split Plot Design

Source	df	MS
Blocks	1	
Valve types	3	V⎫
Whole plot error	3	W⎬
Pulse rates	5	P⎫
Valve types x pulse rates	15	PV⎬
Split plot error	20	S⎭

Harter (1961), Chew (1958, p. 48), and others have doubts that one should always obtain the split plot error by pooling. In biological studies using the "sometimes pooling" technique described by Bozivich et al. (1956) and Bancroft (1968, p. 11), there seems to be good reason to pool "almost always." Some authors have used the sometimes pooling criterion when the split plot treatments are random.

We take the view that if the experimenter has worked in the field of investigation for some time and "knows" the errors involved, he should pool and use the analysis given in Table 7.1.3 if this seems appropriate. If, however, he does not know whether or not blocks interact with the various treatments, we believe he should let the data guide him in this decision. This thinking is consistent with our use of restriction error in the RCBD and that the following linear model is appropriate and allows for the sometimes pooling methods (a numerical example will be given later):

$$y_{ijk} = \mu + B_i + \delta_{(i)} + V_j + BV_{ij} + \omega_{(ij)} + P_k$$

$$+ BP_{ik} + VP_{jk} + BVP_{ijk} + \varepsilon_{(ijk)} \tag{7.1.3}$$

$$i = 1, 2 \quad j = 1, 2, 3, 4 \quad k = 1, 2, \ldots, 6$$

where

y_{ijk} = maximum flow gradient (mmHg) obtained from the ith block, j^{th} valve type, and k^{th} pulse rate

μ = overall mean

B_i = effect of the ith block (random), NID $(0, \sigma_B^2)$

$\delta_{(i)}$ = first restriction error zero df, NID $(0, \sigma_\delta^2)$

V_j = effect of jth valve type (fixed)

BV_{ij} = effect of the interaction of the ith block with the jth valve type, NID $(0, \sigma_{BV}^2)$

$\omega_{(ij)}$ = second restriction error, zero df, NID $(0, \sigma_\omega^2)$ (cf Kempthorne, 1952, p. 375, the η_{ij} of Eq. (13) is similar to this $\omega_{(ij)}$)

P_k = effect of kth pulse rate

BP_{ik} = effect of the interaction of the ith block and the kth pulse rate, NID $(0, \sigma_{BP}^2)$

VP_{jk} = effect of the interaction of the jth valve type with the kth pulse rate

BVP_{ijk} = effect of the interaction of the ith block with the jth valve type with the kth pulse rate, NID $(0, \sigma_{BVP}^2)$

$\varepsilon_{(ijk)}$ = within error, zero df, NID $(0, \sigma^2)$

The corresponding analysis of variance is given in Table 7.1.4. We believe that the controversy on pooling or not pooling BP and BVP may be resolved by a sometimes pooling procedure for the given problem, unless the experimenter wishes to specify his model before running his experiment.

TABLE 7.1.4

ANOVA Using Equation (7.1.3)

Source	df	EMS
Blocks (B_i)	1	$\sigma^2 + 6\sigma_\omega^2 + 24\sigma_\delta^2 + 24\sigma_B^2$
First restriction error ($\delta_{(i)}$)	0	$\sigma^2 + 6\sigma_\omega^2 + 24\sigma_\delta^2$
Valve types (V_j)	3	$\sigma^2 + 6\sigma_\omega^2 + 6\sigma_{BV}^2 + 12\phi(V)$
Interaction (BV_{ij})	3	$\sigma^2 + 6\sigma_\omega^2 + 6\sigma_{BV}^2$
Second restriction error ($\omega_{(ij)}$)	0	$\sigma^2 + 6\sigma_\omega^2$
Pulse rates (P_k)	5	$\sigma^2 + 4\sigma_{BP}^2 + 8\phi(P)$
Interaction (BP_{ik})	5	$\sigma^2 + 4\sigma_{BP}^2$
Interaction (VP_{jk})	15	$\sigma^2 + \sigma_{BVP}^2 + 2\phi(VP)$
Interaction (BVP_{ijk})	15	$\sigma^2 + \sigma_{BVP}^2$
Within error ($\varepsilon_{(ijk)}$)	0	σ^2

To give further explanation for our position in analyzing split plot data in this manner, let us compare this analysis step by step with the nested factorial (Table 6.2.2). The premise we use is that there is a possibility of a block by treatment (BV) interaction if a split plot design is used. This source has 3 df in Table 7.1.4. There is no BV interaction, of course, if the nested factorial design is used. Then the blocks and BV are not separable as in occurrences in valve types with 4 df in Table 6.2.2.

If it is agreed that a block by treatment interaction may exist, and this seems quite logical when block means are of interest as in many engineering experiments, then there certainly may be a block by treatment interaction at another stage in the design. For example BP may exist, in which case we can see no reason for the possibility that BVP cannot exist separate from BP.

All this is not to say, in many agricultural experiments where block effects and block error are not separable or the experimenter does not want to separate them, that he should not run his split plot analysis as given in Table 7.1.3. In this case the experimenter has used a model that he "knows" is correct when he began his experiment.

Our experience in various other fields is that experimenters do not have this information and may gain information by using the flexible Eq. (7.1.3) and test this model using the analysis in Table 7.1.4. The use of the restriction errors at the various stages allows the investigator to think about various sources of variation that are covered up when only the analysis in Table 7.1.3 is carried out.

Example 7.1: Using the actual data from Beeson (1965), the analysis turned out to be as that given in Table 7.1.5. When the correct tests are made, only one source, pulse rates, is significant at the $\alpha = 0.05$ level. However, the mean squares of the interactions with blocks at the various stages are about the same size, indicating that the actual variance components with blocks are possibly zero [from Eq. (7.1.3)]. Using a certain criterion for pooling, the more liberal "sometime poolers" may even obtain the analysis of Table 7.1.6. This shows that all effects and the interaction are significant; however, this last analysis is valid only if the errors are poolable and pooling is correct. In general, for a correct analysis, all restrictions on randomization must be considered before pooling procedures are undertaken.

TABLE 7.1.5

ANOVA for Data Using Eq. (7.1.3)

Source	df	MS
Blocks (B_i)	1	17.52
First restriction error $(\delta_{(i)})$	0	
Valve types (V_j)	3	420.41
Interaction (BV_{ij})	3	77.19
Second restriction error $(\omega_{(ij)})$	0	
Pulse rates (P_k)	5	427.64*
Interaction (BP_{ik})	5	51.97
Interaction (VP_{jk})	15	133.79
Interaction (BVP_{ijk})	15	56.84
Within error $(\varepsilon_{(ijk)})$	0	

*Significance at $\alpha = 0.05$.

TABLE 7.1.6

ANOVA After Obtaining "Pooled Error" from
All Sources with Blocks from Table 7.1.5

Source	df	MS
Valve types	3	420.41**
Pulse rates	5	427.64**
Valve types x pulse rates	15	133.79*
Pooled error	24	56.73

**Significance at $\alpha = 0.01$.

*Significance at $\alpha = 0.05$.

Frequently experimenters in industry do not recognize a split plot design and analyze it as a factorial experiment CRD. An example occurred in the steel industry in which the effect of temperature and orientation on strength of alloys was investigated using four furnaces in the laboratory. Each furnace had a different temperature and within each furnace the investigator randomly placed (orientation 1) two samples from each of three alloys and had an aligned arrangement (orientation 2) of two other samples from the same three alloys. In other words, he put 12 samples of alloys in each furnace, six randomly oriented and six aligned arranged. He analyzed the experiment as a completely randomized design using the between samples within the temperature, alloy, orientation source for the error. The analysis he presented for the factorial experiment is given in Table 7.1.7.

TABLE 7.1.7

ANOVA for the CRD Factorial

Source	df	EMS
Temperatures (T)	3	$\sigma^2 + 12\phi(T)$
Alloys (A)	2	$\sigma^2 + 16\phi(A)$
T x A	6	$\sigma^2 + 4\phi(TA)$
Orientations (O)	1	$\sigma^2 + 24\phi(O)$
T x O	3	$\sigma^2 + 6\phi(TO)$
A x O	2	$\sigma^2 + 8\phi(AO)$
T x A x O	6	$\sigma^2 + 2\phi(TAO)$
Between samples within (T-A-O) Cell	24	σ^2
Total	47	

It was shown that temperatures had a major effect and the best temperature from this analysis was used in production. Some time later in the plant production, it was shown that the new temperature used in this experiment had no significant effect over the old one on strength of the alloys and the experimenter could not understand why his experiment had given him the wrong results.

If he had analyzed the experiment as he had designed it (as a split plot), he would have had no error to test temperatures. Since alloy-orientation combinations are completely randomized in temperatures (furnaces), a more appropriate model for the analysis of the data from this designed experiment is:

$$y_{ijk\ell} = \mu + T_i + \delta_{(i)} + A_j + TA_{ij} + O_k + TO_{ik}$$
$$+ AO_{jk} + TAO_{ijk} + \varepsilon_{(ijk)\ell} \tag{7.1.4}$$

where

$$i = 1, 2, 3, 4 \quad j = 1, 2, 3 \quad k = 1, 2 \quad \ell = 1, 2$$

and all effects and interactions are defined in the usual manner. The corresponding analysis of variance is given in Table 7.1.8.

TABLE 7.1.8

ANOVA Using Equation (7.1.4)

Source	df	EMS
Temperatures and/or furnaces	3	$\sigma^2 + 12\sigma_\delta^2 + 12\phi(T)$
Whole plot error	0	$\sigma^2 + 12\sigma_\delta^2$
Alloys	2	$\sigma^2 + 16\phi(A)$
T x A	6	$\sigma^2 + 4\phi(TA)$
Orientation	1	$\sigma^2 + 24\phi(O)$
T x O	3	$\sigma^2 + 6\phi(TO)$
A x O	2	$\sigma^2 + 8\phi(AO)$
T x A x O	6	$\sigma^2 + 2\phi(TAO)$
Split plot error	24	σ^2
Total	47	

Since the location of the 12 samples in each furnace was at random, there was a completely randomized design in each furnace and the split plot error came from between samples within the temperature, alloy, orientation combination in each furnace and is appropriate for testing the effects and interactions from alloys through the T x A x O source. The temperatures were not replicated in this experiment and were completely confounded with furnaces, which formed the whole plot. There was a restriction on randomization in that once a temperature was chosen for a furnace the alloys and orientation level had to go in that furnace. There was no overall randomization of the combinations of temperature, alloys, and orientation for all 48 samples. Hence this is a split plot design and has no whole plot error estimate.

To obtain a whole plot error estimate it would be necessary for the experimenter to repeat the whole experiment at least once, possibly more, or take only two temperatures and obtain a repeat of the experiment using the remaining two furnaces. With furnaces random and temperatures fixed a random assignment of temperatures to the furnaces would give a RCBD at the whole plot level. There would be no test on furnaces but the interacting source would test temperatures as explained in Chapter 5. Again, the research worker must determine his inference space before he decides how he should run the experiment.

Example 7.2: In order to illustrate the calculational procedure for the analysis of data from a split plot design, consider the model presented in Eq. (7.1.4). The measured variable in this case was alloy strength. The coded values for such an experiment are shown below. The data are set up to correspond to ANOVA table presented in Table 7.1.8.

	Temperature											
	675			700			725			750		
	Alloy			Alloy			Alloy			Alloy		
	1	2	3	1	2	3	1	2	3	1	2	3
Orientation 1	2	15	20	27	35	48	47	55	62	23	33	25
	19	28	26	40	39	55	55	63	58	34	39	17
2	15	25	25	48	48	55	48	63	68	28	33	14
	23	31	33	55	62	64	60	68	62	37	38	12

The ANOVA for these data is as follows:

Source	df	SS	MS	EMS	F
Temperature (T)	3	10865.83	3621.94	$\sigma^2 + 12\sigma_\delta^2 + 12\phi(T)$	None
Whole plot error	0			$\sigma^2 + 12\sigma_\delta^2$	
Alloys (A)	2	434.29	217.15	$\sigma^2 + 16\phi(A)$	5.49*
T x A	6	1240.55	206.76	$\sigma^2 + 4\phi(TA)$	5.23*
Orientation (O)	1	468.75	468.75	$\sigma^2 + 24\phi(O)$	11.85*
T x O	3	400.42	133.47	$\sigma^2 + 6\phi(TO)$	3.38*
A x O	2	74.63	37.31	$\sigma^2 + 8\phi(AO)$	0.94
T x A x O	6	75.20	12.53	$\sigma^2 + 2\phi(TAO)$	0.32
Split plot error	24	949.00	39.54	σ^2	
Total	47	14508.67			

*Significant at the 0.05 level

The sum of squares calculations are as follows (note that the calculation is identical to an ordinary three factor experiment for a completely randomized design):

$$\sum_i \sum_j \sum_k \sum_\ell y_{ijk\ell} = T_{....} = 1880 \qquad CT = \frac{1880^2}{48} = 73633.33$$

$$\sum_i \sum_j \sum_k \sum_\ell y^2_{ijk\ell} = 88142$$

SS total = $88142 - CT = 14508.67$

$$SS\ temp = \frac{1}{12} \sum_i T^2_{i...} - CT = \frac{1}{12} (262^2 + 576^2 + 709^2 + 333^2)$$

$$- CT$$

$$= 84,499.16 - CT$$

$$= 10865.83$$

$$SS\ alloys = \frac{1}{16} \sum_j T^2_{.j..} - CT = \frac{1}{16} (561^2 + 675^2 + 644^2) - CT$$

$$= 74067.62 - CT$$

$$= 434.29$$

$$SS\ orientation = \frac{1}{24} \sum_k T^2_{..k.} - CT = \frac{1}{24} (865^2 + 1015^2) - CT$$

$$= 74,102.08 - CT$$

$$= 468.75$$

The interaction sum of squares calculations are carried out by first forming the three 2-way tables of totals as follows:

		Alloy		
		1	2	3
	675	59	99	104
	700	170	184	222
Temp	725	210	249	250
	750	122	143	68

		Orientation	
		1	2
	675	110	152
	700	244	332
Temp	725	340	369
	750	171	162

		Orientation	
		1	2
	1	247	314
	2	307	368
Alloy	3	311	333

Now,

$$SS \ T \times A = \frac{1}{4} \sum_i \sum_j T^2_{ij\cdot\cdot} - CT - SS \ A - SS \ T$$

$$= \frac{1}{4} (59^2 + 170^2 + \ldots + 250^2 + 68^2) - CT$$

$$- SS \ A - SS \ T$$

$$= 86,174 - CT - SS \ A - SS \ T$$

$$= 1240.55$$

$$SS \ T \times 0 = \frac{1}{6} \sum_i \sum_k T^2_{i\cdot k\cdot} - CT - SS \ T - SS0$$

$$= \frac{1}{6} (110^2 + 244^2 + \ldots + 162^2) - CT - SS \ T - SS \ 0$$

$$= 85368.33 - CT - SS \ T - SS \ 0$$

$$= 400.42$$

$$SS \ A \times 0 = \frac{1}{8} \sum_j \sum_k T^2_{\cdot jk\cdot} - CT - SS \ A - SS \ 0$$

$$= 74,611.00 - CT - SS \ A - SS \ 0$$

$$= 74.63$$

As before, the three factor interaction sum of squares is given by

$$SS \ T \times A \times 0 = \frac{1}{2} \sum_i \sum_j \sum_k T^2_{ijk\cdot} - CT - SS \ T - SS \ A - SS \ 0 - SS \ TxA$$

$$- SS \ T \times 0 - SS \ A \times 0$$

$$= \frac{1}{2} (21^2 + 38^2 + 43^2 + \ldots + 26^2) - CT$$

$$- SS \ T - \cdots - SS \ A \times 0$$

$$= 87,193 - CT - SS \ T - \cdots - SS \ A \times 0$$

$$= 75.20$$

$$SS \ error = SS \ total - SS \ T - SS \ A - \cdots - SS \ T \times A \times 0$$

$$\text{SS error} = \sum_i \sum_j \sum_k \sum_\ell y^2_{ijk\ell} - \frac{1}{2} \sum_i \sum_j \sum_k T^2_{ijk}.$$

$$= 88{,}142.00 - 87193.00$$

$$= 949.00$$

Since there are only two observations per cell for this problem, the error sum of squares may be calculated by summing the square of the differences of the two observations in each cell and dividing by two, i.e.,

$$\text{SS error} = \frac{1}{2} (17^2 + 8^2 + 13^2 + \ldots + 2^2)$$

$$= \frac{1}{2} (1898.00)$$

$$= 949.00$$

This last relationship may be explained by letting y_1 and y_2 be any two observations in any one cell. Then this cell contributes $(y_1 - \bar{y})^2 + (y_2 - \bar{y})^2$ to the error sum of squares where

$\bar{y} = \dfrac{y_1 + y_2}{2}$, the cell mean. Note that

$$(y_1 - \bar{y})^2 + (y_2 - \bar{y})^2 = (y_1 - \frac{y_1 + y_2}{2})^2 + (y_2 - \frac{y_1 + y_2}{2})^2$$

$$= (\frac{2y_1 - y_1 - y_2}{2})^2 + (\frac{2y_2 - y_1 - y_2}{2})^2$$

$$= \frac{(y_1 - y_2)^2}{4} + \frac{(y_2 - y_1)^2}{4}$$

$$= \frac{(y_1 - y_2)^2}{2}$$

$$= \text{difference squared over 2}$$

Summing over all cells gives the sum of squares for error for all cells.

Problem 7.1.1. An agronomist investigated the effects of three different perennial legumes on weight gain in cattle. A large field was divided into four parts. Each random part (replicate) had three fenced-in areas one for each perennial legume.

Thirty six steers of comparable age, weight and breed were allowed to graze the 12 pastures with three randomly selected steers per pasture (replication-legume combination) for a fixed amount of time. Other food and water was kept as nearly equal as possible. The gain in weight for the grazing period of time was recorded on each steer the first year.

The next year the same pastures were used again with 36 different, but similar, steers. The same type randomization of the cattle to the pastures was carried out. This experiment was continued for a total of five years.

(a) Sketch the design of the experiment.

(b) Show an appropriate model, the corresponding analysis and the tests on legumes, years and their interaction.

(c) How would you improve this experiment using the same number of animals?

Problem 7.1.2. A research worker agreed to investigate four types of heart valves. For the research, a mechanical apparatus with a water storage and circulatory system with a pulsing pump was constructed. Water was to be used to simulate the blood and a pump attached to an electric motor was to simulate the heart. The motor could vary the pulse rate from 0 to 220 beats per minute and it was decided to use the six levels: 60, 80, 100, 120, 140, and 160 beats per minute in the experiment.

The variable to be analyzed was a function of a measured efficiency variable. The experimental unit is a valve set in the equipment. When more than one pulse rate is used on a given valve there may be correlation of errors.

The purpose of the experiment was to select the best valve type for all pulse rates and/or the best valve for particular pulse rates.

Using the information above, design three different experiments with conditions stated in a, b, and c, and show their analyses:

(a) Three experimental units (valves) per treatment combination with no restriction on randomization.

(b) Allow three valves per valve type or a total of twelve valves to be used in the experiment. Choose one of the twelve valves at random, insert it in the system and run all six pulse rates at random. Next, choose one of the remaining eleven valves at random. Remove the first valve and insert the second one chosen. Run all six pulse rates rerandomized. Continue in this manner until all four valve types have been used three separate times (using a different valve each time, completely randomly chosen) with the six pulse rates at random each time.

(c) Over a six-months period, choose three days at random.

For the first randomly chosen day, choose a valve type at
random. Run all six pulse rates at random on the chosen valve.
Next choose one of the remaining three valve types at random.
Remove the first valve and insert the second randomly chosen
type. Run all six pulse rates rerandomized. Continue in the
same manner for the remaining two valve types. At this stage
only one valve has been used for each type.
        Repeat the experiment as above for the other two randomly
chosen days, using different valves for each type.
        (d) How would you prefer running the experiment [not
necessarily one of (a), (b) or (c)] to get the widest
inferences and the best tests using a maximum of seventy two
valves and/or observations? Why? Comment on costs. Consider
correlated errors.

7.2 SPLIT-SPLIT PLOT DESIGNS

Restrictions on randomization of the treatments with the
experimental units may be carried on at any number of stages in a
split plot type experiment. Each of these stages may be described
in terms of splits in the design, and a design with two stages of
restrictions on randomization is often called a split-split plot.

An example of such a design occurred in a company dealing with
metals. Three chemical formulations were examined using six heats
(two heats for each chemistry with only one randomly chosen ingot
for each heat). From each ingot three wheels of metal were cut,
one near the top of the ingot, one near the center and one near the
bottom.

From each wheel six specimens were cut at random. Three of
these specimens were placed in one environmental condition and the
other three were placed in another environmental condition. This
procedure was carried out for all combinations of heats and wheels.

In this case the heats (or ingots representing heats) acted as
whole plots, the wheels acted as sub plots because the ingots were
examined at particular spots from top to bottom (wheels). Finally
the sub-sub plots were represented by the two conditions (three
specimens placed in the same environmental condition).

A reasonable model for the analysis of this experiment is:

$$Y_{ijk\ell m} = \mu + C_i + H_{(i)j} + \delta_{(ij)} + W_k + CW_{ik} + HW_{(i)jk}$$

$$+ \omega_{(ijk)} + c_\ell + Cc_{i\ell} + Hc_{(i)j\ell} + Wc_{k\ell} + CWc_{ik\ell} \quad (7.2.1)$$

$$+ HWc_{(i)jk\ell} + \varepsilon_{(ijk\ell)m}$$

$i = 1, 2, 3 \quad j = 1, 2 \quad k = 1, 2, 3 \quad \ell = 1, 2 \quad m = 1, 2, 3$

where

$Y_{ijk\ell m}$ = variable to be analyzed

μ = overall mean

C_i = effect of the ith chemistry (fixed)

$H_{(i)j}$ = effect of the jth heat (random) in the ith chemistry

$\delta_{(ij)}$ = first restriction error

W_k = effect of the kth wheel height (fixed)

CW_{ik} = effect of the interaction of the ith chemistry with the kth wheel height

$HW_{(i)jk}$ = effect of the interaction of the jth heat in the ith chemistry with the kth wheel height

$\omega_{(ijk)}$ = second restriction error

c_ℓ = effect of the ℓth condition (fixed)

$Cc_{i\ell}$ = effect of the interaction of the ith chemistry with the ℓth condition

$Hc_{(i)j\ell}$ = effect of the interaction of the jth heat in the ith chemistry with the ℓth condition

$Wc_{k\ell}$ = effect of the interaction of the kth wheel height with the ℓth condition

$CWc_{ik\ell}$ = effect of the interaction of the ith chemistry with the kth wheel height with the ℓth condition

$HWc_{(i)jk\ell}$ = effect of the interaction of the jth heat in the ith chemistry with the kth wheel height with the ℓth condition

$\varepsilon_{(ijk\ell)m}$ = within error of specimens within conditions, wheels and heats in chemistries

The corresponding analysis of variance is given in Table 7.2.1.

TABLE 7.2.1

ANOVA Using Equation (7.2.1)

Source	df	EMS
Whole Plot		
Chemistry (C)	2	$\sigma^2 + 6\sigma_\omega^2 + 18\sigma_\delta^2 + 18\sigma_H^2 + 36\phi(C)$
Heats and/or ingots in C(H)	3	$\sigma^2 + 6\sigma_\omega^2 + 18\sigma_\delta^2 + 18\sigma_H^2$
First restriction error	0	$\sigma^2 + 6\sigma_\omega^2 + 18\sigma_\delta^2$
Split Plot		
Wheel heights (W)	2	$\sigma^2 + 6\sigma_\omega^2 + 6\sigma_{HW}^2 + 36\phi(W)$
C x W	4	$\sigma^2 + 6\sigma_\omega^2 + 6\sigma_{HW}^2 + 12\phi(CW)$
H x W	6	$\sigma^2 + 6\sigma_\omega^2 + 6\sigma_{HW}^2$
Second restriction error	0	$\sigma^2 + 6\sigma_\omega^2$
Split-Split Plot		
Conditions (c)	1	$\sigma^2 + 9\sigma_{Hc}^2 + 54\phi(c)$
C x c	2	$\sigma^2 + 9\sigma_{Hc}^2 + 18\phi(Cc)$
H x c	3	$\sigma^2 + 9\sigma_{Hc}^2$
W x c	2	$\sigma^2 + 3\sigma_{HWc}^2 + 18\phi(Wc)$
C x W x c	4	$\sigma^2 + 3\sigma_{HWc}^2 + 6\phi(CWc)$
H x W x c	6	$\sigma^2 + 3\sigma_{HWc}^2$
Within error	72	σ^2
Total	107	

The tests are obvious from the expected mean squares. It is
interesting to notice, however, that the information obtained by
taking three specimens per condition is not necessary. It would
have been almost as informative to take only two specimens per
condition.

> Problem 7.2.1. Would it have improved the experiment above
> with the ANOVA given in Table 7.2.1 to have taken more than
> one ingot per heat and taken fewer heats? Explain carefully.

To reduce the cost of the experiment, the investigators would
have liked to have run only one heat per chemistry, which would have
given no degrees of freedom for testing chemistries. Of course, this
would have given no information on the most important factor. To
examine the efficiency of the design one could look at the
combinations of the number of heats per chemistry and the number of
specimens per condition with the corresponding number of degrees of
freedom for the errors.

Consider the number of heats per chemistry as follows:

Number of heats per Chemistry	Degrees of freedom for testing chemistries
1	0
2	3
3	6
4	9

When considering efficiency there is not much value in running the
experiment if only one heat is used per chemistry. There is a 100%
gain in degrees of freedom for error going from two to three heats
per chemistry, but only 50% gain in going from three to four. The
increase in efficiency of the design decreases as the number of heats
per chemistry increases from there on.

Next, consider the number of specimens per condition. If only
one were used there would be zero degrees of freedom for the within
error, which may not make much difference.

It seems there is almost no advantage in taking more than one specimen per condition if the variances are homogeneous, but usually experimenters would like a few repeats in cells to check equipment and operators.

Overall recommendations for the design of the experiment:

(a) If possible use three heats per chemistry and one specimen within conditions, with a few repeats. Of course four heats would be better but the experimenters were not allowed that much expense.

(b) If (a) is too expensive, use two heats per chemistry.

(c) If two heats per chemistry is too expensive and a test for chemistry is desired, there is not much reason to run the experiment.

Problem 7.2.2. In a heat paper battery experiment the variable to be analyzed was the energy output of 25 small pieces of paper cut from a large square sheet of heat paper. The sheets of heat paper were made up of combustible material spread over a sheet of paper as evenly as the equipment would allow. The experimental units (small circular pieces) were cut out of the sheets in the following manner:

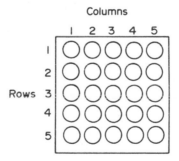

The active ingredients are most important to the experimenter and the effects of rows and columns in describing the response surface of the sheets are less important. Each sheet was selected at random for a run through the equipment and was made up of one combination of three active ingredients (A at 4 levels, B at 3 levels and C at 2 levels). This made a total of 24 sheets (4 x 3 x 2) for one experiment.

(a) Show the design of the experiment, emphasizing the randomization and inference space.

(b) Show the model and analysis of variance for your experiment.

(c) Explain what results can be obtained (include the response surface for the sheets).

(d) Did the design allow for the most important effects to be estimated with the smallest variance? Comment.

(e) When the actual experiment was run and the data analyzed, the error mean square within the sheets of heat paper was 10 times larger than the error mean square between the sheets. What would you suggest doing next?

Problem 7.2.3. An experiment was conducted on cold rolled steel, and was repeated a week later. Two heats were run each week and the same two cold rolled methods were used for each of the four heats. All interaction with heats and weeks (both considered random and no interest to the experimenter except to provide the inference base) were impossible because there must be different heats nested in the different weeks.

Hence there were eight coils of steel, four from each rolling method, and each of these coils was cut in two. One-half of each coil was run at one milling speed and the other half at another speed. Each half coil was then examined at two places; one place was at the end away from the cut that divided the coil and the other was 50 feet from the end. A schematic diagram of a half coil is shown in Fig. 7.1. Two random samples of steel were taken at each of these two places on each half coil to allow measurements of many variables. Using a variable to be analyzed associated with magnetic properties the analysis of variance (only the source and mean squares) for this problem is shown in Table 7.2.2. These results were obtained from a computer output that did not take into account the nesting quality of heats.

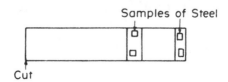

Samples of Steel

Cut

Fig. 7.1. A schematic diagram of a half coil.

(a) Show a diagram of the whole experiment.

(b) Assuming all factors are fixed except heats and weeks, analyze the data appropriately showing the model and correct ANOVA using the mean square data and degrees of freedom. (The df and SS for all nested sources may be obtained in a fashion similar to that presented in Example 6.1).

(c) Run the appropriate tests of significance and show the standard errors necessary to run individual comparisons where necessary. Explain the results.

TABLE 7.2.2

ANOVA for Problem 7.2.3

Source	MS
Weeks (W)	4.4
Heats (H)	34.7
H x W	20.6
Cold rolled methods (C)	116.8
C x H	1.4
C x W	0.4
C x H x W	11.4
Mill speed (S)	3.1
S x H	23.6
S x W	17.9
S x H x W	2.5
S x C	9.1
S x C x H	0.3
S x C x W	12.8
S x C x H x W	2.0
Positions (P) (ends or 50 feet)	23.9
P x H	0.7
P x W	6.3
P x H x W	0.2
P x C	3.2
P x C x H	9.5
P x C x W	1.0
P x C x H x W	3.4
P x S	23.4
P x S x H	0.5
P x S x W	0.2
P x S x H x W	0.2
P x S x C	2.8
P x S x C x H	3.8
P x S x C x W	1.2
P x S x C x H x W	4.7
Within cells	2.6

(d) Explain the role of the two experimental units for each treatment combination. How would you have designed the experiment? Why?

7.3 REGRESSION ANALYSES FROM SPLIT PLOT DATA

In Chapter 2 it was made clear that there could be only one error in a regression analysis. This implies that a CRD is necessary for obtaining the data. Of course, in actual practice most of the regression analyses are run on unstructured data. That is, the design of the experiment is ill-conceived or the investigator merely uses data on hand rather than controlling levels of factors. In general, however the regression analysis assumes there is only one error and the design is completely randomized.

Many times experimenters want to use a regression analysis after the ANOVA has been run on the split plot data. If the various errors in the analysis are not poolable the whole plot effects and interactions regression coefficients will have variances that are too small and the split plot main effects and interactions regression coefficients will have variances that are too large. A reference for this concept is Daniel and Wood (1971, p. 59) on nested data. In effect the regression error from split plot data or any data coming from a design with restriction on randomization such as RCBD and nested factorials, is a diluted average of the appropriate errors.

One proposal for correct regression analyses of split plot data is to use the means of the whole plot data with a separate regression analysis and individual observation data for the split plot part. Thus forcing two regression analyses for split plot data. This, of course, is not very satisfactory and it would be much better if the errors could be pooled in the ANOVA before the regression analysis is run.

Problem 7.3.1. A research organization ran the following experiment to examine the compressive strength (y) of concrete when steel fibers were used in the mix.
Three factors were used in the experiment.

(a)   Mix types: (1)   sand with cement; (2)   3/8 inch
rock with cement; (3)   3/4 inch rock with cement.
(b)   Percent volume of fibers: 0, 1, 1.5, 2, 2.5.
(c)   Size of cylinders: 3" x 6", 4" x 8", 6" x 12".
The experiment was conducted in the following manner:

Day 1:   One batch of mix (sand with cement) was made up and
        split into 5 parts.
           The first part had   0% fibers added,
           the second part had   1% fibers added,
           the third part had   1.5% fibers added,
           the fourth part had   2% fibers added,
        and the fifth part had 2.5% fibers added.

        Each part was made into 6 cylinders, two each for the
        three sizes: 3" x 6", 4" x 8", 6" x 12".

Day 2:   Repeat of Day 1.

Day 3:   The batch of mix was made of 3/8" rock with cement and
        the same procedure was followed for the other two
        factors as in Days 1 and 2.

Day 4:   Repeat of Day 3.

Day 5:   The batch of mix was made of 3/4" rock with cement and
        the same procedure as above was followed for the other
        two factors.

Day 6:   Repeat of Day 5.

Each day's cylinders were tested at random after each day's
cylinders set for a given time.

    Answer the following:
(a)   Show the layout of the experiment.   If the three factors
are fixed and days are random, describe the inference space.
(b)   Describe briefly two weaknesses at the design stage and
what these could cause in the interpretation of the analysis.
(c)   Write down the model assuming the weaknesses have not
prevented an analysis of data, show the limits on each index
and identify each symbol.
(d)   Show the ANOVA (source, df and EMS) and indicate
appropriate tests using arrows.   Indicate where pooling of
errors is likely.
(e)   If the source and df printout from the computer are as
shown in the following tabulation, show what sources and df
must be pooled and where the restriction errors are to get the
correct analysis in (d).

Source	df
Mixes (M)	2
Days (D)	1
M x D	2
Fibers (F)	4
MF	8
DF	4
MDF	8
S	2
MS	4
DS	2
MDS	4
FS	8
MFS	16
DFS	8
MDFS	16
within	90

(f)   If it turned out that the main effects were all significant
and no interactions were significant show what cautions you
would make in the interpretations and what further analyses you
would do on each.

(g)   What further analyses would be appropriate if the
interaction of mixes x fibers turned out to be significant?
Consider regression analyses.

(h)   Consider the whole experimental procedure used and
recommend a better design explaining at each stage why you
think your design is better. (Keep the cost of running the
experiment in mind, so you do not run a completely randomized
design each day.)

7.4 REFERENCES

Anderson, R. L. and Bancroft, T. A. Statistical Theory in Research,
McGraw-Hill, New York, 1952, pp. 344-351.

Bancroft, T. A. Topics in Intermediate Statistical Methods, Iowa
State University Press, Ames, Iowa, 1968.

Beeson, J. Unpublished M.S. Thesis, Purdue University Library,
West Lafayette, Indiana, 1965.

Bennett, C. A. and Franklin, N. L. Statistical Methods (In Chemistry
and Chemical Industry), Wiley, New York, 1954, pp. 539-545.

Bozivich, H., Bancroft, T. A. and Hartley, H. O. Ann. Math. Statist.
27:1017 (1956).

Bliss, C. I. Statistics in Biology, McGraw-Hill, New York, 1960,
Section 12.2.

Chew, V. (Editor) Experimental Designs in Industry, Wiley, New York, 1958, pp. 48-52.

Cochran, W. G. and Cox, G. M. Experimental Designs, 2nd ed., Wiley, New York, 1957, Chapter 7.

Daniel, C. and Wood, F. S. Fitting Equations to Data, Wiley, New York, 1971.

Federer, W. T. Experimental Design (Theory and Application), Macmillan, New York, 1955, Chapter X.

Fisher, R. A. The Design of Experiments, 8th ed., Hafner, New York, 1966.

Harter, H. L. Biometrics 17:144 (1961).

Kempthorne, O. Design and Analysis of Experiments, Wiley, New York, 1952, Chapter 19. Distributed by Krieger Pub. Co., Huntington, New York.

Ostle, B. Statistics in Research, 2nd ed., Iowa State University Press, Ames, Iowa, 1963.

Snedecor, G. W. and Cochran, W. G. Statistical Methods, 6th ed., Iowa State University Press, Ames, Iowa, 1967.

Steel, R. G. D. and Torrie, J. H. Principles and Procedures of Statistics, McGraw-Hill, New York, 1960, Chapter 12.

Winer, B. J. Statistical Principles in Experimental Design, 2nd ed., McGraw-Hill, New York, 1971.

Yates, F. Fifth Berkeley Symp. Vol. IV, Biology and Problems of Health, 777 (1967).

Chapter 8

LATIN SQUARE TYPE DESIGNS

The completely randomized design (CRD), the randomized complete
block design (RCBD) and the split plot designs are used more
frequently in experimentation than the designs in this chapter. The
Latin square (LS) and other designs similar to it are used by some
research workers, but are misused by others more frequently. Hence
in this chapter we will try to point out the type of experimental
problems in which these designs are appropriate and the mistakes
that experimenters may make if these designs are not handled
correctly.

8.1 LATIN SQUARE

In the RCBD each block is large enough to contain all of the t
treatments. In the Latin square design "square" refers to the
existance of t^2 experimental units, each of the t treatments are
assigned in a special way, as described below, to t of the
experimental units. The t treatments are labeled with Latin
letters. Hence the name Latin square.

The actual arrangement of the t treatments onto the t^2
experimental units may be handled by drawing a square where the
sides of the square are called rows and columns and the restrictions
on the drawn square are such that each treatment must be represented
once and only once in each row and once and only once in each column.
The assignment of treatments to the experimental units is done at
random as long as the two restrictions are observed. Thus the term

random square is used to refer to a specific design.   The
randomization procedure and tables of many squares are given by
Fisher and Yates (1963, pp. 22-25 and 86-89).

In agronomy field experiments (where the inference is confined
to a specific field) frequently the experimenters find that there
is a fertility trend in the direction perpendicular to the blocks
used in a randomized complete block design.   To account for the
effect of this additional extraneous variable and remove it from
the error in the model, blocks may be set up as

Columns (Blocks)

Fertility Trend ————————————————————————→

		1	2		c
	1	1	2	. . . .	t
	2	2		. . . .	
Rows (Blocks) :	:	.	.	. . . .	.
	r	t	.	. . . .	

In this case t plots are required for blocking in both directions.
This naturally forces the number of rows (r) and columns (c) to be
equal to the number of plots (t) and the t treatments are assigned
at random to each row and column combinations as discussed above.
The t treatments are sometimes given Latin letters, A, B, ..., to
distinguish them from the numbers in the rows and columns and this
constructed square is called a Latin square.

There are two restrictions inherent in this randomization scheme
and these restrictions are recognized as rows and columns in the
analysis (comparable to blocks in the RCBD).   Using the same thought
process as was used on the RCBD with blocks fixed and assuming all
interactions are zero, the model could be written as:

$$y_{ijk} = \mu + R_i + \delta_{(i)} + C_j + \eta_{(j)} + T_k + \varepsilon_{(ijk)} \qquad (8.1.1)$$

$$i = j = k = 1, 2, \ldots, t$$

where

y_{ijk} = variable to be analyzed from the ith row j^{th} column and k^{th} treatment

μ = overall mean

R_i = effect of the ith row

$\delta_{(i)}$ = row error, due to restriction on randomization (Anderson, 1970) of treatments in the ith row

C_j = effect of the jth column

$\eta_{(j)}$ = column error, due to restriction on randomization of treatments in the jth column

T_k = effect of the kth treatment

$\varepsilon_{(ijk)}$ = experimental error from the ith row, j^{th} column and k^{th} treatment estimated by the residual.

The corresponding ANOVA is shown in Table 8.1.1.

TABLE 8.1.1

ANOVA Using Equation (8.1.1)

Source	df	EMS
Rows (R)	$t - 1$	$\sigma^2 + t\sigma_\delta^2 + t\phi(R)$
Row restriction error	0	$\sigma^2 + t\sigma_\delta^2$
Columns	$t - 1$	$\sigma^2 + t\sigma_\eta^2 + t\phi(C)$
Column restriction error	0	$\sigma^2 + t\sigma_\eta^2$
Treatments	$t - 1$	$\sigma^2 + t\phi(T)$
Experimental error (residual)	$(t - 1)(t - 2)$	σ^2
Total	$t^2 - 1$	

It should be noted that for the Latin square design, since it
is a type of fractional replication (to be discussed in Chapter 10),
the algorithm to obtain the EMS given in Chapter 1 cannot be used.
One must go through a proof of the EMS for each source. References
for deriving EMS are Anderson and Bancroft (1952), Wilk and
Kempthorne (1957), and Edwards (1962).

Since the agricultural experimenter is not interested in the
effects of rows and columns (essentially he knows they are large),
the analysis satisfies him (which indicates the design is
satisfactory). As shown in the above analysis, the variation
remaining (after removal of rows, columns and treatments variation)
was used as the experimental error. Under the usual experimental
conditions and if no interactions exist among these three factors
(rows, columns, and treatments) the test is quite valid. For many
experiments in agriculture the requirements for valid tests may
usually be met, but for many experiments in engineering, social
sciences and other research areas they are not met and the results
could be interpreted erroneously if the experimenter merely uses
this design without understanding the assumptions.

To illustrate some of the difficulties an experimenter may
encounter using this design, consider an experiment on explosive
switches. There were five fixed treatments, that is, five levels
of packing pressures (10,000, 15,000, 20,000, 25,000, and 30,000
psi), five fixed machines (which act as rows) to make the switches
and five fixed men (which act as columns) to fire them. Hence there
are 25 combinations of machines and men with a treatment in each
cell. The design of a selected random square is given in Table 8.1.2.

The 25 switches were fired in a completely random order. If we
do not assume the interactions are zero and if all treatment
combinations were run as a CRD, the model would be

TABLE 8.1.2

Explosive Switch Latin Square

Machines (M)	Men (m)				
	1	2	3	4	5
1	20,000	15,000	10,000	25,000	30,000
2	15,000	25,000	30,000	20,000	10,000
3	10,000	30,000	20,000	15,000	25,000
4	25,000	10,000	15,000	30,000	20,000
5	30,000	20,000	25,000	10,000	15,000

$$y_{ijk} = \mu + M_i + m_j + Mm_{ij} + T_k + MT_{ik} + mT_{jk}$$
$$+ MmT_{ijk} + \varepsilon_{(ijk)}$$

(8.1.2)

where

y_{ijk} = the firing time of the switch made by the ith machine, by the jth man with the kth packing pressure

μ = the overall mean

M_i = the effect of the ith machine

m_j = the effect of the jth man

T_k = the effect of the kth packing pressure

$\vdots$

and all interactions, plus

$\varepsilon_{(ijk)}$ = error within the $(ijk)^{th}$ cell.

The ANOVA for the part of the factorial experiment actually run from Table 8.1.2 and using a CRD as given for Eq. (8.1.2), is given in Table 8.1.3.

TABLE 8.1.3

ANOVA Using Equation (8.1.2)

Source	df	EMS
Machine (M)	4	$\sigma^2 + \phi(MmT) + \phi(mT) + 5\phi(M)$
Men (m)	4	$\sigma^2 + \phi(MmT) + \phi(MT) + 5\phi(m)$
Packing pressure (T)	4	$\sigma^2 + \phi(MmT) + \phi(Mm) + 5\phi(T)$
Residual	12	$\sigma^2 + \phi(MmT) + \phi(Mm) + \phi(MT) + \phi(mT)$
Within error	0	σ^2

There is no restriction error in this analysis because the design is a CRD, taking only one-fifth of the 125 possible treatment combinations. One should note that there has been a restriction utilized in selecting 25 treatment combinations but for the analysis given in Table 8.1.3 it is assumed that the restriction error is negligible. References for developing the EMS are Wilk and Kempthorne (1957) and Edwards (1962, p. 312). This design is not, however, a usual fractional factorial. This concept will be clarified in Chapter 10.

In the randomized complete block design, allowance was made in the model for the possible interaction of the treatments and the blocks. In the Latin square, no such allowance is usually made for the tests, and if there happens to be an interaction present, the estimates of the effects will be biased. For example, in Table 8.1.3, if men interact with machines, part of that effect goes into the packing pressure effect which in turn may appear large when in fact it is not.

Whether rows, columns and treatments are random or fixed does not make much difference because all influences of the interactions must be zero before the treatments can be analyzed properly.

Experimenters have used this design to reduce the number of experimental units from t^3 to t^2 without understanding that they were assuming the interactions were zero. When one looks at the model and compares it with the analysis it is immediately known, but the consequences of this assumption are not always appreciated. To obtain a separate estimate of the interactions in one term and still keep the number of observations rather small, Youden and Hunter (1955) suggested repeating t of the combinations on the diagonal.

The Latin square design is most valuable to those experimenters who use only restrictions on randomization (not true factors of interest to the experimenter) for the rows and columns and know there are no interactions present. The design is also useful in screening for major effects when all three factors are of interest to the experimenter if the assumption is made that even if interactions are present, they are small relative to the main effects. Hence if the main effect mean square is large, the experimenter concludes it is due to the main effect more than the interactions, and if an effect appears to be small it is concluded that it has little influence on the variable being analyzed and would not be used in future experimentation (see Cochran and Cox, 1957).

The Latin square design is not too useful for fewer than four treatments nor more than eight (Kempthorne, 1952).

Example 8.1: In order to illustrate the calculational technique for data from a Latin square design consider the coded data given below which represent the firing time for the design given in Table 8.1.2. In order to eliminate confusion in the table of observations, the packing pressures 10,000, 15,000, 20,000, 25,000, and 30,000 psi are represented by the respective Latin letters A, B, C, D, and E.

	Men										
Machines	1		2		3		4		5		Total
1	C	25	B	17	A	13	D	32	E	42	129
2	B	2	D	26	E	35	C	8	A	8	79
3	A	6	E	43	C	16	B	14	D	25	104
4	D	27	A	9	B	17	E	31	C	19	103
5	E	31	C	22	D	21	A	2	B	13	89
Total		91		117		102		87		107	504

The treatment totals are obtained as follows:

A	B	C	D	E
13	17	25	32	42
8	2	8	26	35
6	14	16	25	43
9	17	19	27	31
2	13	22	21	31
38	63	90	131	182

The resulting ANOVA is shown in the following tabulation.

Source	df	SS	MS	F
Columns (Men)	4	117.76	29.44	
Rows (Machines)	4	284.96	71.24	
Pressures	4	2598.96	649.74	42.44
Residual	12	183.68	15.31	
Total	24	3185.36		

where

$$\sum_i \sum_j y_{ijk} = 504 \qquad CT = \frac{504^2}{25} = 10,160.64$$

$$\sum_i \sum_j y_{ijk}^2 = 13,346$$

SS total = 13,346 - CT = 3185.36

$$SS \text{ columns} = \frac{1}{5} (91^2 + 117^2 + 102^2 + 87^2 + 107^2) - CT$$

$$= 10278.4 - CT$$

$$= 117.76$$

$$SS \text{ rows} = \frac{1}{5} (129^2 + 79^2 + 104^2 + 103^2 + 89^2) - CT$$

$$= 284.96$$

$$SS \text{ pressures} = \frac{1}{5} (38^2 + 63^2 + 90^2 + 131^2 + 182^2) - CT$$

$$= 2598.96$$

$$SS \text{ residual} = SS \text{ total} - SS \text{ columns} - SS \text{ rows} - SS \text{ pressure}$$

$$= 3185.36 - 117.76 - 284.96 - 2598.96$$

$$= 183.68$$

8.2 ASSOCIATED DESIGNS

An associated design is the Graeco-Latin square which adds one more restriction on randomization and uses Greek letters within the t^2 cells along with the Latin letters in such a manner that no combination of Greek-Latin letters are repeated (Hicks, 1973, pp. 79-80). In this case four factors can be investigated at once, there may be restrictions on randomization, but if all combinations were considered there is a reduction from t^4 to t^2 experimental units. Another associated design is the Youden square (a form of an incomplete Latin square design) which allows rectangular arrangements (Hicks, 1973, Section 5.4). In addition, there are systematic squares such as the Knut Vik, that have been used by experimenters, but the experimental error is in question on these designs even if there are no interactions (Kempthorne, 1952, p. 330; Fisher, 1966, pp. 76-80; and Yates, 1964, pp. 318-319).

A very useful design for experimenters confronted with the problem that some of the experimental units are better than others or time causes a change in the units is the cross over design. Cochran and Cox (1957, pp. 127-141) show how a cross over design is a special type of the Latin square design.

Example 8.2: An illustration of the cross over design occurred in a problem for a drug manufacturing company. Two formulas of a drug were given to 18 human beings selected at random from a population to which the drug would be given. Nine random subjects were given formula one and the other nine were given formula two. The amount of the drug in the blood of each of the 18 subjects was measured at specified intervals of time and the cumulative amount was recorded.

After 7 days of "washout" period in which the drug supposedly was excreted, another blood sample was taken and all 18 subjects had zero amount of the drug in their blood samples. The experimenter then had the group of nine who took formula one take formula two and vice versa for the other group. This part of the experiment is called the "cross over" part of the design.

A model for the analysis of this experiment is:

$$y_{ijk} = \mu + S_i + \delta_{(i)} + O_j + \eta_{(j)} + F_k + \epsilon_{(ijk)}$$

$$i = 1, 2, \ldots, 18 \qquad j = 1, 2 \qquad k = 1, 2$$

where

y_{ijk} = cumulative amount of drug in the blood of the ith subject after 8 hours given order j and formula k.

All other terms are comparable to those of Eq. (8.1.1).

The data from this experiment are given in the following tabulation.

	Subjects								
Order	1	2	3	4	5	6	7	8	9
	Formula One								(351.7)
1 (Initial) 7 day "washout"	51.9	35.1	38.6	36.1	34.6	39.7	37.8	38.8	39.1
	Formula Two								(376.1)
2 (Crossover)	43.5	45.4	35.4	43.7	49.8	39.9	41.4	37.5	39.5
Total	95.4	80.5	74.0	79.8	84.4	79.6	79.2	76.3	78.6

	Subjects								
Order	10	11	12	13	14	15	16	17	18
	Formula Two								(332.0)
1 (Initial) 7 day "washout"	50.8	41.1	39.1	35.7	33.7	31.2	34.3	31.8	34.3
	Formula One								(316.6)
2 (Crossover)	44.2	33.4	32.7	33.1	33.4	27.1	33.1	39.5	40.1
Total	95.0	74.5	71.8	68.8	67.1	58.3	67.4	71.3	74.4

$$\text{SS order} = \frac{(\text{Initial})^2 + (\text{Crossover})^2}{18} - \left[\frac{(\text{Grand total})^2}{36} = (\text{CT})\right]$$

$$= \frac{(683.7)^2 + (692.7)^2}{18} - \frac{(1376.4)^2}{36} = \underline{2.25}$$

$$\text{SS formula} = \frac{(\text{Formula one})^2 + (\text{Formula two})^2}{18} - (\text{CT})$$

$$= \frac{(668.3)^2 + (708.1)^2}{18} - 52,624.36 = \underline{44.00}$$

$$\text{SS subjects} = \frac{(95.4)^2 + (80.5)^2 + \dots + (74.4)^2}{2}$$

$$- 52,624.36 = \underline{717.79}$$

$$\text{SS total} = (51.9)^2 + (35.1)^2 + \ldots + (40.1)^2$$

$$- 52,624.36 = \underline{1,093.98}$$

SS error = SS total - SS order - SS formulas - SS subjects

$$= 1093.98 - 2.25 - 44.00 - 717.79 = \underline{329.94}$$

The corresponding ANOVA is tabulated in the following.

Source	df	SS	MS	EMS
Subjects (S_i)	17	717.79	42.22	$\sigma^2 + 2\sigma_\delta^2 + 2\sigma_S^2$
Subject restriction error $(\delta_{(i)})$	0			$\sigma^2 + 2\sigma_\delta^2$
Order (O_j)	1	2.25	2.25	$\sigma^2 + 18\sigma_\eta^2 + 18\phi(0)$
Order restriction error $(\eta_{(j)})$	0			$\sigma^2 + 18\sigma_\eta^2$
Formulas (F_k)	1	44.00	44.00	$\sigma^2 + 18\phi(F)$
Residual $(\epsilon_{(ijk)})$	16	329.94	20.62	σ^2
Total	35	1,093.98		

To find the smallest error from these data we may test for order

$$H_o : \sigma_\eta^2 + \phi(0) = 0$$

$$F_{1,16} = \frac{2.25}{20.62} < 1$$

Hence accept H_o at $\alpha > 0.25$ and pool mean squares, which actually results in a RCBD where subjects act as blocks and formulas are treatments. The error mean square with 17 df is:

$$\text{Error mean square} = \frac{(1)(2.25) + (16)(20.62)}{17} = \underline{19.54}$$

Subjects are random and no interest other than to provide the inference space. The test on formulas is

$$F_{1,17} = \frac{44.00}{19.54} = 2.25$$

Tabled $F_{1,17}$ $\begin{array}{l} \alpha = 0.10 = 3.03 \\ \alpha = 0.25 = 1.42 \end{array}$

Hence the effect of formulas is not significant at $\alpha = 0.10$ level.

Since the tabled $F_{1,\infty}(\alpha = 0.10) = 2.71$, the sample size (number of subjects) could be increased indefinitely with these results (F = 2.25) and there would not be a significant effect due to formulas at the $\alpha = 0.10$ level.

Problem 8.2.1. In a problem on processing of plastics, yield strength of the material (coded for analysis purposes) was of interest. The factors used in the experiment were:

Preconditioning temperatures 1, 2, 3, 4
Preconditioning time   1, 2, 3, 4
Materials   a, b, c, d

The experimental layout and coded yield strengths are given in the following tabulation.

Times	Temperatures 1		2		3		4	
1	a	9.4	b	13.1	c	12.4	d	30.2
2	b	22.6	c	20.9	d	33.7	a	13.7
3	c	9.6	d	34.6	a	4.6	b	13.8
4	d	29.7	a	10.2	b	15.2	c	18.7

(a) Analyze the data (include the model, analysis of variance, means of levels of all factors and standard errors). Be sure to show your EMS.
(b) Discuss the design of the experiment and assumptions involved for this design.
(c) Draw conclusions from the experiment.
(d) Assuming design is LS, determine the relative efficiency of this Latin square design with the RCBD where times are blocks. Also determine the relative efficiency of this Latin square design with the CRD (refer to Section 5.4).
(e) Based on your results of part (d), how would you design future experiments of the same type.

Problem 8.2.2. In a study comparable to that of Example 8.2 except that blood samples were not taken after the 7 days to check the "washout" period. The data are given in the following tabulation.

Order	Subjects								
	1	2	3	4	5	6	7	8	9
	Formula Two								
1 Initial	51.9	35.1	38.6	36.1	34.6	39.7	37.8	38.8	39.1
	Formula One								
2 Crossover	58.5	60.4	50.4	58.7	64.8	54.9	56.4	52.5	54.5

Order	Subjects								
	10	11	12	13	14	15	16	17	18
	Formula One								
1 Initial	50.8	41.1	39.1	35.7	33.7	31.2	34.3	31.8	34.3
	Formula Two								
2 Crossover	62.2	51.4	50.7	51.1	51.4	45.1	51.1	57.5	58.1

(a) Analyze the data appropriately; showing the model and ANOVA with df, SS, MS and EMS.
(b) Interpret the results.
(c) Does this design result in a RCBD? Explain.

8.3 REFERENCES

Anderson, R. L. and Bancroft, T. A. Statistical Theory in Research, McGraw-Hill, New York, 1952.

Anderson, V. L. Biometrics 26:255 (1970).

Cochran, W. G. and Cox, G. M. Experimental Designs, 2nd ed., Wiley, New York, 1957.

Edwards, A. L. Experimental Design in Psychological Research, Holt, Rinehart and Winston, New York, 1962, pp. 311-313.

Fisher, R. A. The Design of Experiments, 8th ed., Hafner, New York, 1966.

Fisher, R. A. and Yates, F. Statistical Tables for Biological, Agricultural and Medical Research, 6th ed., Hafner, New York, 1963.

Hicks, C. R. Fundamental Concepts of Design of Experiments, 2nd ed., Holt, Rinehart and Winston, New York, 1973.

Kempthorne, O. The Design and Analysis of Experiments, Wiley, New York, 1952. Distributed by Krieger Pub. Co., Huntington, New York.

Wilk, M. G. and Kempthorne, O. JASA 52:218 (1957).

Yates, F. Biometrics 20:307 (1964).

Youden, W. J. and Hunter, J. S. Biometrics 11:399 (1955).

Chapter 9

2^n FACTORIAL EXPERIMENTS (COMPLETE AND INCOMPLETE BLOCKS)

In Chapter 4, Section 4.2, a general discussion of factorial experiments was presented. Whereas the meaning of factorial experiments is that all possible combinations of the levels of the factors exist in the experiment, this in no way describes the design of the experiment as dictated by the randomization of the combinations onto the experimental units. Hence, "factorial" refers only to the type of experiment and not to the design or layout. In this chapter the special case in which all factors of a factorial experiment have only two levels is described and the design is assumed to be completely randomized within each block.

In many experimental situations, the investigator would like to examine the effects and interactions of many variables (n factors) simultaneously on a dependent variable. The 2^n factorial experiments provide information on all possible two factor interactions if the experiment is designed properly. In many cases the investigator will follow an experiment of this type with another experiment with more levels on a subset of the original factors in order to obtain more detailed information on the most important factors. In order to investigate non-linear trends one must have more than two levels of the factors.

In the past it was felt that two leveled factorials were good only for exploratory work by some research workers. Recently it seems that the philosophy of experimenters has changed in this regard. Many investigators believe it is better to introduce many factors each with a high and low level into the experiment rather

225

than choosing arbitrarily a few factors and run many levels on each.
The reasoning for this whole approach seems to be that experimenters
who encounter many factors (independent variables) in their
investigations would prefer to have experimental, along with their
theoretical, evidence to discard independent variables rather than
using the theoretical approach only. The reason for this approach
is that the actual theory may be unknown but assumed based on past
experience. Thus, with the availability of rather efficient designs
of experiments involving two-leveled factorials this "theoretical
only" approach does not seem very scientific. In any case the
inference space must be carefully considered.

Other designs based on two-leveled factorials, including
composite designs, will be discussed in Chapter 13. This chapter
will deal mainly with techniques for this special case of two-leveled
factorials assuming CRD, and not discuss the randomization procedures
in detail as in former chapters.

9.1 COMPLETE BLOCKS

This section includes designs in which all the combinations of
the levels of the factors (treatment combinations) in the experiment
are assumed to be handled under similar conditions called a block.
Usually if no repeats of the treatment combination occurs and the
treatment combinations are completely randomized within a block,
the higher factor interactions are frequently assumed zero and the
resulting mean square is used as the error for testing the main
effects and lower factor interactions. As discussed in Chapter 4,
however, the usual approach is to repeat a few treatment combinations
on other experimental units and use the appropriate differences as
an estimate of error. References for all this material are
Kempthorne (1952, Chapter 13); Hicks (1973, Chapter 7), and John
(1971, Chapter 7).

9.1.1 The 2^2 Factorial

This is the simplest of factorial experiments. It is hardly ever used in actual experimentation but it is used here to demonstrate many basic notions of two-leveled factorials.

Consider an example in metals. The variable to be analyzed is y, heat loss per unit of time, and the treatments are factor a, hardness of metals (soft, hard) and factor b, metals (alloy, steel). The interests are in finding out the effects of hardness of metals over alloy and steel and of metals over both soft and hard on heat loss per unit of time (main effects) or the effects of each of the two factors over the levels of the other factor; plus the effect of certain combinations of hardness and metals on heat loss per unit of time (interaction of the two factors).

In the layout of the experiment small letters indicate the treatment combinations and yields for the corresponding treatment combinations depending on context, and in the analysis capital letters indicate the main and interaction effects. In the 2 x 2 example a_0 indicates soft and a_1, hard; b_0 indicates alloy and b_1 steel. Hence $a_0 b_0$ indicates the treatment combination of soft alloy, and so on. There are various ways to portray the treatment combinations compared to $a_0 b_0$ and the remaining treatment combinations. Two other corresponding notations commonly used are as follows:

Notations

1	2	3
$a_0 b_0$	(1)	00
$a_1 b_0$	a	10
$a_0 b_1$	b	01
$a_1 b_1$	ab	11

The association is obvious; however, it should be recognized that notation 2 can be used only for two-leveled factorials (either

presence or absence concept), while notation 3 is general. That is,
if there are three levels one merely uses "2" for level three, and
so on for any number of levels.

The data may be organized as follows.

	b		
a	0	1	Total
0	00 or (1)	01 or b	(1) + b
1	10 or a	11 or ab	a + ab
Total	(1) + a	b + ab	(1) + a+ b + ab

where notation 3 is a matrix indicator. In all instances the
notation for the yields of the treatment combinations and the
treatment combinations are not distinguishable but it is hoped that
there will be no confusion for analysis purposes.

The analysis of variance is given in Table 9.1.1.

TABLE 9.1.1

ANOVA Using a 2^2 Factorial CRD

Source	df	SS
A	1	$[a + ab - (1) - b]^2/4$
B	1	$[b + ab - (1) - a]^2/4$
AB	1	$[ab + (1) - a - b]^2/4$

The total effects are defined as

$$[A] = [ab - b + a - (1)]$$
$$[B] = [ab + b - a - (1)]$$
$$[AB] = [ab - b - a + (1)]$$

See Kempthorne (1952, Section 13.2.1). Notice the correspondence
of the total effects and the sums of squares in the analysis of

variance. Really these are merely contrasts (defined in Section 1.4.1) or comparisons and they are orthogonal. The sum of squares for each constrast is given by the square of the contrast divided by the sum of the squares of the coefficients within the contrast, e.g., SS A = $[A]^2/[(1)^2 + (1)^2 + (-1)^2 + (-1)^2]$. Note that the denominator is $\sum_{i=1}^{n} c_i^2$ where the coefficient, c_i, is either (-1) or $(+1)$. Many times investigators are interested in average effects, e.g., A = $(1/2)[ab - b + a - (1)]$. Kempthorne (1952, p. 236) explains average effects in detail.

One way to display the interrelationship of treatment combinations and effects is given in Table 9.1.2.

TABLE 9.1.2

Treatment Combinations and Effects in 2^2

Effect	Treatment combination				Average effect divisor
	(1)	a	b	ab	
A	-	+	-	+	2
B	-	-	+	+	2
AB	+	-	-	+	2
Mean	+	+	+	+	4

Problem 9.1.1. (a) Show that A, B, and AB are orthogonal comparisons.
    (b) Show that the sums of squares calculated using the usual analysis method is equivalent to that given in Table 9.1.1.
    (c) Could the comparison method of obtaining the sums of squares be used if there were three levels? Explain.

9.1.2 The 2^3 Factorial

The eight treatment combinations for the 2^3 factorial using factors a, b, and c are: (1), a, b, ab, c, ac, bc, abc. Notice how the treatment combinations are generated by multiplication. As in the 2^2 factorial the relationship between the effects and treatment combinations for the 2^3 factorial are shown in Table 9.1.3.

TABLE 9.1.3

Treatment Combinations and Effects in 2^3

Effect	Treatment combination								Divisor
	(1)	a	b	ab	c	ac	bc	abc	
A	−	+	−	+	−	+	−	+	4
B	−	−	+	+	−	−	+	+	4
AB	+	−	−	+	+	−	−	+	4
C	−	−	−	−	+	+	+	+	4
AC	+	−	+	−	−	+	−	+	4
BC	+	+	−	−	−	−	+	+	4
ABC	−	+	+	−	+	−	−	+	4
Mean	+	+	+	+	+	+	+	+	8

As before all main effects have a plus (+) sign when the small letter is present in the treatment combination and a minus (−) when absent. To get the signs for the interactions multiply the corresponding main effect rows. Multiplication of like signs is defined to give a positive sign and multiplication of unlike signs is defined to give a negative sign. The analysis of variance for a 2^3 factorial experiment is given in Table 9.1.4.

TABLE 9.1.4

ANOVA Using 2^3 Factorial

Source	df
A	1
B	1
AB	1
C	1
AC	1
BC	1
ABC	1
Total	7

9.1.3 The 2^n Factorial

The procedure for writing down the signs of the treatment combinations that go into each main effect and interaction for the 2^n factorial may be extended from the 2^3 factorial. There is another method for writing these down, however, from the average effects as follows:

$$A = \frac{1}{2^{n-1}} [(a-1)(b+1)(c+1)(d+1)(e+1) \ldots]$$

$$AB = \frac{1}{2^{n-1}} [(a-1)(b-1)(c+1)(d+1)(e+1) \ldots]$$

$$ABD = \frac{1}{2^{n-1}} [(a-1)(b-1)(c+1)(d-1)(e+1) \ldots]$$

and so on, where the products inside the square brackets will produce the treatment combinations with the correct sign to obtain the designated effect or comparison. A reference for this method is Kempthorne (1952, p. 246).

Analysis of variance for 2^n factorial is given in Table 9.1.5.

TABLE 9.1.5

ANOVA Using 2^n Factorial

Source	df
Main effects	n
2 factor interactions	$\dfrac{n(n-1)}{2}$
3 factor interactions	$\dfrac{n(n-1)(n-2)}{3!}$
.	
.	
.	
n factor interaction	1
Total	$2^n - 1$

Problem 9.1.2. Write out the treatment combinations with
appropriate signs for the interactions ABC, ABD, and ABCD in
a 2^4 factorial.

Example 9.1: An experiment which illustrates the 2^n factorial
(completely randomized in each replication) arose in the finishing
of metal strips in a metallurgical process. The measured response
was a score for smoothness of the surface finish where a small
reading was desirable and a large number indicated roughness. The
factors (two levels of each) which were included for this experiment
were as follows.

Factor		Levels	
Solution temperature	(T)	Low	High
Solution concentration	(C)	Low	High
Roll size	(R)	1	2
Roll tension	(F)	Low	High
Replication (blocks)	(B)	I	II

If all the assumptions for an ANOVA are met, the ANOVA in this
case is similar to that given in Table 9.1.5 except that one must
account for the restriction on randomization in each replicate. In
addition one will have a measurement of error from the block by
treatment interactions.

The data for this experiment are as follows:

$(y = $ smoothness score$)$.

Treatment combination				Replication	
t	c	r	f	I	II
L	L	1	L	18	16
L	L	1	H	10	12
L	L	2	L	10	12
L	L	2	H	8	19
L	H	1	L	9	7
L	H	1	H	9	7
L	H	2	L	8	10
L	H	2	H	14	10
H	L	1	L	12	16
H	L	1	H	21	14
H	L	2	L	15	15
H	L	2	H	21	21
H	H	1	L	17	15
H	H	1	H	24	18
H	H	2	L	4	4
H	H	2	H	13	13
Total				213	209

The calculation of the sums of squares may be carried out following the usual procedure using blocks as a fifth factor, if one has a computer program available. However, if the calculations are to be done by hand the method of using orthogonal contrasts as described in Table 9.1.1 will be easier. To illustrate this equivalence, consider the sum of squares for replicates. The normal method of calculation yields

$$SS\ B = \frac{(213)^2 + (209)^2}{16} - \frac{(213 + 209)^2}{32} = 0.5$$

The replicate contrast is $[B] = 209 - 213 = -4$ and $SS\ B = (-4)^2/32$ $= 0.5$. The main effect contrast is easily displayed by recalling that the low level of the factor is associated with a negative sign and the high level with a positive, e.g.,

$$[T] = 12 + 16 + 21 + 14 + \ldots + 13 + 13 - 18 - 16 - 10 - \ldots$$
$$- 14 - 10$$
$$= 243 - 179 = 64$$

and

$$\text{SS } T = (64)^2/32 = 128.0$$

In order to illustrate the calculation of the two factor contrasts and sums of squares the detailed calculation for the TC interaction is displayed. The two-way table of totals for these two factors is

		C	
		L (-1)	H (+1)
T	L (-1)	(+) 105	(-) 74
	H (+1)	(-) 135	(+) 108

where the -1, -1 cell is obtained by the sum

$$18 + 16 + 10 + 12 + 10 + 12 + 8 + 19 = 105$$

One always has a check on the cell totals as the total for all cells should give the grand total of 422. The appropriate sign for the cell total is obtained by multiplying the column sign times the row sign and placing the result in the corresponding cell. Now

$$[TC] = (105 - 74 - 135 + 108) = 4$$

and

$$\text{SS } TC = (4)^2/32 = 0.5$$

The calculation of the three-way interaction follows in the same fashion except that a three-way table must be formed, e.g., consider the TCF interaction. The three-way table is shown here.

T	C	F L(-1)	H(+1)
	L(-1)	(-)(18+16+10+12) = -56	(+)(10+12+8+19) = +49
L(-1)	H(+1)	(+)(9+7+8+10) = +34	(-)(9+7+14+10) = -40
	L(-1)	(+)(12+16+15+15) = +58	(-)(21+14+21+21) = -77
H(+1)	H(+1)	(-)(17+15+4+4) = -40	(+)(24+18+13+13) = +68

Note that the sum of the values of the cells prior to the sign alteration checks out to a total of 422. The TCF contrast is given by

$$[TCF] = -56 + 49 + 34 - 40 + 58 - 77 - 40 + 68$$

$$= -4$$

and

$$SS\ TCF = (-4)^2/32 = 0.5$$

The calculation for all the remaining effects may be carried out in a similar fashion. The resulting ANOVA table is as follows:

Source	df	MS	F(Based on error)
Replicates (B)	1	0.5	
Restriction error	0		
Treatments	15		
T 1		128.000	15.4*
C 1		105.125	12.6*
TC 1		0.500	< 1
R 1		24.500	2.9
TR 1		36.125	4.3
CR 1		32.000	3.8
TRC 1		136.125	16.4*
F 1		66.125	8.0*
TF 1		72.000	8.7*
CF 1		15.125	1.8
TCF 1		0.500	< 1
RF 1		40.500	4.9*
TRF 1		3.125	< 1
CRF 1		2.000	< 1
TCRF 1		6.125	< 1
B x treatments	15	8.833	
[†]Error(Reps and B x Trt. or within treatments)	16	8.312	

*Significant at the $\alpha = 0.05$ level.

The significance of the three factor interaction TRC should motivate one to set up a table of means in order to visualize the physical relationship among these factors. A Newman-Keuls test may be considered to explore the individual means.

[†]Since the mean square for replicates is so small the design may be considered CRD and the error is within treatments.

Example 9.2: The material presented in Example 9.1 was
analyzed using the method of orthogonal contrasts. The calculational
procedure, even though it is easily understandable, is somewhat time
consuming. A faster technique for doing hand calculations is Yates'
(1937) method. Another reference for this procedure is Davies 1971.
This procedure, which only applies to 2^n experiments, is illustrated
by utilizing the data of Example 9.1. The calculations are given
below. Replicates I and II are added together and the analysis is
carried out as if the experiment was a 2^4. The explanation of the
method of calculation follows.

The first step in this procedure, which is illustrated below,
is to list the treatment combinations in the systematic fashion
using notation 2 described in Section 9.1.1. The appropriate
observation is listed next to the treatment combination. In this
example it is the total of r = 2 observations. This analysis could
have been carried out by listing twice as many treatment combinations,
taking the individual replications into account, and using r = 1.
The first half of the values in column (1) are obtained by adding
together successive pairs of observations from the previous column
e.g., 56 = 34 + 22 and 49 = 22 + 27. The last half of the values
in column (1) are obatined by subtracting the first member of a pair
from the second member in the observation column, e.g., -12 = 22 - 34
and 5 = 27 - 22. Column (2) is then generated from column (1) in the
same fashion that column (1) was generated from the observation
column. This process is continued until column n is generated from
column n - 1 for a 2^n factorial. The calculation of the average
effect and sum of squares columns is self-explanatory. The source
column is written down by changing the small Latin letter in the
treatment combination column to a capital Latin letter.

The ANOVA shown in Example 9.1 could now be generated by merely
calculating the replicate and total sum of squares.

Source	Treatment combination	Observation sum of r=2 values	(1)	(2)	(3)	Total Effect (4)	Average effect $(4)/2^4-1 r$	Sum of squares $(4)^2/2^4 r$
	(1)	34	56	105	179	422		
F	f	22	49	74	243	46	2.875	66.125
R	r	22	32	135	-1	-28	-1.750	24.500
RF	rf	27	42	108	47	36	2.250	40.500
C	c	16	63	-7	3	-58	-3.625	105.125
CF	cf	16	72	6	-31	22	1.375	15.125
CR	cr	18	74	19	23	-32	-2.000	32.000
CRF	crf	24	34	28	13	-8	-0.500	2.000
T	t	28	-12	-7	-31	64	4.000	128.000
TF	tf	35	5	10	-27	48	3.000	72.000
TR	tr	30	0	9	13	-34	-2.125	36.125
TRF	trf	42	6	-40	9	-10	-0.625	3.125
TC	tc	32	7	17	17	4	0.250	0.500
TCF	tcf	42	12	6	-49	-4	-0.250	0.500
TCR	tcr	8	10	5	-11	-66	-4.125	136.125
TCRF	tcrf	26	18	8	3	14	0.875	6.125

Problem 9.1.3. Analyze the data of Example 9.1 as if it were a 2^5 experiment using Yates procedure or a digital computer. Calculate the error mean square by pooling the replicate and all fifteen (replicate by treatment sums of squares). This error sum of squares should agree with that shown in Example 9.1.

Problem 9.1.4. A steel company was interested in causes of weld breaks in its plant and set up an experiment that had the following factors and coded levels:

	Factor	Levels	
		Low	High
(a)	Tap Setting	1	2
(b)	Time of Weld	5	7
(c)	Gage bar setting	3	4
(d)	Upset (real)	L	H
(e)	Upset cycles	1	3

When welding steel in this company two strips to be welded are placed in position so the edges to be welded are facing each other. The thickness of the strips is measured and a setting to control the voltage is determined. This setting is called "tap setting." The strips are then clamped in dies. The two strips are manually brought together until the distance between them is small enough to permit an electric arc to begin (caused by the different polarity of the dies on each strip). The start of the arc is termed the "start of the flash time" and initiates the following automatic cycle. The arc or flash continues for a given time, and is followed by a heat cycle. During the two cycles the strips are automatically traversed towards one another and are literally melted together. The final cycle, termed the "upset," forcibly jams the strips together and finishes the cycle. The force or "upset (real)" in this experiment was set at L (low) and at H (high). The number of the "upset cycles" was either 1 or 3.

The distance the strips move toward each other during the flash and heat cycles is determined by a bar inserted in the proper place on the welder, with various thicknesses of this bar chosen as a function of the strip gage. The thickness chosen, and the resultant distance of the strip traverse during the weld cycle is termed and determined by the "gage bar setting." The total time from start of flash through upset is termed the "time of weld." Everything from the start of flash through upset is automatic, and determined by various cycle times and welder settings.

A replicated CRD factorial experiment (really a RCBD) was set up for a certain grade and gage of steel. Assume that replications are random and all five factors are fixed. The ultimate strength of the weld was measured and the following coded data were obtained:

	Replication I			Replication II	
	a b c d e	Strength (coded)		a b c d e	Strength (coded)
1.	1 7 3 H 1	62	3.	2 7 4 L 1	42
2.	2 7 4 L 3	45	6.	2 7 4 H 3	39
3.	2 7 4 L 1	50	10.	2 5 4 L 3	62
4.	2 5 3 L 1	49	16.	1 7 4 L 3	48
5.	1 5 3 H 3	60	21.	1 5 4 H 1	47
6.	2 7 4 H 3	61	29.	1 7 3 H 3	61
7.	2 7 4 H 1	60	26.	2 5 3 H 3	57
8.	2 5 4 H 3	56	19.	1 5 4 L 3	61
9.	1 5 3 H 1	59	13.	2 5 3 L 3	49
10.	2 5 4 L 3	61	2.	2 7 4 L 3	60
11.	1 7 4 H 1	52	8.	2 5 4 H 3	43
12.	1 7 3 L 3	60	15.	1 7 3 L 1	61
13.	2 5 3 L 3	61	18.	2 5 4 H 1	41
14.	2 5 4 L 1	60	1.	1 7 3 H 1	62
15.	1 7 3 L 1	62	20.	2 5 3 H 1	61
16.	1 7 4 L 3	47	28.	1 7 4 H 3	45
17.	2 7 3 L 3	58	31.	2 7 3 L 1	56
18.	2 5 4 H 1	60	25.	1 5 3 L 1	59
19.	1 5 4 L 3	57	17.	2 7 3 L 3	61
20.	2 5 3 H 1	61	7.	2 7 4 H 1	61
21.	1 5 4 H 1	61	4.	2 5 3 L 1	49
22.	2 7 3 H 1	62	9.	1 5 3 H 1	60
23.	2 7 3 H 3	57	22.	2 7 3 H 1	56
24.	1 7 4 L 1	44	27.	1 5 4 L 1	60
25.	1 5 3 L 1	60	12.	1 7 3 L 3	61
26.	2 5 3 H 3	61	14.	2 5 4 L 1	46
27.	1 5 4 L 1	58	23.	2 7 3 H 3	61
28.	1 7 4 H 3	56	30.	1 5 4 H 3	61
29.	1 7 3 H 3	60	32.	1 5 3 L 3	60
30.	1 5 4 H 3	61	24.	1 7 4 L 1	30
31.	2 7 3 L 1	53	5.	1 5 3 H 3	60
32.	1 5 3 L 3	61	11.	1 7 4 H 1	50

The usual ANOVA for this type of experiment (refer to Section 5.2) is:

Source	df
Replications (R)	1
Restriction error (δ)	0
Treatments (T)	31
Pooled error (R x T)	31

(a)   Using models, show what assumptions are required to use the pooled error term in the above ANOVA.

(b)   Analyze the data with the pooled error, but split up the 31 df for treatments appropriately.   Then set up tables of means with the appropriate standard errors to allow Newman-Keuls' tests where needed.

(c)   Discuss the inference space.

(d)   Later in the investigation it was determined that high carbon steel, impurities and grade differences were the major causes of weld breaks.   What, then, did the results of this experiment contribute to solving the weld break problem?

9.2   INCOMPLETE BLOCK

If the total number of treatment combinations in a factorial experiment is too large to be handled under similar conditions or in one block, more blocks with only a fraction of the total number of treatment combinations is used in each block.   Assuming that the treatment combinations that are chosen (purposely) for each block are randomly assigned to the experimental units within the blocks, we have what is called a randomized incomplete block design.

The main problem is to assign certain sets of treatment combinations to blocks so that the effect or interactions confounded with blocks is not important to the experimenter.   It is immediately obvious that one must lose information on as many effects and/or interactions as there are degrees of freedom for blocks.

9.2.1   The 2^2 Factorial

Let us assume that only two treatment combinations of the four in a 2^2 factorial can be run on a given day.   If main effects are more important than the interaction and there is an estimate of the error variance σ^2, we should set up the design as

Day 1   (1)   ab

Day 2   a   b

To compare Day 1 vs Day 2 we could use

$$[(1) + ab - a - b]$$

which is exactly the comparison used to estimate the AB interaction effect. Hence Days are completely confounded with AB. This leaves us with complete or full information on the two main effects A and B, but no information on AB.

The analysis of variance for this incomplete block design is given in Table 9.2.1.

TABLE 9.2.1

ANOVA Using 2 Blocks of 2 in a 2^2 Factorial

Source	df
AB and/or days	1
Restriction error	0
A	1
B	1
Total	3

Problem 9.2.1. Show the corresponding design if A is confounded with days.

Kempthorne (1952, p. 244) shows the variance of the effects and interactions.

9.2.2 The 2^3 Factorial

Assume we can run only 4 observations of the 8 under similarly controlled conditions, how can we design a 2^3 experiment? Usually the three factor interaction is least important so we will confound ABC with blocks. Using the general approach of putting the four treatment combinations with (+) sign in one block and the four with (−) sign in the other block we get the experiment

Block 1 (−)	Block 2 (+)
(1)	a
ab	b
ac	c
bc	abc

Note: Once we have block one we can take any remaining treatment combination, say a, and multiply the treatment combinations in block 1 by a to generate block 2:

$$a \cdot (1) = a$$
$$a \cdot ab = a^2b = b$$
$$a \cdot ac = a^2c = c$$
$$a \cdot bc = abc$$

The basis for this generation is given by Kempthorne (1952) and Winer (1971, p. 606). Since there are only two levels possible in this design and we define the levels as 0 and 1, the design has a basis of modulo 2. Modulo 2 means dividing the number obtained (here 2) by 2 and using the remainder (here 0) as the appropriate number. Hence $a^2b = a^0b = b$, and so on.

The analysis of variance for 2^3 in blocks of 4 is given in Table 9.2.2.

TABLE 9.2.2

ANOVA Using 2^3 Factorial in 2 Blocks of 4 Each

Source	df
ABC and/or Blocks	1
Restriction error	0
A	1
B	1
AB	1
C	1
AC	1
BC	1
Total	7

There is full information on all main effects and two factor interactions but we lose the information on ABC completely if there is a difference between blocks. Of course, one must have information on the error variance for this design to be of practical value.

Consider the case of four blocks of two. There must be three degrees of freedom confounded with blocks. It can be shown that if A is confounded with blocks and B is confounded with blocks, then AB is also confounded with blocks. This makes up one possible design. Another possibility is if AB and BC are confounded with blocks then AB^2C or AC is also confounded, since $AB^2C = AB^0C = A \cdot 1 \cdot C = AC$, modulo 2.

BC	AB	
	−	+
−	(b), (ac)	(ab), (c)
+	(a), (bc)	(1), (abc)

Consider the interaction of this table, i.e., [the block with (b) and (ac) plus the block with (1) and (abc)] minus [the block with (a) and (bc) plus the block with (ab) and (c)] is the same as

$$[AC] = [(1) - a + b - ab - c + ac - bc + abc]$$

Hence AC is confounded with the blocks for the interaction. This notion holds in general and AC is an example of what is called generalized interaction. A good reference in addition to Kempthorne (1952) is Davies (1971, Chapter 9).

9.2.3 General Approach

Let us switch to a more general approach which can be used on other factorials but will be demonstrated on a 2^5 factorial [Kempthorne (1952) explains the procedure].

Consider a 2^5 in four blocks of eight. There must be three degrees of freedom confounded with blocks. One procedure is to confound ABD and ACE. It follows that the generalized interaction is ABD x ACE = BCDE. Hence these three interactions are confounded with blocks. Let all treatment combinations that satisfy the following equations called defining equations

$$ABD : x_1 + x_2 + x_4 = 0, \text{ mod } 2$$
$$ACE : x_1 + x_3 + x_5 = 0, \text{ mod } 2,$$

be in block 1. This block is called the intrablock subgroup since it contains the treatment combination 00000 or control. In order to determine the remaining composition of the intrablock subgroup which gives rise to this confounding scheme one writes the two defining equations

$$x_1 + x_2 + x_4 = 0 \text{ mod } 2$$
$$x_1 + x_3 + x_5 = 0 \text{ mod } 2$$

where the presence of x_i in one of these equations indicates the presence of the ith factor in the interaction term which is being confounded e.g., for ABD we include x_1, x_2, and x_4, respectively. The values of each x_i are either 0 or 1 depending on the level of factor which corresponds to x_i. Using the models and the above procedure we arrive at the same result as expressed in block 1 that

follows in Table 9.2.3. Blocks 2, 3, and 4 are obtained in the
usual multiplication manner described previously.

TABLE 9.2.3

A Design of a 2^5 Factorial
Experiment in 4 Blocks of 8

a b c d e		ABD: 0 ACE: 0 Block 1	1 1 Block 2	1 0 Block 3	0 1 Block 4
0 0 0 0 0	=	(1)	a	b	c
0 0 1 0 1	=	ce	ace	bce	e
0 1 0 1 0	=	bd	abd	d	bcd
0 1 1 1 1	=	bcde	abcde	cde	bde
1 0 0 1 1	=	ade	de	abde	acde
1 0 1 1 0	=	acd	cd	abcd	ad
1 1 0 0 1	=	abe	be	ae	abce
1 1 1 0 0	=	abc	bc	ac	ab

The analysis of variance for the design given in Table 9.2.3 is
given in Table 9.2.4.

TABLE 9.2.4

ANOVA Using the Design in Table 9.2.3

Source	df
Blocks and/or (ABD, ACE, BCDE)	3
Restriction error	0
Main effects	5
2 factor interactions	10
Error	13
(assume 3 factor and higher interactions are zero)	
Total	31

In many experiments the blocks must be small to retain homogeneity of the experimental units but sometimes there is enough experimental material that many replications of the whole experiment may be made. In this case the experimenter would like to know how best to use the experimental material he has. In most instances, where information on main effects and/or interactions is lost between blocks, partial information is desired on these effects and/or interactions and, if possible, equal information is quite desirable.

Let us consider an example of a design that has equal information on all main effects and interactions which is called a balanced incomplete block design (BIBD). There are two factors each at two levels and three replicates are possible, but there can be blocks of two experimental units only. If information on main effects and interactions are of equal interest, the design in Table 9.2.5 seems best.

TABLE 9.2.5

BIBD of a 2^2 Factorial Using 3
Replications of 2 Blocks of 2

Replication		
I	II	III
Block 1	Block 3	Block 5
(1) b	(1) a	(1) ab
Block 2	Block 4	Block 6
a ab	b ab	a b

The corresponding analysis of variance is given in Table 9.2.6.

TABLE 9.2.6

ANOVA of the BIBD in Table 9.2.5

	df	
Replications	2	
Blocks in Replications	3	
Restriction error	0	
A	1	Two-thirds information on
B	1	each. The sum of squares are computed from those
AB	1	replicates in which the
Error	3	effect or interaction is not confounded.
Total	11	

Of course any multiple of three replicates will give equal information on the main effects and interactions if the design is "balanced" as that given for the three replicates.

(Note: In all systems of confounding, care must be taken to assure the experimenter that the effects or interactions of least interest are the ones confounded with blocks. It is not necessary that the effects or interactions to be confounded are small, only that they are of least interest.)

One must remember that the requirement for a negligible effect or interaction is necessary for using that source of variation for the estimate of the error but not for confounding purposes. This idea is sometimes confusing to the novice in designing experiments, but is never a difficulty if the reasons are understood.

Example 9.3: An example from the area of Animal Science concerning the gain in weight of lambs over a two month period will illustrate the necessary calculations required for computing the sums of squares from data for a BIBD design. The design is the same as that shown in Table 9.2.5. The use of 6 pairs of twin lambs provide the 12 experimental units and each set of twins constitute the blocks. The two factors A and B are low and high

amounts of vitamin A and protein respectively. The coded data for
these results follow:

Replication

I	II	III
Block 1	Block 3	Block 5
(1) = 6 b = 12	(1) = 6 a = 7	(1) = 6 ab = 12
Block 2	Block 4	Block 6
a = 8 ab = 17	b = 13 ab = 17	a = 9 b = 10

The calculations of the various sums of squares is shown below.
The treatment sum of squares must be modified so that, e.g., the
sum of squares for A is calculated by using only the replicates
that do not have A confounded with blocks, i.e., replicates II and
III.

$$\Sigma \; y = 123$$

$$\Sigma \; y^2 = 1437$$

$$CT = \frac{(123)^2}{12} = 1260.75$$

$$TSS = 1437 - 1260.75 = 176.25$$

$$SS \; reps = \frac{(43)^2 + (43)^2 + (37)^2}{4} - CT$$

$$= 1266.75 - 1260.75 = 6.00$$

$$SS \; blocks \; in \; reps = \frac{18^2 + 25^2 + 13^2 + 30^2 + 18^2 + 19^2}{2} - 1266.75$$

$$= 1351.50 - 1266.75$$

$$= 84.75$$

$$SS\ A = \frac{(7 + 17 + 12 + 9)^2 + (6 + 13 + 6 + 10)^2}{4} - \frac{(45 + 35)^2}{8}$$

$$= 812.5 - 800 = 12.5$$

$$SS\ B = \frac{(12 + 17 + 12 + 10)^2 + (6 + 8 + 6 + 9)^2}{4} - \frac{(51 + 29)^2}{8}$$

$$= 860.5 - 800 = 60.5$$

$$SS\ AB = \frac{(6 + 17 + 6 + 17)^2 + (8 + 12 + 13 + 7)}{4} - \frac{(46 + 40)^2}{8}$$

$$= 929 - 924.50 = 4.50$$

$$SS\ error = 176.25 - 6.00 - 84.75 - 12.50 - 60.50 - 4.50$$

$$= 8.00$$

The resulting ANOVA table is as follows:

Source	df	SS	MS	F
Replication	2	6.00		
Blocks in reps	3	84.75		
Restriction error	0			
A	1	12.50	12.50	4.68
B	1	60.50	60.50	22.66*
AB	1	4.50	4.50	1.69
Error	3	8.00	2.67	
Total	11	176.25		

*Significant at $\alpha = 0.05$

Problem 9.2.2. Repeat the exercise shown in Example 9.3 using only replicates I and II. Comment on the results of this problem as compared to the results of Example 9.3.

Problem 9.2.3. Consider the effects of lubrication, and other factors on life of gears. The factors to consider are weight of load: light and heavy; type of load: shock and steady; speed of operation: low and high; and application of lubrication: spray and bath.

If the experimenter can run only eight treatment combinations under similar conditions at one time, design the most efficient experiment, including a suggested number of replications, and show your analysis of variance. Assume three factor and higher interactions are negligible and that the most interest is on the main effects and two factor interactions.

9.3   REFERENCES

Davies, O. L. (Ed.) Design and Analysis of Industrial Experiments, 2nd ed., Oliver and Boyd, Edinburgh, 1971.

Hicks, C. R. Fundamental Concepts of Design of Experiments, 2nd ed., Holt, Rinehart and Winston, New York, 1973.

John, P. W. M. Statistical Design and Analysis of Experiments, Macmillan, New York, 1971.

Kempthorne, O. Design and Analysis of Experiments, Wiley, New York, 1952. Distributed by Krieger Pub. Co., Huntington, N.Y.

Winer, B. J. Statistical Principles in Experimental Design, 2nd ed., McGraw-Hill, New York, 1971.

Yates, F. Design and Analysis of Factorial Experiments. Imperial Bureau of Soil Sciences, Harpenden, England, 1937.

Chapter 10

FRACTIONAL FACTORIAL EXPERIMENTS FOR TWO-LEVELED FACTORS

In almost all experiments the investigator would like to reduce the number of observations required for a complete factorial. If certain assumptions can be met the use of fractional factorials is a most efficient technique to reduce the number of observations and still obtain the desired information. The usual fractional factorial is still orthogonal, which means that certain effects are estimated independently of one another; however, fewer effects can be estimated and these may be biased. The usual procedure given by many authors including Kempthorne (1952, Chapter 20), Hicks (1973, Chapter 15, Davies (1971, Chapter 10), and Peng (1967, Chapter 7) is the one that will be discussed in this chapter.

The usual fractions given in textbooks are powers of 1/2 for two-leveled factorials. These will be described briefly. Statisticians have developed a set of other fractions such as 3/4 which have experimental possibilities because of their flexibility (see John, 1961, 1966, and 1971, Chapter 7).

This chapter will cover two-level factors only, but one of the methods of fractional replication given here can be applied to any prime number equal leveled factorial such as 3, 5, 7, and so on. Besides, if a nonprime number such as 4, 8, 9, and so on occur then the experimenter can convert the factorial into a 2^2, 2^3, 3^2, and so on, which can be fractionated easily (see John, 1961 and 1966).

10.1   THE 1/2 REPLICATION

The method of obtaining a 1/2 fractional replicate is merely to choose one of the two blocks that are generated by the techniques of Section 9.2.   There are certain procedures available that will show quickly whether the proposed design is very good or not.   The techniques discussed here can easily be applied to higher leveled factorials and will be shown in a Chapter 11.

Using the incomplete block design from the 2^3 factorial for demonstration purposes, let us assume that the experimenter has three factors (a, b and c) each at 2 levels and only four observations are to be used in an experiment.   Hence, we have a 1/2 replication of 2^3 factorial.   If we assume the interactions are zero what is the best design?

Let us go back to Table 9.1.3 of main effects and interactions to see how the treatment combinations are used in a fractional factorial.   This table is reproduced as Table 10.1.1 for convenience purposes.

TABLE 10.1.1

Treatment Combinations and Effects
in a 1/2 Replication of a 2^3

| Effect | Treatment combination | | | | | | | |
	(1)	a	b	ab	c	ac	bc	abc
μ	+	+	+	+	+	+	+	+
A	−	+	−	+	−	+	−	+
B	−	−	+	+	−	−	+	+
AB	+	−	−	+	+	−	−	+
C	−	−	−	−	+	+	+	+
AC	+	−	+	−	−	+	−	+
BC	+	+	−	−	−	−	+	+
ABC	−	+	+	−	+	−	−	+

The randomized incomplete block design from Chapter 9 for 2 blocks
of 4 when the interaction ABC is confounded with blocks is

(-)	(+)
Block 1	Block 2
(1)	a
ab	b
ac	c
bc	abc

If we throw away Block 1, we would have our required four
observations. What do the contrasts or comparisons look like?
(Note Table 10.1.1 and select only the treatment combinations
a, b, c, and abc which are reproduced in Table 10.1.2.)

TABLE 10.1.2

Treatment Combinations and Effects
for the 1/2 Fractional Replication

Effect	Treatment combination			
	a	b	c	abc
A	+	-	-	+
BC	+	-	-	+
B	-	+	-	+
AC	-	+	-	+
C	-	-	+	+
AB	-	-	+	+
μ	+	+	+	+
ABC	+	+	+	+

Hence, the effect for A will be calculated the same as the BC effect;
similarly for B and AC; C and AB; and μ and ABC. This feature is
the basis for saying A is an alias of BC, and so on to μ is an alias
of ABC, because in this particular fractional replication whenever we
measure the effect of A we also measure the effect of BC and so on.

Now that we see what a fractional replicated experiment looks like and we assume the treatment combinations are completely randomized onto the experimental units for the design, let us develop a procedure that will allow us to examine the value of the design quickly.

Given we want a 1/2 replicate of 2^3, assuming interactions are zero, let us try to design an experiment with the most information. Since ABC is the highest factor interaction, in most cases, it should be confounded with blocks and this automatically (notice previous development) makes the mean, μ, an alias of ABC. Using our modulus 2 concept we may write the defining relation or identity relationship (Kempthorne, 1952, p. 395) or defining contrast (Davies, 1971, p. 393) as:

$$\mu = I = ABC$$

Where I means the identity and all effects and interactions may be multiplied by it to show the results of the design. Thus we would have the following display of confounded effects:

$$A = A^2BC = BC$$
$$B = AB^2C = AC$$
$$C = ABC^2 = AB$$

The analysis of variance is given in Table 10.1.3.

TABLE 10.1.3

ANOVA of 1/2 Fractional Replicated 2^3 Factorial

Source	df
A and/or BC	1
B and/or AC	1
C and/or AB	1
Total	3

It should be pointed out that if the block with the control (1), that is,

$$(1)$$
$$ab$$
$$ac$$
$$bc$$

had been used, the relationship between A and BC would have been negative as given in Table 10.1.4.

TABLE 10.1.4

A vs BC and the Treatment Combination

Effect	Treatment combinations			
	(1)	ab	ac	bc
A	−	+	+	−
BC	+	−	−	+

The design for this experiment would be CR and have aliases

$$I = -ABC \qquad B = -AC$$
$$A = -BC \qquad C = -AB$$

For a study of this feature see Davies (1971, p. 512).

The sign is not important in the fractional replicated experiments per se because the sums of squares erase the sign, but this will be of value to us in Chapter 13 in discussing the steepest ascent methods used to obtain optimum levels of factors because signs are of interest for the regression coefficients.

Example 10.1: In an experiment on asphalt concrete using cylindrical specimens, a civil engineer wanted to determine the effects and interactions of certain factors on a coded index of goodness of the material. The factors and their levels are given in the following tabulation:

	Factor	Level	
		1	2
a.	Aggregate gradation	Fine	Coarse
b.	Compaction temperature	250°F	300°F
c.	Asphalt content	5%	7%
d.	Curing condition	Unwrapped	Wrapped
e.	Curing temperature	45°F	72°F

The experimenter wanted to run only 16 treatment combinations and was willing to assume three factor and higher interactions were zero.   Hence a one-half replicate of a 2^5 factorial was constructed and the 16 treatment combinations were run completely at random.

The identity used was

$$I = ABCDE$$

so that all main effects have aliases of four factor interactions and all two factor interactions have three factor interactions as aliases.   Since there are zero degrees of freedom for error in this experiment, the investigator had to rely on a previous estimate of the error mean square.   This estimate was 200.

The treatment combinations run in the experiment and their yields are the following:

Treatment combination	Yield
(1)	13
ae	54
be	44
ab	49
ce	13
ac	14
bc	18
abce	85
de	41
ad	73
bd	79
abde	17
cd	82
acde	58
bcde	10
abcd	29
Total	679

From these 16 yields we can obtain the SS for all 5 main effects and the 10 two factor interactions. For the interactions, we can set up two-way tables.

AB:

$$B$$

	$-$	$+$
$-$	$149 = 13 + 13 + 41 + 82$ $\ \ \ \ \ (1) + ce + de + cd$	$44 + 18 + 79 + 10 \qquad = 151$ $be + bc + bd + bcde$
$+$	$199 = 54 + 14 + 73 + \ 58$ $ae + ac + ad + acde$	$49 + \ 85 \ + \ 17 \ + \ 29 \ = 180$ $ab + abce + abde + abcd$

A is the left-side label.

$$SS\ AB = \frac{(149 + 180 - 151 - 199)^2}{16} = \frac{(-21)^2}{16} = \underline{27.6}$$

Notice that 149 + 180 + 151 + 199 = 679 (the total)

AC:

$$C$$

	$-$	$+$
$-$	$177 = 13 + 41 + 44 + 79$ $(1) + de + be + bd$	$13 + 82 + 18 + \ 10 \qquad = 123$ $ce + cd + bc + bcde$
$+$	$193 = 54 + 73 + 49 + \ 17$ $ae + ad + ab + abde$	$14 + \ 58 \ + \ 85 \ + \ 29 \ = 186$ $ac + acde + abce + abcd$

A is the left-side label.

$$SS\ AC = \frac{(177 + 186 - 123 - 193)^2}{16} = \underline{138.1}$$

Here, too, 177 + 186 + 123 + 193 = 679.

Problem 10.1.1. Set up the two-way tables for CD and DE and calculate the SS in a comparable manner as shown for AB and AC. Check your answers with the ANOVA in Table 10.1.5.

Of course one can see that the SS for the main effects are easily found from the marginal totals of the two-way tables. For A we obtain

$$SS\ A = \frac{(199 + 180 - 149 - 151)^2}{16} = 390.1$$

for B

$$SS\ B = \frac{(151 + 180 - 149 - 199)^2}{16} = 18.1$$

and for C

$$SS\ C = \frac{(123 + 186 - 177 - 193)^2}{16} = 232.6$$

Problem 10.1.2. Find the SS for D and E using the technique above. Check your answers with the ANOVA in Table 10.1.5.

TABLE 10.1.5

ANOVA for the Asphalt Experiment

Source	df	MS	F
A	1	390.1	1.95
B	1	18.1	< 1
C	1	232.6	1.16
D	1	612.6	3.06
E	1	76.6	< 1
AB	1	27.6	< 1
AC	1	138.1	< 1
AD	1	1387.6	6.94**
AE	1	1105.6	5.53*
BC	1	68.1	< 1
BD	1	3052.6	15.26**
BE	1	0.6	< 1
CD	1	0.1	< 1
CE	1	410.1	2.05
DE	1	3570.1	17.85**
Error	∞	200	

Tabled $F_{1,\infty}(0.05) = 3.84$
$F_{1,\infty}(0.01) = 6.63$

*Significant at 0.05 level.

**Significant at 0.01 level.

For further analyses the experimenter would probably want to look at all two-way tables of means that are significant and consider running Newman-Keuls test.

10.2   A MORE COMPLICATED DESIGN (1/4 REPLICATION)

As the number of factors gets larger, it is more efficient to run an experiment with a smaller fraction. For example, the 2^3 completely randomized design should hardly ever be fractionated and the previous discussion was to show the meaning, not a good design example; however, a 2^9 factorial could be run as a 1/4 replicated experiment.

Let us take a 2^9 experiment and assume we have no information on the experimental error. Then we may try a 1/4 replication or have 2^7 = 128 treatment combinations. Factors a, b, c, d, e, f, g, h, j, are available and one good completely randomized design results if the defining relation or identity is:

I = ABCDEF = DEFGHJ = ABCGHJ where ABCDEF and DEFGHJ are selected interactions called defining effects and their product ABCGHJ is the generalized interaction. A few of the aliases are:

$$A = BCDEF = ADEFGHJ = BCGHJ$$
$$AB = CDEF = ABDEFGHJ = CGHJ$$
$$ABC = DEF = ABCDEFGHJ = GHJ$$

It can be seen that all main effects have five factor interactions or higher as aliases, all two factor interactions have four factor or higher interactions as aliases, and no information can be obtained from three factor interactions because they are aliases of three factor interactions. It would be best if we could assume three factor and higher interactions are zero so the three factor interaction source could be used as error.

The analysis of variance for this experiment is given in Table 10.2.1 assuming three factor interactions and higher orders

are zero.   The details on obtaining the treatment combinations to
be used will not be described here.

TABLE 10.2.1

ANOVA for 1/4 Fractional Replicated 2^9 Factorial

Source	df
Main effects	9
Two factor interactions	36
(Remainder) error	82
Total	127

Another method of analysis is the half normal plots given by
Johnson and Leone (1964, Vol. II, pp. 188-191) and Daniel (1959).
The use of half normal plots provides a graphical method of
displaying the average effect for each degree of freedom and a
method of testing effects for significance.   This method of
significance testing essentially produces the same results as
obtained with the classical ANOVA table approach.   For this reason
we do not reproduce the half normal plot procedure in this text.
One advantage of displaying all effects on a half normal plot is
that it gives the experimenter some feeling for the validity of the
assumption of normality in each cell.   This follows since all of
the nonsignificant effects should be normally distributed and
display a straight line on the half normal plot.

Problem 10.2.1.   Using results of Problem 9.1.3 in Chapter 9,
run a half normal plot analysis (see Johnson and Leone, 1964,
pp. 188-191).

Example 10.2:   Utilizing a 1/4 replicate of a 2^5 to demonstrate
the Yates (1937) method for obtaining total effects and sums of
squares for fractional factorial experiments, we set up the
following identity:

$$I = ABC = CDE = ABDE$$

Using the intrablock subgroup

Treatment combination	Yield
(1)	6
ab	10
de	4
abde	5
bcd	9
acd	3
bce	8
ace	0

we leave it as a problem for the reader to show that

$$I = - ABC = -CDE = ABDE$$

A rule that may be used for determining the signs of the effects in the identity relation is that a minus sign is placed before all effects which have an odd number of letters provided the intrablock subgroup is used for the experiment. If we are interested in all main effects we can obtain the following aliases for the 7 degrees of freedom from the experiment:

$$A = -BC = -ACDE = BDE$$
$$B = -AC = -BCDE = ADE$$
$$C = -AB = -DE \quad = ABCDE$$
$$D = -ABCD = -CE = ABE$$
$$E = -ABCE = -CD = ABD$$
$$AD = -BCD = -ACE = BE$$
$$AE = -BCE = -ACD = BD$$

The procedure for calculating the sum of squares for each of the above aliases is not unique. One method would be that given in Example 10.1 where one would construct a table of values for each of the seven aliases given above. A second method would be write down all 32 treatment combinations of the 2^5 design in the usual order and use zeroes for yields of those treatment combinations not

in the experiment together with the actual yields for those
treatment combinations utilized in the experiment. Yates method
can now be used to determine the effects due each source of
variation. Note that there will be a sources of variation for each
main effect and interaction and it will turn out, for example, that

$$[A] = -[BC] = -[ACDE] = [BDE]$$

All effects will turn out as shown in the above aliases.

The second method described above becomes rather tedious and
too large for many computer programs as n becomes large and the
fraction $(1/2^p)$ gets smaller. Berger (1972) gives another method
of analysis which allows one to use Yates procedure as if there were
only n-p factors in the experiment. The data of this example will
be analyzed using this procedure. In general, one must find a
nonunique set of n-p letters, e.g., a, c, e, ... which will be used
to generate the entire set of 2^{n-p} treatment combinations in an
order such that Yates algorithm can be applied to them in the
normal fashion and the generated effects are properly identified.
The remaining p letters are referred to as dead letters.

The defining effects are used to determine the dead letters.
For this example the defining effects are ABC and CDE. First select
a dead letter from the first defining effect which is preferrably
not in any other defining effect, CDE in this case. Suppose B is
selected, since I = ABC, set B = AC, so that if B should appear in
some future defining effect AC can be substituted for it. The
second dead letter could be either D or E as neither appear in ABC.
Another way of selecting D or E is that C is related to the dead
letter B by the identity relation B = AC, so we do not select C.
If E, say, is selected as the second dead letter then we see that
E = CD. Thus the letters A, C, and D are admissible and B and E
are dead.

The admissible letters are used to develop the following table
which is used to implement Yates method using the above data. The
first column is generated in the usual fashion using the admissible

Admissible set in Yates order	Dead letters	Augmented letters	Yield	1	2	Total effect	Aliased effects
(1)		(1)	6	16	24	45	
a	b	ab	10	8	21	-9	A = BC = BDE*
c	be	cbe	8	9	-4	-5	C = AB = DE
ac	e	ace	0	12	-5	-19	AC = B = ADE
d	e	de	4	4	-8	-3	D = CE = ABE
ad	be	abde	5	-8	3	-1	AD = BCD = ACE = BE
cd	bee = b	bcd	9	1	-12	11	CD = ABD = E
acd	ee = 1	acd	3	-6	-7	5	ACD = BD = AE = BCE

*Fourth and higher order interactions are omitted.

letters. The second column is a group of dead letters developed by
making entries opposite the terms which have a single admissible
letter. These entries are made by using the dead letter identity
relations B = AC and E = CD. Since A appears with the dead letter
B but no other dead letters, b is placed in the dead letter column
opposite the admissible letter a. The same type of reasoning places
b and e opposite the admissible letter c. The dead letter entries
b and be are multiplied together to obtain the entry e opposite the
admissible value ac. Since D appears only with E in the identity
relations an e is placed opposite d and the remaining dead letters
are formed by multiplication.

The dead letters are augmented to the admissible letters to
form the third column. The third column must be the complete list
of treatment combinations, hence the associated yield is placed in
the next column. Yates method is now carried out on the 2^{5-2}
observations in the usual manner. The aliased effect column is
obtained by multiplying the letters in the first column, the
admissible set, times the identity relation

$$I = ABC = CDE = ABDE$$

Notice from the signed identity relation

$$I = -ABC = -CDE = ABDE$$

we obtain

$$[A]   = -9$$
$$[BC] = 9$$
$$[ACDE] = 9$$
$$[BDE] = -9$$

Next we need to show the sums of squares for the main effects
and two factor interactions, AD and AE. These sums of squares can
now be obtained as follows:

$$SS = \frac{(\text{Total effect})^2}{\sum\limits_{i=1}^{n} c_i^2}$$

In this case, $n = 8$ and $c_i = \pm 1$. Hence the denominator is 8 for all cases.

Sums of squares:

$$A = \frac{(-9)^2}{8} = 10.1 \qquad\qquad AD = \frac{(-1)^2}{8} = 0.1$$

$$B = \frac{(19)^2}{8} = 45.1 \qquad\qquad E = \frac{(-11)^2}{8} = 15.1$$

$$C = \frac{(-5)^2}{8} = 3.1 \qquad\qquad AE = \frac{(5)^2}{8} = 3.1$$

$$D = \frac{(-3)^2}{8} = 1.1$$

In general, if one has a digital computer and a good factorial program available he can use zeroes for these treatment combinations not in the experiment and multiply the sums of squares from the computer printout by the denominator of the fraction involved to get the correct sums of squares. In this case the computer would divide by 32 to get the SS. Hence, all seven sums of squares on the computer output need to be multiplied by 4 in order to obtain the correct sums of squares.

Problem 10.2.2. Determine the intrablock subgroup for a 1/8 replication of 9 factors using the identity relationship I = ABEGHJ = ACFGJ = BCEFH = ABCD = CDEGHJ = BDFGJ = ADEFH. Use the Berger (1972) procedure to find the order of the 64 treatment combinations that allows the Yates procedure to be used to identify the corresponding main effects and interactions.

An approach for solving Problem 10.2.2:

(1)   Select 3 dead letters out of the three interactions:

        ABEGHJ; ACFGJ; ABCD   (defining effects)

as follows:

        a)   Select a letter in the first defining effect that is
not in the other two, say E.   Since I = ABEGHJ
$$E = ABGHJ.$$
        b)   Similarly for second defining effect, F = ACGJ
        c)   Similarly for the third, D = ABC

(2)   Hence D, E and F are the dead letters and associate with
       the treatment combination letters. Admissible letters are
       associated with the dead letters to obtain augmented column
       as follows:

Admissible	Dead	Augmented
a	def	adef
b	de	bde
c	df	cdf
g	ef	efg
h	e	eh
j	ef	efj

(3)   Arrange the admissible set of 64 treatment combinations
       in Yates order and generate the corresponding augmented
       treatment combinations.

Problem 10.2.3.   Given a 1/2 replicate of a 2^4 factorial,
        (a)   Discuss the quality of the best design.
        (b)   Set up the identity and aliases.
        (c)   Show your intrablock subgroup.
        (d)   If the yields for the corresponding treatment
combinations are (1) = 6; ab = 10; ac = 4, bc = 5, ad = 9,
bd = 3; cd = 8 and abcd = 0, use the Yates method and the
Berger procedure to find the best ANOVA assuming the error
mean square is known to be 5.0.

10.3 BLOCKING WITHIN THE FRACTIONAL REPLICATION

In many fractional replicated experiments there are too many treatment combinations to be handled under homogeneous conditions so that blocking is required, even within the fraction.

Consider a 1/4 replicate of a 2^8 or 64 treatment combinations in four blocks of 16. One design is established from the identity relationship.

$$I = ABCDE = ABFGH = CDEFGH$$

From the identity, I, it is seen that all aliases of main effects and two factor interactions are three factor or higher interactions. Then, finding that all aliases of ACF and BDG are four factors and higher and their generalized interaction, ABCDFG, has three factor or higher interactions confounded with it, we can use these as our blocking basis. These aliases are:

$$ACF = BDEF = BCGH = ADEGH$$
$$BDG = ACEG = ADFH = BCEFH$$
$$ABCDFG = EFG = CDH = ABEH$$

To write out the blocks we may use the four defining equations (see Kempthorne, 1952, Chapter 20).

From I:
$$ABCDE : X_1 + X_2 + X_3 + X_4 + X_5 \qquad\qquad = 0 \qquad mod\ 2$$
$$ABFGH : X_1 + X_2 \qquad\qquad + X_6 + X_7 + X_8 = 0 \qquad mod\ 2$$

From Block's:
$$ACF : X_1 \qquad + X_3 \qquad + X_6 \qquad\qquad = 0, 1, mod\ 2$$
$$BDG : \qquad X_2 \qquad + X_4 \qquad + X_7 \qquad = 0, 1, mod\ 2$$

and using levels such that the equations are all set equal to zero gives the intrablock subgroup. Since we will deal only with the intrablock subgroup of the 1/4 replicate, the levels of ABCDE and ABFGH will always be zero, but the levels of the ACF and BDG will be 0 and 1 to make up the four blocks of 16 in the 1/4 replicate. In other words, the four blocks may be represented as:

```
ABCDE : 0              ACF : 0
ABFGH : 0              BDG : 0
```

```
ABCDE : 0              ACF : 0
ABFGH : 0              BDG : 1
```

```
ABCDE : 0              ACF : 1
ABFGH : 0              BDG : 0
```

```
ABCDE : 0              ACF : 1
ABFGH : 0              BDG : 1
```

The analysis of variance for this entire experiment of 64 observations is given in Table 10.3.1.

TABLE 10.3.1

ANOVA for 1/4 Fractional Replicated 2^8 in Blocks of 16

Source	df
Blocks and/or (ACF, BDG, EFG, CDH, BDEF, BCGH, ADEGH, ACEG, ADFH, BCEFH, ABCDFG, ABEH)	3
Main effects	8
Two factor interactions	28
Error (assuming all 3 factors and higher interactions are zero)	24
Total	63

Designs of experiments, including blocking in the incomplete blocks of fractional replications for 2^n factorials are given in the booklet: "Fractional Factorial Experiment Designs for Factors at Two Levels," issued by National Bureau of Standards Applied Mathematics Series 48 (1957).

Problem 10.3.1.   A company was making hat shells out of foam
plastic by molding it and spraying the molded hat shells with
latex.   There were six spray guns set so that they would hit
different regions of the hat.
    The problem facing the experimenter was to adjust the
latex flow rate and the air pressure on each gun in order to
produce a uniformly coated hat.   A lower and upper level for
each latex flow rate and a lower and upper level for each air
pressure was selected on each gun.   He also had two different
foams, one sparkly and the other dull.   Hence there were 13
factors each at 2 levels for the experiment.
    He could run only 35 treatment combinations per day
(there were day to day differences) and wanted to know the
minimum number of days he would need to run the experiment to
get information on main effects and two factor interactions
assuming that three factor and higher interactions were zero.
    (a)   Design a good experiment, that is, show the identity
and its aliases for his conditions (do not write out the
treatment combinations).
    (b)   Show your analysis of variance and your assumptions.

10.4   SOME OTHER VERSIONS AND DEVELOPMENTS IN FRACTIONAL FACTORIALS

10.4.1   Parallel Fractional Replicates

    If many responses or yields (y_i, i = 1, ..., k) are measured
on the same experimental unit and one factor is known not to have
an effect on one of the yields, one can utilize this information in
knowing that certain main effects and interactions are not present
for that yield.   Then if another factor is known not to have an
effect on another yield, the same experimental units may be analyzed
another way assuming other main effects and interactions are zero
for this new variable.

    Daniel (1960) uses an example on flavor (y_1) and texture (y_2)
in a new food experiment in which factor a is oil, b is wheat
product and c is branlike material.   Factor a does not influence
y_2, b may influence both y_1 and y_2 and c does not influence y_1.

    Given that a 1/2 replicate of the 2^3 was run, the aliases are:

$$I = ABC$$
$$A = BC$$
$$B = AC$$
$$C = AB$$

The analysis of variance for y_1 (knowing factor c has no influence) is given in Table 10.4.1.

TABLE 10.4.1

ANOVA of y_1

Source	df
A	1
B	1
AB	1

The analysis of variance for y_2 (knowing a has no influence) is given in Table 10.4.2.

TABLE 10.4.2

ANOVA of y_2

Source	df
B	1
C	1
BC	1

These analyses are run from the same experiment.

More general type designs are given in the paper by Daniel (1960).

10.4.2 The 2^{n-p} Fractional Designs

Box and Hunter (1961, pp. 317-319) introduce the fractional factorial notions and application areas of the 2^{n-p}, where n = number of factors and $1/2^p$ = the fraction. These types of designs, classified according to the degree of fractionation, are given the following names:

(a) Resolution III: No main effect is confounded with another main effect but main effects are confounded with two factor interactions, and two factor interactions are confounded with one another. An example is 2^{3-1}.

(b) Resolution IV: No main effect is confounded with any other main effect or two factor interaction, but two factor interactions are confounded with one another. An example is 2^{4-1}.

(c) Resolution V: No main effect or two factor interaction is confounded with any other main effect or two factor interaction, but two factor interactions are confounded with three factor interactions. An example is 2^{5-1}.

Resolutions III and IV designs are discussed in great detail in this paper. The list of references at the end of the paper is most appropriate here.

Box and Hunter (1961, p. 449) discuss the Resolution V designs in detail.

10.4.3 Symmetrical and Asymmetrical Fractional Factorial Plans

Addelman (1962, p. 47) gives the following synopsis: "Procedures for constructing plans which permit uncorrelated estimation of all main effects and some specified two factor interaction effects are developed for symmetrical factorial arrangements. These plans can then be adjusted to yield plans with similar properties for asymmetrical factorial experiments."

These plans are for three-level factorials, as well as two-level, and follow up on the idea of the Resolution V type designs. The references at the end of Addelman's paper are most relevant material.

10.4.4 The 3/4 Fractional Factorial

Consider a 2^4 factorial. If a 1/2 replicate were run on the 2^4, ordinarily one would use the identity relationship

$$I = ABCD$$

This would cause complete confounding of all two factor interactions with other two factor interactions. One method of eliminating this problem of confounding two factor interactions is with a 3/4 fractional factorial which is described below.

Using four blocks of four treatment combinations each and confounding BD, ABC and ACD with blocks we would have the following blocks:

I	II	III	IV
(1)	a	b	d
ac	c	abc	acd
bcd	abcd	cd	bc
abd	bd	ad	ab

A three quarters (3/4) replicate may be obtained by combining three half-replicates: blocks II and III; blocks II and IV; and blocks III and IV. Notice that block I is not used in the experiment. Hence the name 3/4 replicate.

The half-replicate made up of blocks
(a) II and III, have the identity relationship of I = ABC
(b) II and IV, have the identity relationship of I = ACD
(c) III and IV, have the identity relationship of I = BD

If three and four factor interactions are zero, main effects and two factor interactions can be estimated from the 3/4

replicated experiment. The aliases and blocks to be used for
estimating them are given in Table 10.4.3. It must be understood
in the actual experiment that all 12 treatment combinations given
in blocks II, III, and IV are run completely randomized in one
block of the 12 treatment combinations. Since there are no degrees
of freedom remaining for estimating error it must be estimated from
an outside source. The reference for this design is John (1961),
which also provides a 3/4 replicate for a 2^5 factorial to obtain 8
degrees of freedom for error assuming three factor and higher
interactions are zero.

TABLE 10.4.3

Aliases for the 3/4 Replicate
and Blocks Estimating the Aliases

A	=	ABD	in	III	and	IV	
B	=	ABCD	in	II	and	IV	
C	=	BCD	in	III	and	IV	
D	=	ABCD	in	II	and	III	
AB	=	BCD	in	II	and	IV	
AC	=	ABCD	in	III	and	IV	
AD	=	BCD	in	II	and	III	
BC	=	ABD	in	II	and	IV	
BD	=	ACD	in	II	and	III	Use the mean of these
	or	ABC	in	II	and	IV	two estimates for BD
CD	=	ABD	in	II	and	III	

John (1966) generalizes the results on 3/4 replication by
augmenting the usual fractions to improve some estimates. John
(1971, Chapter 7), is an excellent overall reference for these
designs.

Addelman (1961) showed plans for $3/2^n$ replicates (irregular
fraction plans) some of which permit the estimation of main effects
and two factor interactions with fewer trials then is required with
an orthogonal plan. Although there is some correlation between some

of the estimates there is a procedure given to estimate the main
effects and certain interactions. With present-day regression
computer programs these designs should be considered for the special
type problems.

10.4.5 Case When Some Three Factor Interactions Cannot Be Assumed
Negligible

If in a 1/4 replicate of a 2^8 the identity relationship:

$$I = ABCDE = DEFGH = ABCFGH$$

is used, the following three factor interactions would be confounded
with four factor interactions or higher:

ADF	BEF
ADG	BEG
ADH	BEH
AEF	CDF
AEG	CDG
AEH	CDH
BDF	CEF
BDG	CEG
BDH	CEH

If the experimenter knew before the experiment was conducted that
he could not assume some three factor interactions were negligible,
he should force the factors to be labeled so that these questionable
three factor interactions would be among the 18 above. If he did
not force the factors to be handled in this manner there may be some
three factor interaction effects mixed up with the two factor
interactions and wrong conclusions regarding the importance of these
two factor interactions could be drawn. A reference for this
situation occurred in an experiment described by Hadley et al (1969).
Figure 10.4.1 pictorially displays an example of an interpretable
three factor interaction from Hadley et al. (1969).

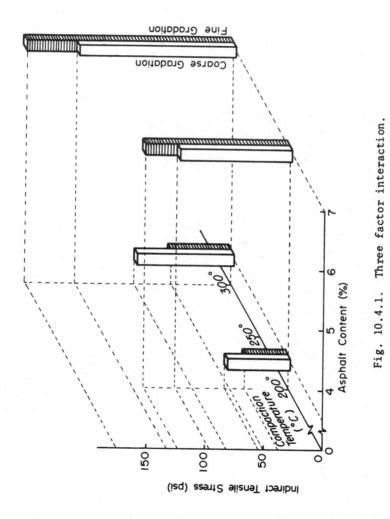

Fig. 10.4.1.   Three factor interaction.

In general, statisticians and experimenters should be more
aware of the possibilities that three factor interactions may be
important to research workers than is usually assumed by almost all
statisticians working in the design of experiments area. If there
is doubt that certain three factor interactions are zero, this
information should be utilized in the design of the experiment by
confounding all main effects, all two factor interactions and three
factor interactions with four factor interactions and higher. It
may turn out that the interpretation of the three factor interactions
is the most essential result. In some cases experimenters have
even interpreted four factor interactions by changing the model and
using the treatment combinations of all four factors simultaneously
as the treatments, thereby ignoring the original main effects,
two factor interactions and three factor interactions. A good
reference for this concept is Marascuilo, L. A. and Levin, J. R.
(1970).

10.5 GENERAL 2^n PROBLEMS

10.5.1. In a 2^6 factorial experiment an investigator could
handle only 17 treatment combinations under one environmental
condition, and he had enough money to run 34 treatment
combinations for the whole experiment. What is a good design
to use? Show your analysis of variance (including source and
degrees of freedom) and state your assumptions on interactions
being negligible.

10.5.2. In an experiment investigating wear and life of gears,
there were 15 factors each at two levels. Let us assume four
factor and higher interactions are negligible. Design (without
blocking within the fraction) a good experiment for this
condition. Show the analysis of your experiment.

10.5.3. In a 2^{11} factorial, what is the minimum number of
observations required to get information on main effects and
two factor interactions if we assume three factor interactions
and higher are zero and we use the maximum number of blocks.
Show the identity.

10.5.4. If, in a 2^9 factorial experiment a research worker did
not want to assume any three factor interactions were negligible,

design a fractional factorial experiment with the smallest
fraction and show your analysis if:
  (a)   three factor interactions are of no interest
  (b)   three factor interactions are of interest
  (c)   comment about the two designs and their analyses

10.5.5. Find a reasonably good design for a 3/4 replicate of
a 2^6 factorial with no blocking in the fraction. State your
assumptions on three factor interactions. Show your analysis
of variance. Compare your results with a 1/2 replicate of a 2^6
factorial with no blocking in the fraction.

10.6 REFERENCES

Addelman, S. Technometrics 3:479 (1961).

Addelman, S. Technometrics 4:47 (1962).

Berger, P. D. Technometrics 14:971 (1972).

Box, G. E. P. and Hunter, J. S. Technometrics 3:311 (August 1961);
  ibid., 3:449 (November 1961).

Daniel, C. Technometrics 1:311 (1959).

Daniel, C. Technometrics 2:263 (1960).

Davies, O. L. (Ed.) Design and Analysis of Industrial Experiments,
  2nd ed., Oliver and Boyd, Edinburgh, 1971.

Hadley, W. O., Hudson, W. R., Kennedy, T. W., and Anderson, V. L.
  Association of Asphalt Paving Technology 1:1 (1969).

Hicks, C. R. Fundamental Concepts of Design of Experiments, 2nd ed.,
  Holt, Rinehart and Winston, New York, 1973.

John, P. W. M. Biometrics 17: 319 (1961).

John, P. W. M. Technometrics 8:469 (1966).

John, P. W. M. Statistical Design and Analysis of Experiments,
  Macmillan, New York, 1971.

Johnson, N. L. and Leone, F. C. Statistics and Experimental Design,
  Vol. II, Wiley, New York, 1964.

Kempthorne, O. Design and Analysis of Experiments, Wiley, New York,
  1952. Distributed by Krieger Pub. Co., Huntington, New York.

Marascuilo, L. A. and Levin, J. R., Am. Ed. Res. J. 7:397 (1970).

Peng, K. C. The Design and Analysis of Scientific Experiments
  Addison-Wesley, Reading, Massachusetts, 1967.

Statistical Engineering Laboratory, "Fractional Factorial Experiment
  Designs for Factors at Two Levels", U. S. Dept. of Com. Nat'l. Bur.
  of Standards, Applied Math. Series 48, (1957).

Yates, F. Design and Analysis of Factorial Experiments, Imperial
   Bureau of Soil Sciences, Harpenden, England, 1937.

Chapter 11

THREE-LEVEL FACTORIAL EXPERIMENTS

One must recognize that the 2^n factorial experiment is a
special case of the general equal prime-number leveled factorial
experiments. In the quantitative two-leveled factorial no mention
is made of investigating for trends because one cannot examine the
concept of deviations from linear. In even the equally spaced
quantitative 3^2 factorial (which is the simplest case with more
than two levels for each factor in the quantitative equal prime
number leveled factorial), one can examine the results for
departures from the linear concept. The reader may want to refer
to Chapters 1 and 2 to reacquaint himself with the usefulness of
quantitative levels and model development. In general, it will be
seen that this chapter on three-leveled factorials will form the
basis for easily extended procedures not only on equally spaced
quantitative equal prime-number leveled factorials but also on the
qualitative equal prime-number leveled factorials. Also, this
chapter deals only with CRD in each block discussed.

When only two levels occur in each factor there is only 1
degree of freedom for each main effect and interaction. For any
other case the interaction degrees of freedom are larger than those
of the main effects. For example, in the simplest case of a 3^2,
the number of degrees of freedom for the two factor interaction is
four, whereas the degrees of freedom for each main effect is only 2.

Since the interactions have the same number of degrees of
freedom as the main effects in the two-leveled factorials, complete
flexibility in using any interaction or main effect in confounding

systems is possible there. In all equal prime-number leveled
factorials this type of flexibility is desired. The procedure,
then, for the factorials with the number of levels greater than two
is to split the interactions into orthogonal pieces with the same
number of degrees of freedom as the main effects. In this case two
of the degrees of freedom for the interaction may be used to
estimate the interaction and the other 2 degrees of freedom can then
be used for confounding purposes.

As indicated in the introduction of Chapter 10 the present
method of experimentation is to include many factors with only two
levels each early in the investigation. Later, however, many levels
of few factors may be used to obtain a thorough description of the
response surface associated with the few factors. Since three
levels really is in between these two extremes, designs of this
nature do not seem to be used as often. There are occasions when
three-leveled designs are useful and a general analysis for them is
given by Margolin (1967), Kempthorne (1952), Davies (1971), and
Winer (1971).

In this chapter a short description of how these designs may
be set up, especially for incomplete blocks and fractional designs,
is given. A thorough description is given by Kempthorne (1952) and
Davies (1971).

11.1   CONFOUNDING IN A 3^n SYSTEM

The system of confounding that was used in the 2^n factorials
given in Chapter 9 will be continued here. To get the pieces of
interactions to have the same number of degrees of freedom as the
main effects we will use the system given by Kempthorne (1952,
Chapter 6) and Hicks (1973). A breakdown of the interactions that
follows the 2^n approach given in Chapter 9 is adaptable to incomplete
blocks and fractional factorial experiments for three-leveled
factorial experiments.

11.1.1 The 3^2 Factorial

Following the method given by Kempthorne (1952, p. 293) and Hicks (1973, Chapter 9), the 3^2 factorial may be examined as follows:

<u>Main effects:</u>   2 degrees of freedom each on margins

$$
\begin{array}{c c c c c}
 & & b & & \\
 & 0 & 1 & 2 & \\
\hline
0 & 00 & 01 & 02 \longrightarrow & X_1 = 0,\ \text{mod } 3 \\
\hline
1 & 10 & 11 & 12 \longrightarrow & X_1 = 1,\ \text{mod } 3 \\
\hline
2 & 20 & 21 & 22 \longrightarrow & X_1 = 2,\ \text{mod } 3 \\
\hline
\end{array}
$$

a

$$X_2 = 0 \qquad X_2 = 1 \qquad X_2 = 2$$

$$\text{mod } 3 \qquad \text{mod } 3 \qquad \text{mod } 3$$

<u>Interactions:</u>   (a)   2 degrees of freedom for right diagonals or AB^2

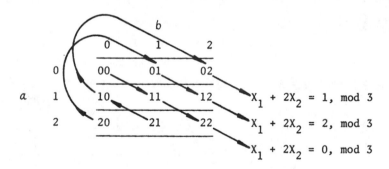

$$X_1 + 2X_2 = 1,\ \text{mod } 3$$
$$X_1 + 2X_2 = 2,\ \text{mod } 3$$
$$X_1 + 2X_2 = 0,\ \text{mod } 3$$

(b)   2 degrees of freedom for left diagonals or AB

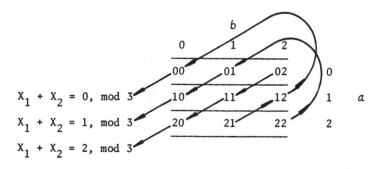

$$X_1 + X_2 = 0,\ \text{mod } 3$$
$$X_1 + X_2 = 1,\ \text{mod } 3$$
$$X_1 + X_2 = 2,\ \text{mod } 3$$

The analysis of variance is given in Table 11.1.1.

TABLE 11.1.1

ANOVA for 3^2 Factorial

Source	df		
A	2		
B	2	$\begin{Bmatrix} AB & (2\ df) \\ AB^2 & (2\ df) \end{Bmatrix}$	$\begin{Bmatrix} \text{This breakdown of the} \\ \text{interaction has no} \\ \text{experimental meaning.} \end{Bmatrix}$
AB	4		

Interactions can be split up in the above manner so that the parts have the same number of degrees of freedom as the main effects and all four pieces (two main effects and the two pieces of interactions) are orthogonal to each other. There is no known experimental value in splitting up AB with 4 df into AB (2 df) and AB^2 (2 df). That is, there is no value in comparing AB and AB^2. The only purpose for this scheme is for use in confounding. Each part of the interaction, AB or AB^2, is equally good for an estimate of the interaction between A and B. There is twice as much information on the AB total interaction from both parts as from either part.

The sums of squares may be described as:

$$SS(A) = \frac{A_0^2 + A_1^2 + A_2^2}{3} - \frac{G^2}{9} = \frac{1}{3} \sum_{i=0}^{2} A_i^2 - \frac{G^2}{9}$$

$$SS(B) = \frac{B_0^2 + B_1^2 + B_2^2}{3} - \frac{G^2}{9} = \frac{1}{3} \sum_{j=0}^{2} B_j^2 - \frac{G^2}{9}$$

$$SS(AB) = \frac{(AB_0)^2 + (AB_1)^2 + (AB_2)^2}{3} - \frac{G^2}{9}$$

$$SS(AB^2) = \frac{(AB_0^2)^2 + (AB_1^2)^2 + (AB_2^2)^2}{3} - \frac{G^2}{9}$$

where

A_i = sum of observations in i^{th} level of a (i = 0, 1, 2)

B_j = sum of observations in j^{th} level of b (j = 0, 1, 2)

G = sum of all observations

AB_0 = total of observations where $X_1 + X_2 = 0$, mod 3 (i.e., 00 + 12 + 21)

AB_1 = total of observations where $X_1 + X_2 = 1$, mod 3 (i.e., 10 + 01 + 22)

AB_2 = total of observations where $X_1 + X_2 = 2$, mod 3 (i.e., 20 + 11 + 02)

AB_0^2 = total of observations where $X_1 + 2X_2 = 0$, mod 3 (i.e., 00 + 11 + 22)

AB_1^2 = total of observations where $X_1 + 2X_2 = 1$, mod 3 (i.e., 02 + 10 + 21)

AB_2^2 = total of observations where $X_1 + 2X_2 = 2$, mod 3 (i.e., 12 + 01 + 20)

Margolin (1967) shows an algorithm for computing effects for three leveled factorials if the levels are equally spaced quantitative. A general discussion using an algorithm for $2^n 3^m$ factorial experiments is also given by Margolin (1967). Of course when n = 0 the 3^m algorithm results and when m = 0 the 2^n algorithm (usually called the Yates method) results.

11.1.2   Other Three-Leveled Factorial Experiments

In a 3^4 factorial experiment:

$(AB^2C^2D)_2$ = total of the observations from the defining equation $X_1 + 2X_2 + 2X_3 + X_4 = 2$, mod 3

$(AC^2)_1$ = total of the observations from the defining equation $X_1 + 2X_3 = 1$, mod 3

Problem 11.1.1. For a 3^4 factorial experiment, show the sum of squares for AB^2CD^2. (Write out the equation, establish the correct treatment combinations for the various totals and indicate the way to obtain the sum of squares.)

For all designs with prime and equal number of levels, the first letter of any part of an interaction may have an exponent of unity. This condition is easily seen for the three-leveled factorials. For example: $AB^2 = A^2B$ since the treatment combinations for each interaction are combined in the same manner as shown below

$$(AB^2)_0 = 00 + 11 + 22 \qquad (A^2B)_0 = 00 + 11 + 22$$
$$(AB^2)_1 = 02 + 10 + 21 \qquad (A^2B)_2 = 02 + 10 + 21$$
$$(AB^2)_2 = 01 + 12 + 20 \qquad (A^2B)_1 = 01 + 12 + 20$$

In order to obtain an exponent of unity for the first letter of an interaction part in a 3^n factorial, square the interaction part. For example:

(a) $(A^2B)^2 = A^4B^2$ $[A^4, \bmod 3 = A]$. Hence $A^2B = AB^2$

(b) $ABC \times AC^2D = A^2BC^3D = A^2BD = (A^2BD)^2 = A^4B^2D^2 = AB^2D^2$
since $(C^3, \bmod 3) = C^0 = 1$

11.1.3 Method of Confounding

Consider what can be confounded in a 3^n factorial experiment. Of the $(3^n - 1)$ total degrees of freedom there are 2 degrees of freedom for each main effect and each piece of interaction, therefore the number of possible pieces (including the main effects) to confound are $(3^n - 1)/2$. In a 3^3 factorial experiment Table 11.1.2 shows there are 13 possible pieces to confound.

TABLE 11.1.2

ANOVA of a 3^3 Factorial Experiment

Source	Pieces that can be confounded	df
A	A	2
B	B	2
AB	AB	2
	AB^2	2
C	C	2
AC	AC	2
	AC^2	2
BC	BC	2
	BC^2	2
ABC	ABC	2
	ABC^2	2
	AB^2C	2
	AB^2C^2	2

In general there has to be a one-to-one correspondence between the degrees of freedom for blocks and the degrees of freedom of effects and/or interactions confounded. One way of summarizing this is the following:

Number of df confounded with blocks	Main effect and/or part of the interactions confounded (2 df each)
2	X
8	X Y XY XY^2
	Once X and Y are confounded XY and XY^2 will also be confounded
26	X Y XY XY^2 Z XZ XZ^2 YZ YZ^2 XYZ XY^2Z XYZ^2 XY^2Z^2
	Once X, Y, and Z are confounded the other effects are also confounded

$\vdots$

and so on

To show the method of handling the generalized interaction concept for three levels consider the example:

	$\underline{X}$	$\underline{Y}$	$\underline{XY}$	$\underline{XY^2}$
(a)	AB	BC	AB^2C	$AB^3C^2 = AC^2$
(b)	AB	AC	$A^2BC = AB^2C^2$	BC^2

One explanation of this is:

   (a) Since modulo 3 is used here, $AB^3C^2 = AB^0C^2 = A1C^2 = AC^2$.

   (b) Since the first letter in an interaction part may be with exponent 1, $(A^2BC) = (A^2BC)^2 = A^4B^2C^2 = AB^2C^2$. The squaring merely provides the exponent (mod 3) to be 1. It is interesting that as this method of dividing interactions into orthogonal parts is used on larger prime-numbered leveled factorials, the same procedure of forcing the first letter to have exponent 1 is used but, of course, the squaring goes to cubing and so on.

Let us consider a 3^3 factorial experiment or 27 observations in nine blocks of three. There must be eight degrees of freedom confounded with blocks. Show a design that provides full information on main effects and as much information on two factor interactions as possible. Try:

	$\underline{X}$	$\underline{Y}$	$\underline{XY}$	$\underline{XY^2}$
(a)	ABC	AB^2C	$A^2C^2 = AC$	B
(b)	ABC	AB^2	$A^2C = AC^2$	$B^5C = BC^2$

Results:

   (a) This design has a main effect, B, confounded with blocks, so, in most cases, it is not satisfactory.

   (b) Only 1/4 of the three factor interaction is confounded, only 1/2 of each of AB, AC and BC is confounded and the three main effects are completely free of confounding. This is usually considered a good design.

For the layout of this design we may use the equations

$$ABC : X_1 + X_2 + X_3 = 0, 1, 2 \mod 3$$
$$AB^2 : X_1 + 2X_2 = 0, 1, 2 \mod 3$$

The blocks may be formed as follows:

levels of ABC :	0	1	2	0	...	2
levels of AB^2 :	0	0	0	1	...	2
Blocks :	1	2	3	4	...	9

0 0 0	0 0 1	0 0 2	0 2 1	...	2 0 0
1 1 1	1 1 2	1 1 0	1 0 2	...	0 1 1
2 2 2	2 2 0	2 2 1	2 1 0	...	1 2 2

The corresponding analysis of variance is given in Table 11.1.3.

TABLE 11.1.3

ANOVA for 3^3 Factorial in Blocks of 3

Source	df
Blocks and/or ABC, AB^2, AC^2, BC^2	8
Restriction error	0
A	2
B	2
AB	2
C	2
AC	2
BC	2
Error and/or AB^2C, ABC^2, AB^2C^2	6
Total	26

In general in designing incomplete block experiments, one should confound the interaction of least interest, and use the interactions which are assumed zero for error.

11.2 SYSTEM OF CONFOUNDING

11.2.1 A 3^2 Factorial Experiment in Blocks of 3

(a) An example of complete confounding:

One may confound AB (2 df) with three blocks. The equation $X_1 + X_2 = 0, 1, 2$; mod 3 is appropriate to indicate the confounding procedure:

Block 1	Block 2	Block 3
$X_1 + X_2 = 0$	$X_1 + X_2 = 1$	$X_1 + X_2 = 2$

0 0	1 0	2 0
1 2	2 2	0 2
2 1	0 1	1 1

The corresponding analysis of variance is given in Table 11.2.1.

TABLE 11.2.1

ANOVA for 3^2 in Blocks of 3^a

Source	df
Blocks and/or AB	2
Restriction error	0
A	2
B	2
AB (actually AB^2)	4 - 2 = 2

[a] AB (actually AB^2) is the estimate of the whole interaction AB coming from the part, AB^2.

(b) An example of partial confounding and balanced incomplete blocks design (BIBD).

One may consider the replications of the experiment above. Confound A in replication I, B in replication II, AB in replication III, and AB^2 in replication IV. The corresponding analysis of

variance for the data from this design is given in Table 11.2.2.
Refer to Example 9.3 for the method of calculation of SS.

TABLE 11.2.2

ANOVA of 4 Replications of 3^2 in Blocks of 3

Source	df			Information
Replications (R)	3			
Blocks in R	8			
Restriction error	0			
A	2		3/4	Balanced
B	2		3/4	information
AB	4	AB	3/4	on all effects
		AB^2	3/4	
Error	16			

Problem 11.2.1. Given the following data:

Replications

I

B1 1	00 = 2 01 = 4 02 = 3
B1 2	10 = 3 11 = 6 12 = 5
B1 3	20 = 7 21 = 6 22 = 5

II

B1 4	00 = 3 10 = 5 20 = 6
B1 5	01 = 5 11 = 7 21 = 8
B1 6	02 = 2 12 = 7 22 = 3

III

B1 7	00 = 1 12 = 5 21 = 9
B1 8	02 = 3 11 = 6 20 = 6
B1 9	01 = 4 10 = 7 22 = 4

IV

B1 10	00 = 4 11 = 6 22 = 1
B1 11	01 = 6 12 = 5 20 = 4
B1 12	02 = 7 10 = 8 21 = 7

     (a) Tell what is confounded in replications I, II, III
and IV. What kind of design is this?
     (b) Run ANOVA on these data. Show the Source, df, SS,
MS, and test the effects of A, B, and AB.

(c) Show how you calculated the SS for all sources and give the amount of information on each of A, B, and AB.

Problem 11.2.2. Delete replication IV in Problem 11.2.1 and reanalyze the data. Compare the results of Problem 11.2.1 with the results in this problem.

11.2.2 Two 3^3 Factorial Experiments

(a) If one considers a design in which there are three blocks of nine treatment combinations in each, there will be two degrees of freedom confounded with blocks and either ABC, AB^2C, ABC^2, or AB^2C^2 may be confounded with blocks.

(b) 9 blocks of 3; 8 df confounded with blocks.

If there were nine blocks of three treatment combinations in each block, then one possible design is:

X	Y	XY	XY^2
AB	BC	AB^2C	AC^2

11.2.3 Two 3^4 Factorial Experiments

(a) If there were three blocks of 27 treatment combinations in each block then there would be two degrees of freedom confounded with blocks and either ABCD, AB^2CD, ..., or $AB^2C^2D^2$ may be confounded with blocks.

(b) If there were nine blocks of nine treatment combinations in each block then one possible design is:

X	Y	XY	XY^2
ABC	AC^2D	AB^2D^2	BC^2D^2

and the corresponding analysis of variance is given in Table 11.2.3.

TABLE 11.2.3

ANOVA for a 3^4 Factorial When
ABC and AC^2D Are Confounded

Source	df
Blocks and/or ABC, AC^2D, AB^2D^2, and BC^2D^2	8
Restriction error	
Main effects	8
2-factor interactions	24
3-factor interactions	24
4-factor interactions	16

Problem 11.2.3. In an experiment conducted by a social
scientist, the variable to analyze is the score on a
citizenship attitude scale. The factors in the experiment
are:
    (a) schools in the university called: 0, 1, 2
    (b) home environments are: low = 0, medium = 1,
high = 2
    (c) scholastic indices are: low = 0, medium = 1,
high = 2
Tests could be given by only one instructor in one room. The
room held 45 students and it was desired that there be five
students for each treatment combination. The design allowed
confounding in blocks of nine and a record was kept on the
total of five students for each treatment combination only.
The following scores are totals of five scores each and the
total will be used as the unit in this analysis. The blocks
are different days in which different students took the tests.

Block I				Block II				Block III			
a	b	c	Score	a	b	c	Score	a	b	c	Score
0	0	0	174	0	0	2	193	0	0	1	186
1	0	1	183	1	0	0	207	1	0	2	175
2	0	2	185	2	0	1	205	2	0	0	191
0	1	2	163	0	1	1	164	0	1	0	184
1	1	0	168	1	1	2	176	1	1	1	193
2	1	1	213	2	1	0	216	2	1	2	215
0	2	1	175	0	2	0	187	0	2	2	197
1	2	2	186	1	2	1	168	1	2	0	182
2	2	0	201	2	2	2	221	2	2	1	207

(a) Analyze the experiment and draw conclusions.

(b) Comment on the design and devise an alternative specifying the assumptions involved.

Problem 11.2.4. Design an experiment for blocks of size three in which the treatment combinations are the same as those in Problem 11.2.3 but AB and BC^2 are confounded. Write out the treatment combinations for each block. Discuss this design relative to the design in Problem 11.2.3.

11.3 FRACTIONAL REPLICATIONS

In two-level factorial experiments dealing with fractional replications it is easy to see at a glance whether or not a design is satisfactory or not because there is only one exponent on the letters of the main effects and interactions. This is not true for the three-level case and peculiarities in this case have generalities not visible in the two-level case.

It was noticed that if there were only three blocks in a three-level case the highest factor interaction was confounded with blocks. Using this idea as a basis for the aliases and setting the identity or mean to be completely confounded with a highest factor interaction part, it can be seen that the mean has only one degree of freedom but the part has two. This peculiarity can be handled by recognizing that a reflection or the square of that part will generate aliases with two degrees of freedom each just as the original part will. The result is that there is one degree of freedom for each part, the original and the squared which provide a basis for generation of aliases and a technique to follow for smaller fractional experiments. Kempthorne (1952) Chapter 21 is an excellent reference for the general case.

11.3.1 A 1/3 Replicate of 3^5 Factorial

A 1/3 replicate of a 3^5 factorial experiment could be handled as follows:

$$I = ABCDE = (ABCDE)^2$$
$$A = AB^2C^2D^2E^2 = BCDE$$
$$B = AB^2CDE = ACDE$$
$$AB = ABC^2D^2E^2 = CDE$$

and so on.

The assumption that three factor and higher interactions are zero would give the analysis of variance in Table 11.3.1.

TABLE 11.3.1

ANOVA of 1/3 Replicate of a 3^5 Factorial

Source	df
Main effects (A,B,C,D,E)	10
Two factor interactions	40
Error	30
Total	80

This analysis shows that one obtains full information on the main effects and two factor interactions from the one block (this is the 1/3 fraction).

11.3.2 A General Approach

If a 1/9 replicate of a 3^n were used, we may set up a general approach

$$X \quad Y \quad XY \quad XY^2 \quad (X)^2 \quad (Y)^2 \quad (XY)^2 \quad (XY^2)^2$$

to investigate the aliases. For example, take a 1/9 replicate of a 3^5; one identity relationship is:

$$I = AB^2CD = BCDE^2 = AC^2D^2E^2 = ABE = (AB^2CD)^2 = (BCDE^2)^2 = (AC^2D^2E^2)^2 = (ABE)^2$$
$$A = ABC^2D^2 = ABCDE^2 = ACDE = AB^2E^2 = BC^2D^2 = AB^2C^2D^2E = CDE = BE$$

and so on for 13 main effects and interactions which have aliases.

The reasoning for there being 13 main effects and interactions to
investigate is:

   I has four aliases with two degrees of freedom each or using
   the reflections or squares there are 8 aliases with one degree
   of freedom each. Hence there is a total of nine degrees of
   freedom on line I.

   A has 8 aliases and there are 18 degrees of freedom on line
   A. These can be generated by multiplication of A with I and
   all of the aliases of I. There are 243 or 3^5 degrees of
   freedom to be accounted for

   243 = total df
   - 9 = df confounded on line I
   234 = df for main effects and interactions remaining.

$$\frac{234 \text{ df}}{18 \text{ df for each set of aliases}} = 13, \text{ the number of main effects}$$

and interactions with aliases in addition to the mean (I).

A reference that gives a thorough coverage of fractional
factorials at three levels with blocking of the fraction is Conner
and Zelen (1959).

11.4 A SPECIAL USE OF FRACTIONAL FACTORIALS

In the past few years research workers have recognized the
importance of fractional replicated experiments and have been
willing to assume three factor and higher interactions are usually
smaller than the two factor interactions, but have been hesitant in
assuming that the interactions used for the error are actually zero.
As a check on the latter, experimenters have been running duplicates
on various treatment combinations (usually three or four) to obtain
a "pure" estimate of the error. A test on the residual is usually
made and then a decision can be made on whether or not the average
of the interactions included in the residual are "near enough" zero

to use the residual as an error estimate. In order to maintain the
homogeneity of the variance of the estimates, these additional
observations are not used in the estimates of the effects or in the
analysis of variance.

In addition to utilization of a few observations to estimate
the error, experimenters are frequently finding that they may be
able to utilize information in a sequence of experiments by using
low fractions as blocks of larger fractions, obtaining information
at the completion of each block. This allows flexibility for the
research worker in that he may discard factors at various stages or
keep them in the design as he chooses.

For example, an experimenter finds that he can obtain only 28
observations under homogeneous conditions, but he is willing to
run the experiment at three different times for a total of 84
observations from five factors each at three levels. If he can
obtain information from the first set of 28 observations that would
give him insight into large main effects, he may be willing to
discard a factor or two. How can such a design be set up so that
intermediate analyses will give him some information, yet he will
have a sound overall analysis if he uses the 84 observations as
originally stated?

A 1/9 replicate of a 3^5 requires 27 observations and an
experimenter would have one observation to duplicate one of the
treatment combinations in order to check the error. Three repeats
of a 1/9 replicate would, however, be much less efficient in
estimating main effects and two factor interactions than considering
three blocks of a 1/3 replicate and run one block at a time. To
show this, let us take $I = (ABCDE) = (ABCDE)^2$ for the mean, and
$AB^2C = ACD^2E^2 = BD^2E^2$ to be confounded with blocks within the 1/3
replicate. The overall analysis of variance (using the extra
observation in each block on a chosen treatment combination) is
given in Table 11.4.1.

TABLE 11.4.1

ANOVA for 1/3 Replicate of 3^5 Factorial[a]

Source	df	
Blocks and/or $(AB^2C,\ BD^2E^2,\ ACD^2E^2)$	2	
Restriction error	0	
Main effects	10	
Two factor interactions	40	
Residual (error)	28	May be broken into components if the mean square is much larger than <u>within</u>, then look at the larger and smaller pieces of these interactions separately
Within duplicate error	3	
Total	83	

[a]Within duplicate error (3 df) is used to test residual (error) to check on the assumption that three and higher factor interactions are zero. The extra three observations are not used for the remaining part of the analysis, totaling 80 df.

This is most satisfactory for the overall analysis, but is the information from one of the blocks any value if one were run before the other two?

The defining equations to consider for the 1/3 replicate with 3 blocks are:

$$ABCDE : x_1 + x_2 + x_3 + x_4 + x_5 = 0 \text{ or } 1 \text{ or } 2 \quad \text{mod } 3 : \text{Fraction}$$

$$AB^2C : x_1 + 2x_2 + x_3 = 0, 1, \text{ and } 2 \quad \text{mod } 3 : \text{Blocks}$$

If the experimenter assumes he will use the intrablock subgroup, he will use zero level for ABCDE for all blocks and zero level for AB^2C for the first block. Hence for the first block he has a 1/9 replicate with the identity

$$I = AB^2C \quad = ABCDE \quad = ACD^2E^2 \quad = BD^2E^2$$

$$= (AB^2C)^2 = (ABCDE)^2 = (ACD^2E^2)^2 = (BD^2E^2)^2$$

It follows that

$$A = ABC^2 = AB^2C^2D^2E^2 = AC^2DE = ABD^2E^2 = BC^2 = BCDE = CD^2E^2 = AB^2DE$$

and similarly for the other main effects. In general, if the two factor interactions are considered smaller than the main effects, a conclusion may be made to discard some factors from the next two blocks if the main effects are quite small.

If, however, no conclusion can be reached from the one block regarding discarding factors the experiment can continue efficiently by taking another block and possibly both additional blocks from the three blocks inside the 1/3 replicate of the 3^5 factorial experiment.

This is more efficient than taking three replicates of a 1/9 replicate of the 3^5 factorial.

11.5 SPECIAL LATIN SQUARE DESIGNS AS FRACTIONAL FACTORIALS

The Latin square design is an excellent design to remove sources of variation in two directions when only one set of treatments is of interest. This is the intent of the design.

In many scientific fields the Latin square design has been used as a method of reducing the number of observations when three factors are of interest. The danger in using this design for a fractional replication is that the main effects are confounded with parts of two factor interactions. To demonstrate this let us take a 3 x 3 Latin square:

		Columns					b			
		1	2	3				0	1	2
	1	A	B	C			0	c_0	c_1	c_2
Rows	2	B	C	A	or	a	1	c_1	c_2	c_0
	3	C	A	B			2	c_2	c_0	c_1

The design can be written in terms of a block of nine from a 3^3 factorial experiment,

a b c	
0 0 0	where:
1 0 1	$$X_1 + X_2 + 2X_3 = 0, \text{ mod } 3$$
2 0 2	satisfies the block. Hence
0 1 1	$$I = ABC^2 = (ABC^2)^2$$
1 1 2	
2 1 0	$$A = AB^2C = BC^2$$
0 2 2	$$B = AB^2C^2 = AC^2$$
1 2 0	$$C = AB = ABC$$
2 2 1	$$AB^2 = AC = BC$$

and it can be seen that all main effects are confounded with parts of two factor interactions. If interactions are zero, this fractional factorial is of value to investigate main effects. In large investigations, the fractional setup as a Latin square has advantages as a screening device if the lower order terms express the larger source of variation of the variable being analyzed. In other words, even if there is confounding of the two factor interactions with main effects, the main effects ought to be larger and indicate the major source of variation.

It should be pointed out that the Latin square must have a pattern, such as

$$\begin{bmatrix} A & B & C & D & E \\ B & C & D & E & A \\ C & D & E & A & B \\ D & E & A & B & C \\ E & A & B & C & D \end{bmatrix}$$

before an identity can be written down for the fractional factorial.
In all cases, however, the mixup of the main effects and two factor
interactions exist. In a nonsystematic Latin square the identity
relationship cannot be explicitly shown.

11.6   EXTENSION OF 3^n TO p^n, WHERE p IS A PRIME NUMBER

Consider a 1/5 replicate of a 5^3 factorial experiment for
which the identity relationship is:

$$I = ABC = (ABC)^2 = (ABC)^3 = (ABC)^4$$
$$A = A^2BC = A^3B^2C^2 = A^4B^3C^3 = A^5B^4C^4, \quad \text{mod } 5$$

To get the first letter of each interaction to have an exponent of
one, it is necessary to raise the interaction to appropriate powers.
For example for A we have:

$$A = (A^2BC)^3 = (A^3B^2C^2)^2 = (A^4B^3C^3)^4 = A^0B^4C^4$$
$$= AB^3C^3 = AB^4C^4 = AB^2C^2 = BC$$

and so on for the other aliases.

If the experiment is 1/25 replicate, there would be 24
aliases for I or μ. As in the three level case where the square
was the reflection, here there are squares, cubes, and quartics to
handle in order to develop the whole set of aliases.

This same procedure is expansible to seven levels and so on
for any p, prime number of levels.

When dealing with a large number of levels and factors in a
small fractional replicated experiment, the problem of handling the
aliases is enormous and usually experimenters resort to computers
to develop the block and aliases for the experiment.

Problem 11.6.1. (a) Show that there does not exist a
fractional factorial with a well defined identity with
aliases for the example in Chapter 8 (the 5 x 5 Latin square).
     (b) Show for (a) above the arrangement of the treatments
within the rows and columns that would give a fractional

replication with a well-defined identity. State the identity
for your design and show the correct ANOVA indicating the
assumptions that must be made on interactions.

11.7 REFERENCES

Conner, W. S. and Zelen, M. Fractional Factorial Experiment Designs
for Factors at Three Levels, National Bureau of Standards, Applied
Math. Series 54, 1959.

Davies, O. L. (Ed.) Design and Analysis of Industrial Experiments,
Oliver and Boyd, Edinburgh, 1971.

Hicks, C. R. Fundamental Concepts of Design of Experiments, 2nd ed.,
Holt, Rinehart and Winston, New York, 1973.

Kempthorne, O. Design and Analysis of Experiments, Wiley, New York,
1952. Distributed by Krieger Pub. Co., Huntington, New York.

Margolin, B. H. Technometrics 9:245 (1967).

Winer, B. J. Statistical Principles in Experimental Design, 2nd ed.,
McGraw-Hill, New York, 1971.

Chapter 12

MIXED FACTORIAL EXPERIMENTS AND OTHER INCOMPLETE BLOCK DESIGNS

There are many designs available for factorial experiments
with unequal levels. These designs are not handled the same as the
equal leveled factorials, so special emphasis must be made on these
so called "mixed" factorial completely randomized incomplete block
designs (Kempthorne, 1952, Chapter 18). These are the designs that
are needed for experiments in which there are large numbers (25 or
more) levels of one factor only investigated simultaneously. For
these designs it is sometimes easy to force the number of levels to
be a perfect square and then use a special type "lattice" design to
compare the levels or treatment means. If the number can only be
forced to be one number times another, "rectangular" lattice
designs may be used. Federer (1955) is a very good reference for
lattice designs.

Many other incomplete block designs exist and special plans are
available for some of them in Cochran and Cox (1957). The treatment
of all incomplete block designs can follow a different procedure
from the one presented here but the results are essentially the
same. In all instances in this chapter it is assumed that there is
complete randomization in the blocks considered.

12.1 DESIGNS FOR FACTORIAL EXPERIMENTS WITH THE NUMBER OF
      LEVELS THE SAME PRIME POWER

If a factorial experiment has the number of levels of each of
the factors equal to $k = p^m$, where p is a prime number and m is an

integer greater than one, special incomplete block designs can be used.

12.1.1 One Factor (to Demonstrate Pseudofactor)

Take an example in which there are eight levels of a factor or call these levels treatments. Then one can consider this a 2^3 pseudofactorial where there are three pseudofactors each at two levels and handle the design as previous two-leveled factorials except that the experimenter must eventually discuss the results in terms of the eight treatments. One way of renaming the eight treatments is the following:

Treatment a	a_1	a_2	a_3
1	0	0	0
2	0	0	1
3	0	1	0
4	0	1	1
5	1	0	0
6	1	0	1
7	1	1	0
8	1	1	1

It should be understood that the correspondence between the treatment and the pseudofactor is arbitrary, but once it is decided upon for that design the correspondence must be maintained throughout the experiment.

Remember this subsection is for demonstration of the pseudofactor technique and does not include a design to provide an error estimate.

12.1.2 Two Factors

If there are two factors each with a number of levels equal to
a multiple of prime numbers, another pseudofactorial type design
may be used. Consider two factors a and b each with eight levels,
one can call these factors six pseudofactors each with two levels.
A possible naming method follows:

	Factor 1				Factor 2		
a	a_1	a_2	a_3	b	b_1	b_2	b_3
1	0	0	0	1	0	0	0
2	0	0	1	2	0	0	1
3	0	1	0	3	0	1	0
4	0	1	1	4	0	1	1
5	1	0	0	5	1	0	0
6	1	0	1	6	1	0	1
7	1	1	0	7	1	1	0
8	1	1	1	8	1	1	1

If a complete 8 x 8 factorial were run, the analysis of the
completely randomized factorial and of the pseudofactorial is
given in Table 12.1.1.

From this analysis it is seen that if a main effect of a
pseudofactor is confounded this is no worse than if an interaction
of the same pseudofactor (e.g., A_1A_2) is confounded because all of
the main effects and interactions from the same factor are really
main effects of the factor of interest (e.g., A here).

TABLE 12.1.1

ANOVA for 8 x 8 Factorial Experiment

Original		Pseudofactorial	
Source	df	Source	df
A	7	A_1	1
		A_2	1
		A_3	1
		A_1A_2	1
		A_1A_3	1
		A_2A_3	1
		$A_1A_2A_3$	1
B	7	B_1	1
		B_2	1
		B_3	1
		B_1B_2	1
		B_1B_3	1
		B_2B_3	1
		$B_1B_2B_3$	1
AB	49	A_1B_1	1
		A_1B_2	1
		.	.
		.	.
		.	.
		$A_1A_2A_3B_1B_2B_3$	1
Total	63		63

For the above experiment consider the case in which only 16 treatment combinations can be run under homogeneous conditions but that four blocks may be run.

One possible design (1) is:

X	Y	XY
$A_1A_2A_3$	$B_1B_2B_3$	$A_1A_2A_3B_1B_2B_3$

and the analysis of variance is given in Table 12.1.2.

TABLE 12.1.2

ANOVA for 4 Blocks of 16 Using Design (1)

Source	df	Information
Blocks	3	
Restriction error	0	
A	6	6/7
B	6	6/7
AB	48	48/49

A second design (2) is:

X	Y	XY
A_1B_1	A_2B_2	$A_1A_2B_1B_2$

and the analysis of variance is given in Table 12.1.3.

TABLE 12.1.3

ANOVA for 4 Blocks of 16 Using Design (2)

Source	df	Information
Blocks	3	
Restriction error	0	
A	7	Full
B	7	Full
AB	46	46/49

Conclusion: Usually the second design would be preferred.

Regardless of the design actually used, the pseudolevels must be converted back to the original levels before the experiment can be run. Consider the second design, the two defining equations (if the variables for a_1, a_2, a_3, b_1, b_2, b_3 are X_1, X_2, X_3, X_4, X_5, X_6) are:

$$X_1 + X_4 = 0, 1 \mod 2$$

$$X_2 + X_5 = 0, 1 \mod 2$$

The corresponding intrablock subgroup is:

Pseudofactor						Factor	
X_1	X_2	X_3	X_4	X_5	X_6	a	b
0	0	0	0	0	0	1	1
1	0	0	1	0	0	5	5
0	1	0	0	1	1	3	4
1	1	1	1	1	0	8	7
.						.	
.						.	
.						.	

The experimenter is interested only in the block that contains the treatment combinations in terms of the actual factor a and b. He has no interest in the pseudofactor block. It is merely a "means to an end."

12.2 DESIGNS IN WHICH THE NUMBER OF LEVELS ARE DIFFERENT PRIME NUMBERS

Consider an example of a 2 x 2 x 3 x 3 factorial. Since all the factor's levels are prime numbers, no pseudofactorial is needed. Let the factors be a, b, c, and d in that order.

The analysis of a completely randomized 2 x 2 x 3 x 3 factorial experiment and the corresponding confounding pieces of main effects and interactions is given in Table 12.2.1.

TABLE 12.2.1

ANOVA for 2 x 2 x 3 x 3 Factorial

Source	df	Pieces to Confound
A	1	A
B	1	B
AB	1	AB
C	2	C
AC	2	AC
BC	2	BC
ABC	2	ABC
D	2	D
AD	2	AD
BD	2	BD
ABD	2	ABD
CD	4	CD and CD^2
ACD	4	ACD and ACD^2
BCD	4	BCD and BCD^2
ABCD	4	ABCD and $ABCD^2$
Total	35	

The possible block sizes for this experiment are 2, 3, 4, 6, 9, 12, and 18.

When choosing interactions to confound, all factors must have the same prime number of levels within a defining equation such as X or Y (e.g., the interaction AB may be selected as a defining equation since both a and b have the same number of prime levels) for the procedure to work. This procedure will not work if AD is used as a defining equation. The resulting interaction need not be one of the defining equations to establish the intrablock subgroup but it is necessary to understand which generalized interaction is confounded with blocks in order to analyze the data properly.

12.2.1 Design with 9 Blocks of 4 for the 2 x 2 x 3 x 3

$$\begin{array}{cccc} X & Y & XY & XY^2 \\ C & D & CD & CD^2 \\ df = 2 & 2 & 2 & 2 \end{array}$$

This is a very poor design because all main effects of C and D are lost and so is their interaction.

12.2.2 Design with 6 Blocks of 6 for the 2 x 2 x 3 x 3

$$\begin{array}{ccc} X & Y & XY \\ AB & CD & ABCD \\ df = 1 & 2 & 2 \end{array}$$

This is a fairly good design because all main effects and 1/2 of CD are estimable. The information on AB, however, is lost.

12.2.3 Design with 4 Blocks of 9 for the 2 x 2 x 3 x 3

$$\begin{array}{ccc} X & Y & XY \\ A & B & AB \\ df = 1 & 1 & 1 \end{array}$$

This is a poor design for the same reasons as those given in Section 12.2.1.

12.2.4 Design with 3 Blocks of 12 for the 2 x 2 x 3 x 3

$$\begin{array}{c} X \\ CD \text{ or } CD^2 \\ df = \quad 2 \end{array}$$

This is not as bad a design as the one given in Section 12.2.3 since one obtains half information on CD, and full information on all main effects and the other interactions.

12.2.5 Design with 2 Blocks of 18 for the 2 x 2 x 3 x 3

$$X$$
$$AB$$
$$df = 1$$

This design is not as good as the one given in Section 12.2.4
because all information on AB is lost.

12.2.6 A Partially Balanced Incomplete Block Design
        (Four Replications of 6 Blocks of 6)

	df $\longrightarrow$	1	2	2
		X	Y	XY
Replication	1	AB	C	ABC
Replication	2	A	CD	ACD
Replication	3	B	CD^2	BCD^2
Replication	4	AB	D	ABD

This design provides 3/4 information on all main effects, 1/2
information on AB and 3/4 information on CD (with 4 degrees of
freedom since one obtains 1/2 information (CD^2) in replication 2 and
another 1/2 (CD) in replication 3). In mixed factorial incomplete
block design it is very difficult to obtain balanced designs without
many replications.

12.3 DESIGNS THAT HAVE THE NUMBER OF LEVELS AS
PRODUCTS OF PRIME NUMBERS

In some mixed factorial experiments the factors have the number
of levels that are products of prime numbers and other factors with
prime numbers. Consider the example 2 x 2 x 6 x 6. One can make
this into an experiment that is a $2^4 3^2$ by allowing

$$\begin{array}{cccccc} & & \overbrace{\quad}^{c} & & \overbrace{\quad}^{d} & \\ a & b & c_1 & c_2 & d_1 & d_2 \\ 2 & 2 & 2 & 3 & 2 & 3 \end{array}$$

The pseudofactors could be

c	c_1	c_2	d	d_1	d_2
1	0	0	1	0	0
2	0	1	2	0	1
3	0	2	3	0	2
4	1	0	4	1	0
5	1	1	5	1	1
6	1	2	6	1	2

Various incomplete block designs could result from such a factorial experiment.

Example 12.1: An experiment in industrial engineering was set up to investigate the effects of various factors on reducing user dissatisfaction with microforms. Microforms consist of photographs of hardcopy originals reduced in size for efficient storage. The measured variables include time measurements of reading, comprehension, and subjective preferences.

The factors and their levels include:

	Factors	Levels
(a)	Screen angles from horizontal	0°, 45°, 90°, 105°
(b)	Screen luminance (foot lamberts)	30, 40, 50
(c)	Ambient illumination (foot candles)	20, 40, 60
(d)	Visual tasks	Reading (R), Information Searching (I)
(e)	Type of projection	Front (F), Back (B)

A subject can comfortably perform 24 conditions (treatment combinations) at one sitting. This takes about one and one half

hours. Since there are 4 x 3 x 3 x 2 x 2 = 144 treatment
combinations, there must be six sittings for each subject to
complete the experiment. This will demand six blocks within each
subject (replication). There are twelve random subjects available
for the experiment. The usual assumption that there is no
correlation among errors within a subject is made.

The design of the experiment, then, must be an incomplete
pseudofactorial, and preferably partially balanced to obtain as
much information on all main effects and two factor interactions as
possible. Use the letter notation (a, b, c, d, e) for designating
the factors and the usual capital letters (A, B, C, D, E) to indicate
the effects.

Since the first factor, a, has four levels we will regard it as
a 2^2 pseudofactorial where the pseudofactors a_1 and a_2 each has two
levels as follows:

a	a_1	a_2
0°	0	0
45°	0	1
90°	1	0
105°	1	1

With five degrees of freedom to confound with blocks (sittings)
for each subject, we may construct the following partially balanced
incomplete block design for the 12 subjects:

df $\longrightarrow$	2	1	2	Number of
Condition	X	Y	XY	subjects
I.	BC	$A_1 DE$	$A_1 BCDE$	2
II.	BC	$A_2 DE$	$A_2 BCDE$	4
III.	BC^2	$A_1 DE$	$A_1 BC^2 DE$	2
IV.	BC^2	$A_1 A_2 DE$	$A_1 A_2 BC^2 DE$	4

Hence we can see that we get 1/2 information on BC interaction since the 2 df BC is represented by 1/2 of the subjects and the other 2 df BC^2 is represented by the other six subjects. In a similar manner there is 2/3 information on ADE because A_1DE is calculated using eight subjects, A_2DE is calculated using eight subjects and A_1A_2DE is calculated using eight subjects with full information on all main effects and all other two and three factor interactions. Since we must confound something with two degrees of freedom in order to get 5 df confounded with blocks, the best (highest factor interaction of factors with 3 levels) we can do is the design above.

The reader must understand that for each subject blocks are chosen at random to run first, second and so on. Then, given a block the treatment combinations within that block are chosen at random to find out which will be run first, second and so on for each sitting.

To indicate how the treatment combinations are selected for each block, let us use the condition I : BC, A_1DE, A_1BCDE and the intrablock subgroup. Then, we must solve the following defining equations:

$$BC : \qquad X_3 + X_4 = 0 \quad \mod 3$$
$$A_1DE : \qquad X_1 + X_5 + X_6 = 0 \quad \mod 2$$

The resulting treatment combinations follow:

| | \multicolumn{6}{c}{Psuedoblock} | | \multicolumn{5}{c}{Real or usable block for experimenter} |
	X_1	X_2	X_3	X_4	X_5	X_6		a	b	c	d	e
	a_1	a_2	b	c	d	e						
1.	0	0	0	0	0	0	1.	0°	30	20	R	F
2.	0	1	0	0	0	0	2.	45°	30	20	R	F
3.	0	0	0	0	1	1	3.	0°	30	20	I	B
4.	0	1	0	0	1	1	4.	45°	30	20	I	B
5.	0	0	1	2	0	0	5.	0°	40	60	R	F
6.	0	0	2	1	0	0	6.	0°	50	40	R	F
7.	0	1	1	2	0	0	7.	45°	40	60	R	F
8.	0	1	2	1	0	0	8.	45°	50	40	R	F
9.	0	0	1	2	1	1	9.	0°	40	60	I	B
10.	0	1	1	2	1	1	10.	45°	40	60	I	B
11.	0	0	2	1	1	1	11.	0°	50	40	I	B
12.	0	1	2	1	1	1	12.	45°	50	40	I	B
13.	1	0	0	0	1	0	13.	90°	30	20	I	F
14.	1	0	0	0	0	1	14.	90°	30	20	R	B
15.	1	0	1	2	1	0	15.	90°	40	60	I	F
16.	1	0	1	2	0	1	16.	90°	40	60	R	B
17.	1	0	2	1	1	0	17.	90°	50	40	I	F
18.	1	0	2	1	0	1	18.	90°	50	40	R	B
19.	1	1	0	0	1	0	19.	105°	30	20	I	F
20.	1	1	0	0	0	1	20.	105°	30	20	R	B
21.	1	1	1	2	1	0	21.	105°	40	60	I	F
22.	1	1	1	2	0	1	22.	105°	40	60	R	B
23.	1	1	2	1	1	0	23.	105°	50	40	I	F
24.	1	1	2	1	0	1	24.	105°	50	40	R	B

Assuming that all 24 different blocks have been set up for the whole experiment we next outline the ANOVA showing the source, df and amount of information for each source.

TABLE 12.3.1

ANOVA for Microform Experiment

Source	df	Amount of Information
Subjects	11	
Blocks and/or interactions	60	
Restriction error	0	
A	3	Full
B	2	Full
AB	6	Full
C	2	Full
AC	6	Full
BC	4	Half
ABC	12	Full
D	1	Full
AD	3	Full
BD	2	Full
ABD	6	Full
CD	2	Full
ACD	6	Full
BCD	4	Full
ABCD	12	Full
E	1	Full
AE	3	Full
BE	2	Full
ABE	6	Full
CE	2	Full
ACE	6	Full
BCE	4	Full
ABCE	12	Full
DE	1	Full
ADE	3	2/3
BDE	2	Full
ABDE	6	Full
CDE	2	Full
ACDE	6	Full
BCDE	4	Full
ABCDE	12	5/6
Error (Residual)	1513	
Total	1727	

Problem 12.3.1. (a) Find the usable block in the microfilm experiment if we use the defining equations

$$BC : \qquad X_3 + X_4 = 1, \quad \mod 3$$

$$A_1DE : \quad X_1 + X_5 + X_6 = 0, \quad \mod 2$$

(b) Indicate where correlated errors may exist in this experiment.

Problem 12.3.2. Find the usable block for the intrablock subgroup using Condition III; BC^2, A_1DE, A_1BC^2DE.

Problem 12.3.3. (a) If only 6 subjects were used in the microfilm experiment (Example 12.1), write out the ANOVA showing source, df and amount of information.

(b) Discuss the value of using 12 subjects rather than only 6 and give your recommendations plus your reasoning. Consider the inference space and correlated errors.

Problem 12.3.4. In a farm cardiac study a problem investigating the effects of certain farm tasks upon heart action were studied. The variable measured was oxygen uptake when men carried one or two pails at six different heights using three different loads. Three men were used in the experiment and the complete factorial would have been a 3 x 2 x 6 x 3 experiment.

Each man could perform only six tasks in any one day, only one man could work on any one day, and eighteen days were allowed for the experiment. All two factor interactions are of interest but three factor and higher are assumed zero. Men are random and may be considered replications. The other three factors are fixed.

(a) Design a good experiment identifying the treatment combinations as the experimenter would want them.

(b) Show the analysis including the amount of information on each main effect and two factor interaction for the three fixed factors (Remember men act only as replications and are not of interest per se.)

(c) Describe the randomization procedure.

12.4 DESIGNS IN WHICH THE PSEUDOFACTORS HAVE DIFFERENT POWERS ON THE PRIME NUMBER OF LEVELS

An example is 9 x 8 or

a_1	a_2	b_1	b_2	b_3
3	3	2	2	2

which results in a pseudofactorial experiment of $3^2 2^3$ if incomplete block designs are required.

12.5 DESIGNS IN WHICH THE NUMBER OF LEVELS ARE PRODUCTS OF POWERS OF PRIME NUMBERS

An example of this type design is the case in which a, b, c, and d are each at 6 levels and e, f, g, and h are each at 10 levels. The factorial experiment can be written as

$$6^4 10^4$$

but to handle incomplete block designs it may be broken into

a_1 a_2 b_1 b_2 c_1 c_2 d_1 d_2 e_1 e_2 f_1 f_2 g_1 g_2 h_1 h_2

2 x 3 x 2 x 3 x 2 x 3 x 2 x 3 x 2 x 5 x 2 x 5 x 2 x 5 x 2 x 5

or a pseudofactorial experiment $2^8 3^4 5^4$.

Problem 12.5.1. In a 6^3 experiment, show a good system of confounding blocks of 36.

Problem 12.5.2. Using Problem 8.2.1 as a reference:
    (a) Rearrange the materials a, b, c, d within the rows and columns so that an identity can be found (using pseudofactors) that will be appropriate for all 16 treatment combinations in the experiment. This will show the fractional replication used to obtain that particular Latin square.
    (b) Show the correct analysis for fractional replication in (a). Indicate the confounding.

12.6 FRACTIONAL MIXED FACTORIALS $2^m 3^n$

The designs in the booklet by Conner and Young (1961) are constructed so that all main effects and two factor interactions are estimated under the condition that three factor and higher interactions are zero. A design is constructed for 39 pairs (m, n) included from $(m + n = 5)$ through $(m + n = 10)$ where $(m, n \neq 0)$.

The objective of the designs was to keep the number of required treatment combinations small while retaining as much orthogonality among the estimates as possible.

A brief example shows the basic method of construction of the design: Consider a $2^3 3^2$ experiment with 72 possible treatment combinations. If one desires a 1/2 replicate of this experiment he must confound either one degree of freedom for a three factor interaction or use a nonorthogonal design. If three factor interactions are of interest to the experimenter he would attempt to confound a degree of freedom from the five factor interaction $A_1 A_2 A_3 B_1 B_2$, but as mentioned earlier one cannot use an interaction from factors with different numbers of levels as a defining contrast. A very good design, however, is obtained by structuring a design by working with the two and three level factors separately and then combining the results.

Working with the 2^3 part of the experiment and the $A_1 A_2 A_3$ interaction for confounding purpose we form the defining equations $X_1 + X_2 + X_3 = 0, 1,$ mod 2, which produces the two following groups of treatment combinations:

	S_1				S_2	
a_1	a_2	a_3		a_1	a_2	a_3
0	0	0		1	1	1
1	1	0		1	0	0
1	0	1		0	1	0
0	1	1		0	0	1

Similarly using $B_1 B_2$ for confounding the 3^2 part, $W_1 + W_2 = 0, 1, 2,$ mod 3 (using W's instead of X's to identity the equation for mod 3), the three resulting groups of treatment combinations are

	R_1		R_2		R_3
b_1	b_2	b_1	b_2	b_1	b_2
0	0	1	0	2	0
1	2	0	1	0	2
2	1	2	2	1	1

To form the 36 treatment combinations given in Conner and Young (1961) one may use the following:

	S_1R_1							S_2R_2							S_2R_3				
	a_1	a_2	a_3	b_1	b_2			a_1	a_2	a_3	b_1	b_2			a_1	a_2	a_3	b_1	b_2
1.	0	0	0	0	0		13.	1	1	1	1	0		25.	1	1	1	2	0
2.	0	0	0	1	2		14.	1	1	1	0	1		26.	1	1	1	0	2
3.	0	0	0	2	1		15.	1	1	1	2	2		27.	1	1	1	1	1
4.	1	0	0	0	0		16.	1	0	0	1	0		28.	1	0	0	2	0
$\vdots$	$\vdots$						$\vdots$	$\vdots$						$\vdots$	$\vdots$				
12.	0	1	1	2	1		24.	0	0	1	2	2		36.	0	0	1	1	1

The analysis of an actual example with data is given in Conner and Young (1961, pp. 9-12). The analyses could be programmed on computers, since, of course, the least square estimates of the effects do become difficult to compute. (These computations assume b's are equally spaced quantitative factors and trends are desired.) Regression programs may be used for analyses on the data from such experiments, too.

12.7 LATTICE DESIGNS (PSEUDOFACTORIAL)

The general class of lattice designs are really a special case of the pseudofactorial incomplete block designs with one factor (discussed briefly in Section 12.1 in this chapter). The number of levels of the factor is usually quite large since the most common usage of the design is in comparing varieties in agronomic studies. Of course, the design can be used for various conditions where there are many treatments, all of equal interest, to be compared.

Since there are too many treatments or varieties to be placed in one block, the main problem is to group the treatments in incomplete blocks in such a manner that maximum information is obtained on the average yields. The various groupings actually

make up the various designs. In all designs the interest is in
retaining information on the treatments or varieties from between
and within the incomplete blocks. Elaborate designs (using partial
confounding methods) confounding in different directions and
utilizing information from many sources, have been evolved. Three
of the best sources of information on such designs are in the books
by Federer (1955, Chapters XI and XII), by Kempthorne (1952,
Chapters 22-26), and Cochran and Cox (1957, Chapter 10).

There is a general notion concerning the effects confounded
with blocks that must be stressed. In all previous designs there
were more than one factor concerned, so that higher factor
interactions were of less value to the investigator than lower
factor interactions and main effects. When there is only one
factor (in lattice designs) each treatment is as important as any
other, and the confounding of effects is completely artificial.

What is meant by "artificial" in this case is the effects
A, B, AB, ..., and AB^{k-1} (if there are k^2 treatments) have no real
meaning as effects, also they are equally valuable because all
represent main effects or comparisons of the treatment means.
Hence, in all cases in lattice designs, the pseudomain effects are
usually confounded first, and the designs are built up from these
main effects and simple interactions.

A few of the lattice designs will be discussed briefly, more
to explain the type of design than to give an account of the minute
details of the layout and analysis of each. Details are given so
well by Federer and Kempthorne that there seems no reason to cover
them here.

12.7.1 One-Restrictional Designs

(a) Simple Lattice

This design has one restriction on randomization (that is,
blocking in one direction). Consider the case of k^2 varieties or

treatments in blocks of k varieties. We can consider this
experiment as a pseudofactorial in which pseudofactor a is at k
levels and pseudofactor b is at k levels. If main effect A is
confounded with blocks in replicate one and B is confounded with
blocks in replicate two, there is blocking in only one direction
and the two pseudo main effects are each confounded in the two
replications. This basic design structure is called a simple
lattice.

If there were a repeat in which two more replications of the
experiment were run, in one of the new replications A was confounded
and in the other new replication B was confounded, the design would
be called a double simple lattice. The analysis for such a design
is given in Federer (1955, pp. 319-333).

It can be seen that even in this simple lattice design there
will be information in one replication lost due to blocking, but
that same information will be available in the other replication.
In other words this is a partially confounded type design with
blocks nested in replications and information on every treatment
will be available even if there are large block to block differences.
In addition, it can be seen that if the block to block variation is
very small or negligible the data may be analyzed as a randomized
complete block design where the replications act as the blocks in
this RCBD. For the simple lattice design analyzed as a randomized
complete block design, the analysis is given in Table 12.7.1.

TABLE 12.7.1

ANOVA of a RCBD from a Simple Lattice

Source	df
Replications	1
Restriction error	0
Treatments	$k^2 - 1$
R x T (error)	$k^2 - 1$

Similarly for the double simple lattice the analysis for RCBD is
given in Table 12.7.2

TABLE 12.7.2

ANOVA for a RCBD from a Double Simple Lattice

Source	df
Replications	3
Restriction error	0
Treatments	$k^2 - 1$
R x T (error)	$3(k^2 - 1)$

Federer (1955, p. 324) shows a method to compare the efficiency of
double lattice design to randomized complete blocks using the ratio
of the two effective error variances. A quicker, but cruder,
method (described in Chapter 5) is to use the Fisher's amount of
information method to determine the efficiency of the lattice
design relative to the randomized complete block design. The
appropriate ratio is

$$\frac{(n_1 + 1)(n_2 + 3) \ s_2^2}{(n_1 + 3)(n_2 + 1) \ s_1^2}$$

where

n_1 = degrees of freedom for the error in the lattice

s_1^2 = error mean square for the lattice

n_2 = degrees of freedom for the error in the randomized
        complete block

s_2^2 = error mean square for the randomized complete block.

The techniques of utilizing intrablock information and
interblock information will not be discussed here. Of course, both
Federer (1955) and Kempthorne (1952) discuss methods thoroughly.

(b) Triple Lattice

If three replicates or multiple of three replications can be run in a k^2 experiment, we can confound A, B and AB in the various replications equally.

Of course quadruple lattices, and so on can be used depending on the number of replications and type of confounding in each set.

(c) Balanced Lattice

If k + 1 replications can be taken

$$A, B, AB, AB^2, \ldots, AB^{k-1}$$

can be confounded with blocks and there will be equal information on each pseudo effect and interaction. As a result there is equal information on all treatment or varietal means. This design is an example of a balanced incomplete block design (BIBD).

This completes the designs called one-restrictional, one-dimensional.

(d) Other One-Restrictional

In addition to one-dimensional, there are two-dimensional designs called rectangular lattices (k p) discussed by Federer (1955, pp. 343-348). In this design the number of varieties or treatments are not a square. Further discussion on three dimensional, <u>cubic</u> <u>lattices</u> (k^3), four dimensional (k^4) and so on to n-dimensional (k^n) is given by Federer (1955, p. 348-368) and the references given by Federer (1955) and Kempthorne (1952). This completes the one-restrictional designs.

12.7.2 Two-Restrictional Designs

(a) Lattice Square

This is the first design with two restrictions on randomization in the lattices and compares, quite naturally, to the Latin square for complete blocks.

In this case A is confounded in one direction and B is confounded in another direction for the same replication for k^2 varieties or treatments, Federer (1955, Chapter XII).

(b) Semibalanced Lattice Square

As more replications are introduced into the design, more interactions are confounded with blocks. For example

Effect confounded with	Replication				
	1	2	3	...	(k + 1)/2
Rows	A	AB	AB^3	...	AB^{k-2}
Columns	B	AB^2	AB^4	...	AB^{k-1}

(c) Balanced Lattice Square (BIBD)

Continuation of the replication pattern established by the semi-balanced lattice square by reversing the confoundings in rows and columns give a complete balanced system.

Effect confounded with	Replication		
	$1 \ldots \dfrac{k + 1}{2}$	$(\dfrac{k + 1}{2} + 1)$	$\ldots (k + 1)$
Rows	$A \ldots AB^{k-2}$	B	$\ldots AB^{k-1}$
Columns	$B \ldots AB^{k-1}$	A	$\ldots AB^{k-2}$

A reference for the two-restrictional designs is Federer (1955, pp. 377-412).

12.7.3 Designs with More Than Two Restrictions

A good reference is Federer (1955, pp. 413-414).

12.8 ADDITIONAL INCOMPLETE BLOCK DESIGNS

Designs that have only one factor with more than one level and all the levels of the factor cannot be handled in a single block have been outlined in Chapter 11 of Cochran and Cox (1957). They have plans and analyses for these designs that are most useful for cases not covered in this material. The reference on pages 468 and 469 of Cochran and Cox (1957) are excellent for this type problem.

A calculus for factorials is also available. A recent paper on applications of this concept is given by Kurkjian and Woodall (1972).

12.9 REFERENCES

Anderson, R. L. and Bancroft, T. A. Statistical Theory in Research, McGraw-Hill, New York, 1952.

Cochran, W. G. and Cox, G. M. Experimental Designs, Wiley, New York, 1957.

Conner, W. S. and Young, S. Fractional Factorial Designs for Experiments with Factors at Two and Three Levels, Nat'l. Bureau of Standards, Applied Mathematics Series 58 (1961).

Federer, W. T. Experimental Design Theory and Application, Macmillan, New York, 1955.

Kempthorne, O. Design and Analysis of Experiments, Wiley, New York, 1952. Distributed by Krieger Pub. Co., Huntington, New York.

Kurkjian, B. and Woodall, R., HDL-TR-1596, U. S. Army Materiel Command, Harry Diamond Laboratories, Washington, D. C., 20438 (May 1972).

Chapter 13

RESPONSE SURFACE EXPLORATION

In the preceding chapters the designs had one common characteristic, that of finding out the influences of the various factors on the variable analyzed. In this chapter that characteristic still exists, but the emphasis includes finding that particular treatment combination which causes the maximum (or minimum) response, yield, or variable analyzed. In addition to finding the treatment combination that causes the optimum yield, response surface exploration includes investigation of the response surface near the optimum yield. Since one can determine the minimum of a response surface by finding the maximum of the negative of the actual response surface we will only discuss the concept of maximum. In this chapter it should be recognized that, at most, a second degree polynomial will be investigated. In general, the response surface will be some nonlinear function which may be sufficiently approximated with a quadratic polynomial in a small region near the optimum operating condition.

Not all of the response surface or sometimes called "optimum" designs will be given here and even those given will not be described in depth. The purpose of presenting the designs in this manner is to expose the reader to some of the most useful designs, describe the strengths and weaknesses of each method and provide references if further reading is desired. The approach is to divide the designs into fixed (nonsequential) type, in which the decision of the maximum response is made with a given set of observations, and into sequential, in which the first design is used to provide information on how to set up the next design and so on until the

326

experimenter is satisfied he has reached the maximum or "near enough" maximum response. In addition, an attempt has been made to list the designs within each classification of fixed and sequential from least to most desirable (from a statistician's view).

13.1 FIXED DESIGNS

The notion of fixed factors has been used in factorial experimentation with the understanding that all of the levels of interest of that factor were in the experiment. The fixed or nonsequential designs used to seek the optimum levels of the factors in a given experiment include previously discussed designs and additional designs that have been used successfully by experimenters. The additional designs include those which have fewer observations than factors and mixture designs that require the sum of the levels in a treatment combination to add to a constant.

13.1.1 Random Balance

The random balance design was suggested by Frank E. Satterthwaite (1956, 1959). The design has been used by various industrial organizations since then. The purpose of this design is to use very few observations in an easy manner to find the maximum response and also to try to describe the response surface.

There has been quite a bit of controversy among statisticians concerning the design and analysis of the random balance experiment. To examine the peculiarities of this design, let us consider an example. An experimenter can name 16 factors or independent variables $(X_1, X_2, \ldots, X_{16})$ that may influence the response (y). Let us say that the number of levels of factors varies from 2 to 7 as follows:

Factor	X_1	X_2	X_3	X_4	X_5	X_6	X_7	X_8	X_9	X_{10}	X_{11}	X_{12}	X_{13}	X_{14}	X_{15}	X_{16}
Levels	3	2	7	4	2	3	6	5	2	3	5	6	3	4	5	2

The total number of treatment combinations in the population is the
product of the levels or in this case 653,184,000. Hence the usual
fractional replicated experiment seems impossible, especially when
the experimenter wants to look at only 25 treatment combinations.

The method of obtaining the 25 treatment combinations out of the
possible 653,184,000 by the random balance method follows:

>   for X_1 take a random number between 1 and 3, say 3
>   for X_2 take a random number between 1 and 2, say 1
>   for X_3 take a random number between 1 and 7, say 6
>   and so on, until finally,
>   for X_{16} take a random number between 1 and 2, say 2

Then the first treatment combination for $(X_1, X_2, X_3, \ldots, X_{16})$ to
be used in the experiment is (3, 1, 6, ..., 2). For the next
treatment combination take a random number between 1 and 3, say 2
for X_1. Continue as above to obtain the second treatment combination.
This method is continued for all 25 treatment combinations with the
restriction that the number of treatment combinations containing
each level of each factor be as equal as possible. For example,
consider factor X_{10} each of the three levels will be represented
at least eight times, but one level must be represented nine times;
whereas for factor X_8, each of the five levels must be present in
the experiment five times. This completes the design.

One part of the analysis is to declare that treatment
combination with the highest response the winner. If the number of
treatment combinations used in the experiment is more than the
number of factors, a multiple regression analysis using only the
linear main effects may be used assuming all the other effects and
interactions are zero.

In this case $y = \beta_0 + \beta_1 X_1 + \beta_2 X_2 + \beta_3 X_3 + \ldots + \beta_{16} X_{16} + \varepsilon$.
If the number of treatment combinations used in the experiment is
less than the number of factors, it has been suggested that graphing
of effects and interactions may still give good information. One

experimenter even declared that he could evaluate the higher factor interactions with fewer observations than factors. If one considers the design as a fractional replicated experiment with an unknown identity, he must acknowledge the difficulty in such a claim.

The following statement given by James H. Stapleton in the October, 1959, issue of the American Statistician summarized the feelings of the authors about this design:

> The major error committed by those who advocate the "random balance" approach is in their viewpoint, centering on the results to be expected from a random experiment and not on the results to be expected from the particular experiment chosen. Let us suppose that there are k factors of interest, that a complete factorial would require M experiments, and that n experiments are to be made, with n much smaller than M. The problem is to choose these n experiments from the M possible in some optimum fashion. The solution of the "random balancists" is to choose these n at random under possibly a few restrictions. It is hoped that these n experiments will form an optimum design in the sense that the important effects will be measured as well as possible. Why leave it to chance? The only possible answer to this is that it is felt that somehow chance will come up with a better design than the statistician can find deterministicly. If this is so, and I very much doubt that it is, it only points out the inadequacy of the theory of the design of experiments. Certainly if M is much larger than n, the design determined in a nonrandom manner must include much complete and partial confounding, but so must the random design. As a statistician I feel much better if I have control over this confounding.

13.1.2 Systematic Supersaturated

Supersaturated designs are characterized by there being more factors (q) than treatment combinations or observations (n) and no repeats of any treatment combination so there must be an external error estimate. The effects of certain factors must be assumed predominant and all interactions negligible. These designs look similar to the Plackett and Burman (1946) designs for the 2^q fractional factorials. In fact two of the systematic supersaturated designs described by Booth and Cox (1962) are regular Plackett and Burman designs.

While the Plackett and Burman designs are obtained by systematically cycling the first member of the basic complete factorial, Booth and Cox acquire their two-leveled designs (when n < q) by using an even number of factors (q) and have a computer try all possible treatment combinations. To obtain an admissible design the factors are arrayed as column headings and the complete factorial of treatment combinations can be listed under the headings (one treatment combination per row) as input. The admissible designs require that the number of pluses be equal to the number of minuses in each column so that each effect can be estimated (not necessarily uncorrelated). Hence there are 1/2 n negative ones and 1/2 n positive ones in each column.

If the design were orthogonal all of the sums of cross products of all the pairs of columns would be zero, but since there are fewer treatment combinations (n rows) than factors (q columns) there cannot be complete orthogonality and some of the sums of cross products must not be zero. To obtain the best design Booth and Cox require that the set of treatment combinations (the design) must provide a minimum value of the maximum absolute value of the sum of cross products of all the pairs of columns and if there is more than one design meeting this requirement select that design in which the number of pairs of columns attaining this minimum value for the maximum absolute value of the sum of cross products is a minimum.

An example given by Booth and Cox (1962) for q = 16, n = 12, follows:

Factors (q)

Treatment Combination (n)

	1	2	3	4	5	6	7	8	9	10	11	12	13	14	15	16
1	+	+	+	+	+	+	+	+	+	+	+	−	−	−	−	−
2	+	−	+	+	+	−	−	−	+	−	−	−	−	−	−	−
3	−	+	+	+	−	−	−	+	−	−	+	+	+	−	+	+
4	+	+	+	−	−	−	+	−	−	+	−	−	+	+	+	+
5	+	+	−	−	−	+	−	−	+	−	+	+	+	−	+	−
6	+	−	−	−	+	−	−	+	−	+	+	+	+	+	−	+
7	−	−	−	+	−	−	+	−	+	+	+	+	−	+	+	+
8	−	−	+	−	−	+	−	+	+	+	−	+	−	+	−	+
9	−	+	−	−	+	−	+	+	+	−	−	+	+	+	+	−
10	+	−	−	+	−	+	+	+	−	−	−	−	+	+	−	−
11	−	−	+	−	+	+	+	−	−	−	+	−	−	−	+	+
12	−	+	−	+	+	+	−	−	−	+	−	−	−	−	−	−

Any factor goes in the experiment (−) as often as (+). This is similar to balance in the random balance design. Notice that there are six pluses and six minuses in each column. The sum of products between column 1 and column 2 is zero, but between column 15 and column 16 is 4. If all possible pairs of columns are investigated 67 pairs have zero and the other 53 pairs have 4 for the absolute value of the sum of cross products.

Booth and Cox (1962) provide designs for the following:

| Designs | q | n | $\lvert s\rvert$* | | | | | Number of pairs of columns = $\binom{q}{2}$ |
			0	2	4	6	8	
I (Plackett-Burman)	16	12	67		53			120
II	20	12	75		115			190
III	24	12	101		175			276
IV	24	18		193		83		276
V	30	18		281		154		435
VI	36	18		385		245		630
VII (Plackett-Burman)	30	24	295		81		59	435

*$\lvert s\rvert$ = absolute value of the sum of cross products.

The analyses of the data from an experiment may be to find the optimum treatment combination, i.e., the one that produces the maximum response, or to look at the effects of the factors. As in the random balance design, it is difficult to interpret a regression analysis because of the confounding even of main effects (since $q > n$).

Booth and Cox (1962) compare the systematic supersaturated designs given above with random balance designs by looking at the variance of the measure of nonorthogonality. This variance is obtained from the frequency of the sum of cross products given in the previous table. An example using the design $q = 16$, $n = 12$ follows:

The probability is 67/120 that $\lvert s\rvert$ = 0 and 53/120 that $\lvert s\rvert$ = 4. Booth and Cox (1962) indicate that since the number of -4's and +4's are equally likely over the set of designs from which this particular design was evolved, the E(s) = 0 and the variance of s is $E(s^2) = (53/120)(4^2) + (67/120)(0^2)$. Hence the variance for this design is 7.07.

Further, for the random balance design the experimenter is dealing with finite sampling and the variance of s is $[n^2/(n-1)]$ or for this case the variance is $(12^2/11) = 13.1$. A table summarizing comparisons of the two designs follows:

Design	q	n	Random balance	Systematic supersatured
I	16	12	13.1	7.07
II	20	12	13.1	9.68
III	24	12	13.1	10.1
IV	24	18	19.1	13.6
V	30	18	19.1	15.$\tilde{3}$
VI	36	18	19.1	17.4
VII	30	24	25.0	11.4

This design includes only 2-leveled factors and is not as practical as the random balance design which allows factors with varying levels such as 3, 2, 7, 6, 4.

13.1.3   Random Method

Brooks (1958) describes a procedure for seeking optimum combinations of levels of factors or of independent variables with maximum yield of the dependent variable. The procedure is very similar to the random balance design discussed in 13.1.1 except that the treatment combinations are chosen completely at random. The random method described by Brooks allows the experimenter to choose the sample size, n (number of treatment combinations in the experiment), so that he has probability, S, of obtaining at least one treatment combination which will fall in the optimum subregion. This optimum subregion is that proportion, a, of the factor space (all possible treatment combinations from which n is a random sample) which produces the highest responses.

The probability that at least one of the n trials falls in the optimum subregion is $S = 1 - (1 - a)^n$,

where: $(1 - a)$ = the probability that a given trial will fall outside the optimum subregion.

Solving for n (for certain values of a and S) Brooks obtains the data in Table 13.1.1.

TABLE 13.1.1

Number of Observations (n)

a	S			
	0.80	0.90	0.95	0.99
0.100	16	22	29	44
0.050	32	45	59	90
0.025	64	91	119	182
0.010	161	230	299	459
0.005	322	460	598	919

If the experimenter requires an estimate of experimental error he must take some extra trials at the same points.

The procedure for running the experiment is to set up the factor space or population of factors and levels of each, then select the n combinations of the levels of the factors, that is a treatment combination at random and conduct the experiment at these n points. The analysis is merely to select the treatment or factor combination that gave maximum response and call these conditions optimum. Of course, one could examine the response surface by a regression analysis as described for the random balance design.

The characteristics of this design are:

(1) The number of trials does not depend on the number of factors or levels of each factor.

(2) The number of trials does not depend on the function of the surface.

(3) The design is fixed, not sequential.

Brooks explains a stratified random design in which the factors and/or levels of factors are stratified before the treatment combinations are drawn. Then there is only one treatment combination drawn at random from each stratum.

13.1.4 Mixture Designs

In experimentation dealing with mixtures of ingredients (as in many chemical-type experiments), the sum of the levels of the factors for each treatment combination adds to a constant. For example, in manufacturing a marker flare the ingredients magnesium, sodium nitrate, strontium nitrate, and a binder make up 100% of the ingredients, while in stainless steel many times the amount of iron is fixed so that the sum of the variable ingredients such as carbon, chromium, and others is 100% minus the percentage of the iron.

The design of these mixture experiments is most unusual because the experimenter must restrict the treatment combinations to sum to a constant. This causes an entirely different design approach from the usual factorial procedure.

(1) Simplex Lattice. A mathematical definition of a simplex is a convex hull of any set of $n + 1$ points from a Euclidean n space (E^n) which do not lie on a hyperplane in E^n.

A comprehensive description of the simplex lattice design is given by Scheffé (1958). When the simplex lattice design is used in mixture problems, the responses are measured at the simplex lattice composition points. Polynomial equations having a special correspondence with the simplex lattice points are then used to represent the response surface. This polynomial equation is quite easily determined, the response surfaces can then be drawn and the optimum proportion of components may be determined.

The following tabulation, taken from an article by Gorman and Hinman (1962), relates the number of response points to the number of factors and the number of spacings per factor:

Number of spacings (m) →	Quadratic	Special Cubic	Cubic	Quartic
	2	2	3	4
Number of factors (q)	Number of response points (k)			
3	6	7	10	15
4	10	14	20	35
5	15	25	35	70
6	21	41	56	126
8	36	92	120	330
10	55	175	220	715

In general,

$$k = \frac{(m + q - 1)!}{m!(q - 1)!}$$

where q is the number of factors and m is an integer related to the spacing of points between 0 and 1. There are (m + 1) equally spaced points from 0 to 1. Hence if m = 3, the responses will be measured at the proportions 0, 1/3, 2/3, and 1 for each component. This would be a simplex cubic lattice. The above formula does not hold for the special simplex cubic lattice which is a simplex quadratic lattice with additional center points on the two-dimensional face (s). The number of response points in the special simplex cubic lattices is given by:

$$k = \frac{q(q + 1)}{2} + \frac{q(q - 1)(q - 2)}{6}$$

The special simplex cubic lattice is sometimes referred to as the augmented quadratic.

Graphical aids facilitate the calculations associated with simplex lattice designs and as an aid to see where the responses are measured and what simplex lattice designs look like, the figure below shows typical 3-factor designs.

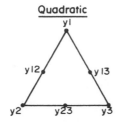

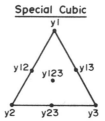

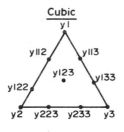

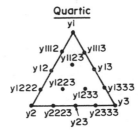

The polynomials associated with each of these 3-component designs are given below.

Quadratic: $y = \gamma_1 X_1 + \gamma_2 X_2 + \gamma_3 X_3 + \gamma_{12} X_1 X_2 + \gamma_{13} X_1 X_3 + \gamma_{23} X_2 X_3$

Special Cubic: $y = \gamma_1 X_1 + \gamma_2 X_2 + \gamma_3 X_3 + \gamma_{12} X_1 X_2 + \gamma_{13} X_1 X_3 + \gamma_{23} X_2 X_3$
$$+ \gamma_{123} X_1 X_2 X_3$$

Cubic: $y = \gamma_1 X_1 + \gamma_2 X_2 + \gamma_3 X_3 + \gamma_{12} X_1 X_2 + \gamma_{13} X_1 X_3 + \gamma_{23} X_2 X_3$
$$+ \eta_{12} X_1 X_2 (X_1 - X_2) + \eta_{13} X_1 X_3 (X_1 - X_3) + \eta_{23} X_2 X_3 (X_2 - X_3)$$
$$+ \gamma_{123} X_1 X_2 X_3$$

Quartic: $y = \gamma_1 X_1 + \gamma_2 X_2 + \gamma_3 X_3 + \gamma_{12} X_1 X_2 + \gamma_{13} X_1 X_3 + \gamma_{23} X_2 X_3$
$$+ \eta_{12} X_1 X_2 (X_1 - X_2) + \eta_{13} X_1 X_3 (X_1 - X_3) + \eta_{23} X_2 X_3 (X_2 - X_3)$$
$$+ \delta_{12} X_1 X_2 (X_1 - X_2)^2 + \delta_{13} X_1 X_3 (X_1 - X_3)^2 + \delta_{23} X_2 X_3 (X_2 - X_3)^2$$
$$+ \nu_{1123} X_1^2 X_2 X_3 + \nu_{1223} X_1 X_2^2 X_3 + \nu_{1233} X_1 X_2 X_3^2$$

where the coefficients are special functions of the usual regression terms. The usual second-order polynomial with three independent variables is

$$y = \beta_0 + \beta_1 X_1 + \beta_2 X_2 + \beta_3 X_3 + \beta_{12} X_1 X_2 + \beta_{13} X_1 X_3 + \beta_{23} X_2 X_3 + \beta_{11} X_1^2$$
$$+ \beta_{22} X_2^2 + \beta_{33} X_3^2$$

where y is the dependent variable and the X_1, X_2, X_3 are the independent variables. It is required that $X_1 + X_2 + X_3 = 1$ in the simplex lattice design. We can form the following four identities:

$$\beta_0 X_1 + \beta_0 X_2 + \beta_0 X_3 = \beta_0$$
$$X_1^2 + X_1 X_2 + X_1 X_3 = X_1$$
$$X_1 X_2 + X_2^2 + X_2 X_3 = X_2$$
$$X_1 X_3 + X_2 X_3 + X_3^2 = X_3$$

Substituting these identities into the general equation for β_0, X_1^2, X_2^2, and X_3^2, we will get, after simplification,

$$y = (\beta_0 + \beta_1 + \beta_{11})X_1 + (\beta_0 + \beta_2 + \beta_{22})X_2 + (\beta_0 + \beta_3 + \beta_{33})X_3$$
$$+ (\beta_{12} - \beta_{11} - \beta_{22})X_1X_2 + (\beta_{13} - \beta_{11} - \beta_{33})X_1X_3$$
$$+ (\beta_{23} - \beta_{22} - \beta_{33})X_2X_3$$

Letting

$$\gamma_1 = \beta_0 + \beta_1 + \beta_{11}$$
$$\gamma_2 = \beta_0 + \beta_2 + \beta_{22}$$
$$\gamma_3 = \beta_0 + \beta_3 + \beta_{33}$$
$$\gamma_{12} = \beta_{12} - \beta_{11} - \beta_{22}$$
$$\gamma_{13} = \beta_{13} - \beta_{11} - \beta_{33}$$

and

$$\gamma_{23} = \beta_{23} - \beta_{22} - \beta_{33}$$

it follows that the quadratic polynomial for the simplex lattice design with three factors is

$$y = \gamma_1 X_1 + \gamma_2 X_2 + \gamma_3 X_3 + \gamma_{12}X_1X_2 + \gamma_{13}X_1X_3 + \gamma_{23}X_2X_3 \quad (13.1.1)$$

<u>Problem 13.1.1.</u> Let $X_1 = 1 - X_2 - X_3$ and use
$$y = \delta_0 + \delta_2 X_2 + \delta_3 X_3 + \delta_{23}X_2X_3 + \delta_{22}X_2^2 + \delta_{33}X_3^2$$
to show that this is the same equation as Eq. (13.1.1).

If the design of the experiment is

y	X_1	X_2	X_3
y_1	1	0	0
y_2	0	1	0
y_3	0	0	1
y_{12}	1/2	1/2	0
y_{13}	1/2	0	1/2
y_{23}	0	1/2	1/2

(The y's correspond to the y's in the diagram for quadratic.)

the least square estimates are

$$\hat{\gamma}_1 = y_1, \ \hat{\gamma}_2 = y_2, \ \hat{\gamma}_3 = y_3, \ \hat{\gamma}_{12} = 4y_{12} - 2y_1 - 2y_2$$

$$\hat{\gamma}_{13} = 4y_{13} - 2y_1 - 2\bar{y}_3 \text{ and } \hat{\gamma}_{23} = 4y_{23} - 2y_2 - 2y_3$$

Using these estimates of the parameters (γ's) in the prediction equation for three independent variables we obtain the predicted value ($\hat{y}$) as

$$\hat{y} = X_1(2X_1 - 1)\bar{y}_1 + X_2(2X_2 - 1)\bar{y}_2 + X_3(2X_3 - 1)\bar{y}_3$$
$$+ 4X_1X_2\bar{y}_{12} + 4X_1X_3\bar{y}_{13} + 4X_2X_3\bar{y}_{23}$$

where X_i (i = 1, 2, 3) is the value of the ith independent variable to be used as a predictor of y; $\bar{y}_i$ or $\bar{y}_{ij}$ (i < j = 1, 2, 3) is the mean of the observations at the designated treatment combination (i or ij). It follows that the variance of the estimate is

$$V(\hat{y}) = \sigma^2 \left[\sum_{i=1}^{3} \frac{[X_i(2X_i-1)]^2}{r_i} + \sum_{(i<j)=1}^{3} \frac{(4X_iX_j)^2}{r_{ij}} \right]$$

where σ^2 = error variance from a previous experiment or estimated
from the within treatment combinations; r_i or r_{ij}
(i < j = 1, 2, 3) = the number of observations at the
designated treatment combination (i or ij).

To test the adequacy of the quadratic model, the center point y_{123} can be estimated from the model of the simplex lattice and observed with the point added to the design and the t-test used if an error variance is available.

After the experimenter obtains the response surface equation, he can examine the contours of the surface. Gorman and Hinman (1962, pp. 474-479) discuss a procedure for doing this. They have examined the surfaces, by the signs and sizes of the interaction γ's for the three factor quadratic model as follows:

Case	γ_{12}	γ_{13}	γ_{23}
A	0	0	0
B	+	+	+
C	+	0	-
D	+	+	0
E	+	+	-

where 0 indicates linearity; (+) indicates synergism; (-) indicates antagonism.

This specific type design in which each factor makes up 100% of one of the treatment combination is easily handled mathematically, but for many experiments it is impossible to have any treatment combination made up completely of one factor. Scheffé (1958, p. 354) discusses a modification that allows one factor to have an upper bound less than 1. In the analysis he uses a "pseudocomponent" (coding of the original variables) which allows the use of a regression equation in terms of the coded variables.

(2) Simplex Centroid. To provide experimental coverage of the response surface in the center of the planes and hyperplanes, Scheffé (1963) introduced the simplex centroid design. One way to visualize the design is to consider the structure of a three factor experiment, next a four factor experiment, etc. The four factor simplex centroid has one center point in each if its four faces (sides) of the design plus one in the center of the tetrahedron. This procedure is extended to any q-factor simplex centroid design.

The center points (centroids for a regular q-factor simplex lattice) are located in the design by finding the mean level for all factors involved. For example, the center point for a three factor simplex or any face of a q-factor simplex is found to be (1/3, 1/3, 1/3), while the center point of a four factor simplex or tetrahedron is (1/4, 1/4, 1/4, 1/4) and so on.

The analysis of a simplex centroid is similar to that of a simplex lattice since the restriction that the sum of the levels of the factors

$$\sum_{i=1}^{q} X_i = 1$$

must be used in the regression analysis for the simplex centroid also. If an experimenter is interested in optimizing the levels of the factors, he will probably use a second degree model. In the case that the simplex lattice design allowed only an incomplete model, the simplex centroid may allow a complete second degree regression model because of the additional center points.

(3) Extreme Vertices. In most experimenter problems dealing with mixtures, some portion of all ingredients are necessary for the experimental unit to work. For example, in a certain type flare some of each ingredient: magnesium, sodium nitrate, strontium nitrate, and binder must be present for the flare to function correctly. In addition, many experimenters will know the lower and upper bounds of each ingredient that they wish to examine before the experiment is designed.

The extreme vertices design (McLean and Anderson, 1966) takes into account certain aspects of the simplex centroid design and allows upper and lower bounds on every factor that are inside the usual simplex boundaries. If one considers q-factors (ingredients) represented by a proportion X_i of the total mixture, then

$$\sum_{i=1}^{q} X_i = 1,$$

and

$$0 \leq a_i \leq X_i \leq b_i \leq 1$$

where $i = 1, \ldots, q$ and a_i and b_i are constraints on the X_i imposed by the experimenter.

Once a_i and b_i are determined for all q-factors the extreme vertices experiment can be designed as follows:

Write down all possible two-level treatment combinations using

the a_i and b_i levels for all but one factor which is left blank,
e.g., $(a_1, b_2, a_3, \underline{\quad}, a_5, b_6)$, for a six factor experiment. This
procedure generates $q \cdot 2^{q-1}$ possible treatment combinations with one
factor's level blank in each.

Then, go through all $q \cdot 2^{q-1}$ possible treatment combinations and
fill in those blanks that are admissible, i.e., that level
(necessarily falling within the constraints of the missing factor)
which will make the sum of the levels for that treatment combination
equal to 1. Each of the admissible treatment combinations is a
vertex; however, some vertices may appear more than once.

The hyperpolyhedron so constructed contains a variety of centroids.
There is one located in each bounding 2-dimensional face,
3-dimensional face, ..., r-dimensional face ($r \leq q - 2$), and the
centroid of the hyperpolyhedron. The latter point being the
treatment combination obtained by averaging all the factor levels
of the existing vertices. The centroids of the 2-dimensional faces
are found by isolating all sets of vertices for which each (q - 3)
factor levels remains constant within a given set and by averaging
the factor levels for each of the three remaining factors. All
remaining centroids are found in a similar fashion using all
vertices which have (q - r - 1) factor levels constant within a set
for an r-dimensional face where $3 \leq r \leq q - 2$. It should be noted
that under the assumptions given above the dimensionality of the
hyperpolyhedron formed by the extreme vertices will always be q - 1.

Example 13.1: In manufacturing one particular type of flare
the chemical constituents are magnesium (x_1), sodium nitrate (x_2),
strontium nitrate (x_3), and binder (x_4). Engineering experience
has indicated that the following constraints on a proportion by
weight basis should be utilized:

$$0.40 \leq x_1 \leq 0.60$$
$$0.10 \leq x_2 \leq 0.50$$
$$0.10 \leq x_3 \leq 0.50$$
$$0.03 \leq x_4 \leq 0.08$$

The problem is to find the treatment combination (x_1, x_2, x_3, x_4) which gives maximum illumination as measured in candles.

The vertices of the polyhedron consisting of all the admissible points of the factor space are found by applying the above procedure. See the following tabulation which indicates eight admissible vertices and six faces. These eight treatment combinations are shown in Fig. 13.1.1.

Treatment combination	x_1	x_2	x_3	x_4	Treatment combination	x_1	x_2	x_3	x_4
	0.40	0.10	0.10	___	(1)	0.40	0.10	0.47	0.03
	0.40	0.10	0.50	___	(2)	0.40	0.10	0.42	0.08
	0.40	0.50	0.10	___		0.40	0.50	___	0.03
	0.40	0.50	0.50	___		0.40	0.50	___	0.08
	0.60	0.10	0.10	___	(3)	0.60	0.10	0.27	0.03
	0.60	0.10	0.50	___	(4)	0.60	0.10	0.22	0.08
	0.60	0.50	0.10	___		0.60	0.50	___	0.03
	0.60	0.50	0.50	___		0.60	0.50	___	0.08
(5)	0.40	0.47	0.10	0.03		___	0.10	0.10	0.03
(6)	0.40	0.42	0.10	0.08		___	0.10	0.10	0.08
	0.40	___	0.50	0.03		___	0.10	0.50	0.03
	0.40	___	0.50	0.08		___	0.10	0.50	0.08
(7)	0.60	0.27	0.10	0.03		___	0.50	0.10	0.03
(8)	0.60	0.22	0.10	0.08		___	0.50	0.10	0.08
	0.60	___	0.50	0.03		___	0.50	0.50	0.03
	0.60	___	0.50	0.08		___	0.50	0.50	0.08

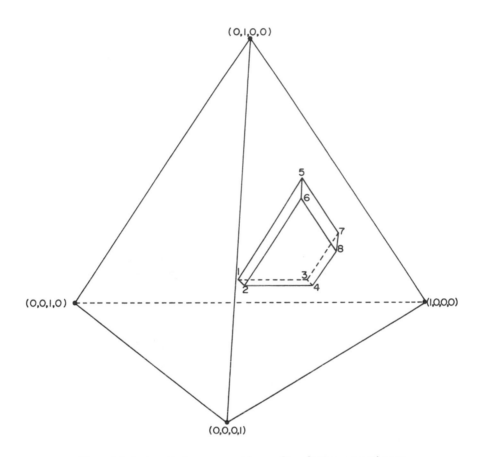

Fig. 13.1.1. Extreme vertices for flare experiment.

In order to complete the design, one must determine the six centroids for each and the centroid for the polyhedron. To do this we must isolate all existing faces. These are made up of at least 3 vertices which have 4-3 or 1 factor level which is constant. For example, vertices (1), (3), (5), and (7) all have the binder (x_4) level equal to 0.03. All possible faces and their associated centroids are as follows:

Treatment combination	Centroid	Treatment combination which form the face
(9)	(0.50, 0.1000, 0.3450, 0.055)	(1), (2), (3), (4)
(10)	(0.50, 0.3450, 0.1000, 0.055)	(5), (6), (7), (8)
(11)	(0.40, 0.2725, 0.2725, 0.055)	(1), (2), (5), (6)
(12)	(0.60, 0.1725, 0.1725, 0.055)	(3), (4), (7), (8)
(13)	(0.50, 0.2350, 0.2350, 0.030)	(1), (3), (5), (7)
(14)	(0.50, 0.2100, 0.2100, 0.080)	(2), (4), (6), (8)

and the final centroid of the polyhedron, of course, comes from the average of all eight treatment combinations and is

(15)   (0.50, 0.2225, 0.2225, 0.055)

Fifteen flares assembled at each of the above treatment combinations produced, respectively, the following amounts of illumination (measured in 1000 candles):

(1)	75	(6)	230	(11)	190
(2)	180	(7)	220	(12)	310
(3)	195	(8)	350	(13)	260
(4)	300	(9)	220	(14)	410
(5)	145	(10)	260	(15)	425

Standard least squares techniques were used on the above data to obtain the complete quadratic model.

$$\hat{y} = -1,558x_1 - 2,351x_2 - 2,426x_3 + 14,372x_4 + 8,300x_1x_2 + 8,076x_1x_3$$
$$- 6,625x_1x_4 + 3,213x_2x_3 - 16,998x_2x_4 - 17,127x_3x_4$$

The squared multiple correlation coefficient (R^2) for this model is 0.9833, with only five degrees of freedom for residual. If only x_1, x_2, x_3, x_1x_2, x_1x_3, x_1x_4, x_2x_3 were used, the R^2 would be 0.9829, with eight degrees of freedom. Since all four variables still appeared in the latter model, the authors decided to retain the full model. The reader should recognize that, as in any model development problem, one must have stopping rules for evolving models from data. The purpose of this example, however, is merely to demonstrate the use of the regression model to determine the optimum treatment combination not to elaborate on model development per se.

In order to consider all the necessary constraints, a quadratic programming routine was used to calculate the optimum treatment combination

$$(0.5233,\ 0.2299,\ 0.1668,\ 0.0800)$$

which is the desired solution to the problem. The predicted value of y for this optimum point is 397.48. It should be noted that this procedure only guarantees an optimum in the case where the response surface is a concave function.

It is quite feasible that one would like to further verify the initial estimate of the optimum condition. This could be done by applying another extreme vertices design to a localized region containing this initial point.

Snee and Marquardt (1974) developed an algorithm for selecting a set of extreme vertices for linear models which tend to minimize the average variance of the estimated regression coefficients. This article also includes an extended list of references on the subject of mixture designs. Snee (1975) also discusses construction of efficient designs for constrained mixture systems in which the response can be described by a quadratic model.

Problem 13.1.2. An experiment on steel alloys was needed to find the optimum combination of carbon, nickel, chromium, manganese and iron to maximize the strength of the alloy (keeping other desirable properties).

Using the extreme vertices designs, design experiments and show approach for analyses for the following conditions.

(a)

	Low	High
Carbon	0.0004	0.0010
Nickel	0.0800	0.1200
Chromium	0.1200	0.2000
Manganese	0.0050	0.0200

keep Iron at 0.7000 (fixed).

(b)   Use same levels of carbon through manganese but use

	Low	High
Iron	0.65	0.75

(Indicate the difficulty in handling this part but do not try
to solve for the centroids).

13.1.5 Rotatable Designs

The rotatable designs require the usual assumptions for
regression and the sum of the levels of the factors for any
treatment combination does not necessarily add to 1. The treatment
combinations are located equidistant from the originally estimated
maximum point, designated by (0, 0, 0, ..., 0), so that the
standard error of the estimate of y is the same for all points on
the circumference of the design. Box and Hunter (1957) have done
extensive work on these designs and Cochran and Cox (1957) have
given many of these designs.

(1) First-Order Design, Two Factors X_1, X_2 and No Interaction
(Nonorthogonal).

The simplest experiment possible to obtain information on both
main effects requires three observations. The rotatable design for
such an experiment may be pictured as:

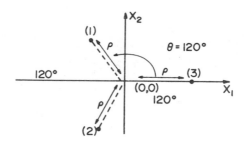

where points (1), (2), and (3) make up the design of the experiment,
   ρ = distance from the center point (0, 0) in standardized units
      which the experimenter believes will give information on
      the response surface; θ = the angle from one point,
      treatment combination to another. (In this case θ = 360°/3
      = 120°), and the resulting design is an equilaterial triangle

The regression equation for this experiment is

$$y = \beta_0 + \beta_1 x_1 + \beta_2 x_2$$

where the usual assumptions apply to the nonorthogonal regression analysis except that there is no estimate of error. To obtain an error estimate, at least one of the three treatment combinations must be repeated in the experiment. Usually for this type of experiment previous knowledge of the error is available.

(2) Second-Order Design, Two Factors with Interaction (Nonorthogonal)

In this rotatable design there must be a minimum of six treatment combinations one located at the center and the arrangement to give equal distance from the center point is a regular pentagon. In this case $\theta = 360°/5 = 72°$ between each point and the distance again can be ρ, the length specified by the experimenter.

The regression equation for this experiment is

$$y = \beta_0 + \beta_1 x_1 + \beta_2 x_2 + \beta_{11} x_1^2 + \beta_{22} x_2^2 + \beta_{12} x_1 x_2$$

with no error estimate from the data unless there are repeated observations for at least one of the treatment combinations. Box and Wilson (1951, pp. 8 and 9), compare pentagonal and factorial. There is little to choose between the two. Quadratic effects are estimated more accurately with pentagonal and interactions less accurately.

(3) General Second-Order Two Factor (Nonorthogonal)

Cochran and Cox (1957, p. 346) recommend that to obtain a general second-order model one may choose n_1 points in the center (0, 0) and n_2 points on the circumference of the circle at distance ρ, and $\theta = 360°/n_2$ apart such that the standard errors of estimate of y are approximately the same. One can locate the treatment combinations by using the following:

Treatment combination	X_1	X_2
1	$\rho \cos \theta$	$\rho \sin \theta$
2	$\rho \cos 2\theta$	$\rho \sin 2\theta$
.	.	.
.	.	.
.	.	.
.	.	.
n_2	$\rho \cos n_2\theta$	$\rho \sin n_2\theta$
n_2+1	0	0
.	.	.
.	.	.
.	.	.
n_2+n_1	0	0

With three or more factors, regular figures can be obtained with additional observations at the center, Box and Hunter (1957).

13.1.6 Fractional Factorials (Orthogonal)

As discussed in previous chapters, fractional factorials allow estimation of certain effects and interactions if other interactions are negligible. Great care is taken to assure the experimenter of exactly which interactions are assumed negligible before the experiment is even laid out. Hence the analysis of variance is clearly understood.

The important feature in using fractional factorial type experiments for finding maximum points or regions is the extraction of the regression model from the data. Consider an example of a 1/2 replicate of a 2^3 factorial experiment in which all factors are considered to have only -1 and +1 levels. There are only four treatment combinations, the two factor interactions and the three factor interaction are assumed negligible. This concept can be expressed through the aliases

$$\mu = I = Z_1 Z_2 Z_3$$
$$Z_1 = Z_2 Z_3$$
$$Z_2 = Z_1 Z_3$$
$$Z_3 = Z_1 Z_2$$

where $Z_1 Z_2 Z_3 = Z_2 Z_3 = Z_1 Z_3 = Z_1 Z_2 = 0$.

Then it follows that the regression equation (with no error estimate from the experiment) is:

$$y = \alpha_0 + \alpha_1 Z_1 + \alpha_2 Z_2 + \alpha_3 Z_3$$

As in previous regression equations dealing with orthogonal experiments it is quite easy to transform back to

$$y = \beta_0 + \beta_1 X_1 + \beta_2 X_2 + \beta_3 X_3$$

using the actual levels of the factors X_1, X_2 and X_3.

Continuing the same example, the design can generally be described as

Z_1	Z_2	Z_3	
-1	-1	-1	
1	1	-1	Same as I = -ABC from fractional replicated factorial experiments
1	-1	1	
-1	1	1	

If the levels of X_1 are really 2 and 14, of X_2 are 15 and 23 and of X_3 are 5 and 27, the design could be written as:

X_1	X_2	X_3
2	15	5
14	23	5
14	15	27
2	23	27

13.1.7 Complete Factorials (Orthogonal)

The complete factorials (as discussed previously) allow estimation of all the effects and interactions with no confounding. Of course one can estimate only linear effects and interactions if there are only two levels for each factor and can estimate up to the quadratic effects and corresponding interactions if there are three levels of each factor and so on. This concept was discussed thoroughly in previous chapters.

Here, as in the section on fractional factorials immediately preceding this section, the important feature in using factorial experiments to investigate the maximum points or regions is to obtain a regression model to describe the response surface. Using the 2^3 complete factorial here, the design may be given by

z_1	z_2	z_3		x_1	x_2	x_3
-1	-1	-1		2	15	5
-1	-1	1		2	15	27
-1	1	-1		2	23	5
-1	1	1		2	23	27
1	-1	-1	or	14	15	5
1	-1	1		14	15	27
1	1	-1		14	23	5
1	1	1		14	23	27

The corresponding regression equations (with no estimate of error) are:

$$y = \alpha_0 + \alpha_1 z_1 + \alpha_2 z_2 + \alpha_3 z_3 + \alpha_{12} z_1 z_2 + \alpha_{13} z_1 z_3 + \alpha_{23} z_2 z_3$$
$$+ \alpha_{123} z_1 z_2 z_3$$

and

$$y = \beta_0 + \beta_1 x_1 + \beta_2 x_2 + \beta_3 x_3 + \beta_{12} x_1 x_2 + \beta_{13} x_1 x_3 + \beta_{23} x_2 x_3$$
$$+ \beta_{123} x_1 x_2 x_3$$

Of course if the three factor interaction is assumed zero, one can estimate the error from that component.

It should be noticed that the two-leveled complete factorial experiment allows estimates of interactions, but not of the squared terms. The three-leveled factorials would allow the estimates of squared terms, but too many observations are required for these estimates relative to number of observations required by the composite design (to be discussed in the next section).

13.1.8 Composite Design

The composite design Box and Wilson, (1951); Myers, (1971, Ch. 7; and Davies, (1971, Ch. 11) has three parts: a basic two-leveled factorial or fractional factorial, an extra point at the center of the entire design and 2k (where k = number of factors) extra points, one at either extreme of each factor and at the center of all other factors. Hence in a composite design with a complete factorial experiment in it there are 2^k + 2k + 1 treatment combinations. Correspondingly, if there was a fractional factorial instead of a complete factorial experiment in the design, the 2^k would be reduced as required. The particular type of composite design depends on the location of the extreme points. If the extreme points are located at the same standardized distance from the center point as the factorial points, the design is called a rotatable composite design (sometimes the word "central" is included in the title of these designs to indicate that there is a center point). If the extremes are located at a distance that makes the squared terms in the model orthogonal to each other, the design is called an orthogonal composite design. Any other location of the extreme points may be used, but the analysis is just a nonorthogonal regression analysis.

The advantage of a composite design over the fractional or complete three-leveled factorial is in the reduction of the number of treatment combinations required to estimate the squared terms in a second-order model. This idea may be summarized in the following tabulation :

Number of Treatment Combinations

Number of factors k	Three-leveled factorial, 3^k	Composite, $2^k + 2k + 1$
2	9	9
3	27	15
4	81	25
5	243	43
5 (1/3 fractional)	81	(1/2 fractional) 27
6	729	77
6 (1/3 fractional)	243	(1/2 fractional) 45

As the number of factors increase in an experiment the savings in number of treatment combinations using the composite design instead of three factor factorial increases rapidly.

Two disadvantages in using the composite design instead of the three leveled factorial are (1) estimating effects with unequal variances (to be shown later), and (2) having fewer degrees of freedom for error. The model for a composite design is:

$$y = \beta_0 + \beta_1 X_1 + \beta_2 X_2 + \cdots + \beta_k X_k + \beta_{12} X_1 X_2 + \cdots +$$
$$\beta_{(k-1)k} X_{k-1} X_k + \beta_{11} X_1^2 + \beta_{22} X_2^2 + \cdots + \beta_{kk} X_k^2 + \varepsilon$$

Consider k = 3, then the composite model is:

$$y = \beta_0 + \beta_1 X_1 + \beta_2 X_2 + \beta_3 X_3 + \beta_{12} X_1 X_2 + \beta_{13} X_1 X_3 + \beta_{23} X_2 X_3$$
$$+ \beta_{11} X_1^2 + \beta_{22} X_2^2 + \beta_{33} X_3^2 + \varepsilon$$

where the error has 5 degrees of freedom, and the corresponding model for the three-leveled factorial is

$$y = \beta_0 + \beta_1 x_1 + \beta_2 x_2 + \beta_3 x_3 + \beta_{12} x_1 x_2 + \beta_{13} x_1 x_3 + \beta_{23} x_2 x_3$$
$$+ \beta_{11} x_1^2 + \beta_{22} x_2^2 + \beta_{33} x_3^2$$
$$+ \beta_{112} x_1^2 x_2 + \beta_{122} x_1 x_2^2 + \beta_{113} x_1^2 x_3 + \beta_{133} x_1 x_3^2 + \beta_{233} x_2 x_3^2$$
$$+ \beta_{223} x_2^2 x_3 + \beta_{1122} x_1^2 x_2^2 + \beta_{1133} x_1^2 x_3^2$$
$$+ \beta_{2233} x_2^2 x_3^2 + \beta_{123} x_1 x_2 x_3 + \beta_{1123} x_1^2 x_2 x_3$$
$$+ \beta_{1223} x_1 x_2^2 x_3 + \beta_{1233} x_1 x_2 x_3^2 + \beta_{11223} x_1^2 x_2^2 x_3$$
$$+ \beta_{11233} x_1^2 x_2 x_3^2 + \beta_{12233} x_1 x_2^2 x_3^2 + \beta_{112233} x_1^2 x_2^2 x_3^2$$

In a second-order model, the terms from $\beta_{112} x_1^2 x_2$ through $\beta_{112233} x_1^2 x_2^2 x_3^2$ are assumed zero and could be put into the error estimates. The factorial model would then allow 17 degrees of freedom for error.

In general for optimum designs, the five degrees of freedom in the composite design for the error estimate is adequate and the composite design is preferred over the three-leveled factorial. In the following two subsections two types of composite designs, rotatable and orthogonal, are described.

(1) Rotatable Composite (Partly Orthogonal)

The rotatable composite design has all points equidistant from the center point. The basic factorial terms are orthogonal to one another and the squared terms are orthogonal to the linear and interaction terms but not to each other. In other words the star points, the 2k points used to estimate the squared terms in the model, do not form an orthogonal design. As a result one may run straight multiple regression analysis on all the data or run an analysis of variance on the factorial part before running the regression analysis. The analysis of variance on the factorial part allows the experimenter to examine the linear main effects and the linear by linear interaction parts before examining the overall model.

Here, as in other rotatable designs, the variance of the estimates of y are approximately the same, but the variance of the estimates of the regression coefficients may be quite different. Of course these latter variances can be obtained by the usual least squares method.

(2)   Orthogonal Composite

This design requires that the distance from the center point to each of the star points be such that all effects and interactions estimated in the second order model are orthogonal to one another. This distance, d, is

$$d = \left\{ \frac{[(F + T)^{1/2} - F^{1/2}]^2 \cdot F}{4n^2} \right\}^{1/4}$$

where F = the number of treatment combination in the factorial run (fractional or full); and T = the number of additional points multiplied by the number of observations per treatment combination (n).

In the three factor case; F = 8, T = 7 if there is only one observation for each treatment combination (n = 1). If follows that

$$d = \left\{ \frac{[(8 + 7)^{1/2} - 8^{1/2}]^2 \ 8}{4} \right\}^{1/4} = 1.215$$

An example using the three factor case follows:

The design or the 15 treatment combinations for a three factor orthogonal composite design is:

Factor (Independent variable)

	z_1	z_2	z_3	Treatment combination
1	-1	-1	-1	(1)
2	1	-1	-1	z_1
3	-1	1	-1	z_2
4	1	1	-1	$z_1 z_2$
5	-1	-1	1	z_3
6	1	-1	1	$z_1 z_3$
7	-1	1	1	$z_2 z_3$
8	1	1	1	$z_1 z_2 z_3$
9	0	0	0	0
10	1.215	0	0	d_1
11	-1.215	0	0	$-d_1$
12	0	1.215	0	d_2
13	0	-1.215	0	$-d_2$
14	0	0	1.215	d_3
15	0	0	-1.215	$-d_3$

The analysis in model form is:

$$y = \alpha_0 z_0 + \alpha_1 z_1 + \alpha_2 z_2 + \alpha_{11} z_{11} + \alpha_{22} z_{22} + \alpha_{12} z_1 z_2 + \alpha_3 z_3$$

$$+ \alpha_{33} z_{33} + \alpha_{13} z_1 z_3 + \alpha_{23} z_2 z_3 + \alpha_{123} z_1 z_2 z_3 + \varepsilon$$

There are four degrees of freedom for error in this design, unless $z_1 z_2 z_3$ is zero; then there are five degrees of freedom. It is impossible to estimate (linear x quadratic) interactions separately because $z_1 z_{22} = z_1 z_{33}$ (note table below). The layout of the effects and interactions is given in Table 13.1.2.

TABLE 13.1.2

Treatment Combination

Effect	1	2	3	4	5	6	7	8	9	10	11	12	13	14	15
→	(1)	z_1	z_2	z_1z_2	z_3	z_1z_3	z_2z_3	$z_1z_2z_3$	0	d_1	$-d_1$	d_2	$-d_2$	d_3	$-d_3$
z_0	+	+	+	+	+	+	+	+	+	+	+	+	+	+	+
z_1	−	+	−	+	−	+	−	+	0	1.215	−1.215	0	0	0	0
z_2	−	−	+	+	−	−	+	+	0	0	0	1.215	−1.215	0	0
z_1z_2	+	−	−	+	+	−	−	+	0	0	0	0	0	0	0
z_3	−	−	−	−	+	+	+	+	0	0	0	0	0	1.215	−1.215
z_1z_3	+	−	+	−	−	+	−	+	0	0	0	0	0	0	0
z_2z_3	+	+	−	−	−	−	+	+	0	0	0	0	0	0	0
$z_1z_2z_3$	−	+	+	−	+	−	−	+	0	0	0	0	0	0	0
z_{11}	0.27	0.27	0.27	0.27	0.27	0.27	0.27	0.27	−0.73	0.746	0.746	−0.73	−0.73	−0.73	−0.73
z_{22}	0.27	0.27	0.27	0.27	0.27	0.27	0.27	0.27	−0.73	−0.73	−0.73	0.746	0.746	−0.73	−0.73
z_{33}	0.27	0.27	0.27	0.27	0.27	0.27	0.27	0.27	−0.73	−0.73	−0.73	−0.73	−0.73	0.746	0.746
*z_1z_{22}	−0.27	0.27	−0.27	0.27	−0.27	0.27	−0.27	0.27	0	−0.887	0.887	0	0	0	0
*z_1z_{33}	−0.27	0.27	−0.27	0.27	−0.27	0.27	−0.27	0.27	0	−0.887	0.887	0	0	0	0

*Note equality and hence are completely confounded.

The levels of each factor (-1.215, -1, 0, 1, 1.215) are not equally spaced and the usual orthogonal polynomials do not work directly; however, if in this special case the quadratic variable

$$Z_{11} = \left(Z_1^2 - \frac{\Sigma \ (Z_1^2)}{15} \right)$$

then the analysis is orthogonal. The same procedure is used for Z_{22} and Z_{33}.

All effects and interactions are orthogonal and are estimated in the usual manner $\hat{\alpha} = \Sigma Zy/\Sigma Z^2$. The regression coefficients in the model are not correlated, and the terms are additive. As can be seen by the table of coefficients, however, the effects and interaction coefficients ($\hat{\alpha}$'s) have different variances because Z^2 are different for certain coefficients. The variance of $\hat{\alpha}$ is

$$V(\hat{\alpha}) = \left(\frac{\sigma^2}{\Sigma Z^2} \right)$$

where σ^2 is the error variance, and Z's are as described in the layout. As examples consider the variance of $\hat{\alpha}_1$:

$$V(\hat{\alpha}_1) = \frac{\sigma^2}{\Sigma Z_1^2} = \frac{\sigma^2}{8 + 2(1.215)^2}$$

and the variance of $\hat{\alpha}_{11}$:

$$V(\hat{\alpha}_{11}) = \frac{\sigma^2}{\Sigma Z_{11}^2} = \frac{\sigma^2}{8(.27)^2 + 5(-.73)^2 + 2(.746)^2}$$

whereas in a factorial experiment the variances of the $\hat{\alpha}$'s are all constant.

If an experimenter is interested in investigating only second-degree polynomial surfaces in a fixed experiment the recommendation is that he consider the two-level factorial plus additional points to obtain a composite design rather than a

complete or fractional three-level factorial. The important reason
for the recommendation is that the reduction in the number of
treatment combinations is quite large, especially when the number
of factors is four or higher. In the composite design, however, the
variances of the coefficients for effects and interactions are not
all equal and should always be considered before deciding on the
exact design of the experiment.

Problem 13.1.3. Suppose an experimenter wanted to estimate
the parameters in the equation

$$y = \beta_0 + \beta_1 X_1 + \beta_2 X_2 + \beta_{12} X_1 X_2 + \beta_{11} X_1^2 + \beta_{22} X_2^2 + \beta_{112} X_1^2 X_2$$

and wanted to use a design that would have the fewest number
of observations, could be nonorthogonal and have the standard
errors of estimate as near equal as possible.
   Set up the design specifying the location of each
experimental unit relative to coordinates (0, 0) of (X_1, X_2).

Problem 13.1.4. In a weapons evaluation section of a large
corporation, an experiment was needed to investigate factors
affecting firing time of a weapon. The factors and their
lower and upper limits of the levels were:

   (1) Primary igniter   5 to 15 mg
   (2) Primary initiator   5 to 15 mg
   (3) Packing pressure   12,000 to 28,000 psi
   (4) Amount of igniter   80 to 120 mg
   (5) Delay column pressure   20,000 to 40,000 psi

Eventually an equation of the response surface would be desired
in which interactions of linear x linear and squared terms may
be important. The number of observations should be held to a
minimum and all effects from a third degree term and higher
are considered negligible.
   Design an orthogonal experiment showing the levels of each
factor to be used and explain why you used this particular
design.

Problem 13.1.5. Show the variance of $\hat{\alpha}_1$, $\hat{\alpha}_{12}$ and $\hat{\alpha}_{22}$ in a
3-factor orthogonal composite design.

13.2 SEQUENTIAL DESIGNS

13.2.1 Univariate Method or One-Factor-at-a-Time

The one factor at a time method of finding an optimum
treatment combination has been used for many years. If there are
only two independent variables, the experimenter selects one
variable to hold constant and decides in which direction to go
after the first observation plus the amount of the jump to make for
the other variable. The first observation is taken at the "guessed
at" maximum. If the response on the second observation was larger
than the first, take the third observation in the same direction
with the same increment jump. If the response on the second
observation is smaller than the first, go in the opposite direction
from the first with the same increment. If the response on the
third observation is larger than the second continue in that
direction; but if smaller, go back to the point for which the
response was the greatest and follow a similar procedure in a
perpendicular direction by keeping the variable constant that was
being varied and change the variable that was held constant. Repeat
this procedure until no appreciable increase in yield results and
call the maximum response point the optimum for those independent
variables.

Brooks (1959) describes a slightly different procedure called
"Univariate," but the method is nearly the same. Using Brooks'
example in which 16 trials are allotted for the experiment, take
four trials equally spaced around the "guessed at" maximum in one
direction, say X_2. Then a curve

$$y = \beta_0 + \beta_2 X_2 + \beta_{22} X_2^2 + \beta_{222} X_2^3$$

is fitted and the point in X_2 that maximized this expression along
the line of the experiment is used as the fixed point to try four
more trials in the X_1 (perpendicular to X_2) direction. This
procedure of fitting is repeated for X_1, and four more trials are
taken in the X_2 direction in a similar manner except that the

increments are decreased between points as described by Brooks
(1958). The final four trials are made in the other direction and
the maximum response for the final four gives the optimum factor
combination. Of course one could fit a polynomial over the surface
examined, but the factor space may not be covered adequately to
provide a very good overall response surface for the limits of X_1
and X_2 used in the experiment.

Brooks (1958) gives a numerical example of the univariate
method in his paper.

13.2.2 Simplex (Sequential)

The sequential application of simplex designs presented here
do not require that the sum of the levels of the factors add to one
for each treatment combination. The number of factors, n, then, is
equal to the dimension of the design. For the two factor experiment
the regular simplex design is an equilateral triangle in two
dimensions. It is not necessary for this design to be regular but
the scaling can be done such that the same unit is used for all
dimensions.

To portray the procedure we will use a two dimensional design
centered around a "guessed at" maximum and call this point the
centroid (0, 0). The adjacent points in our design will be one
unit apart. The layout for this design is as follows.

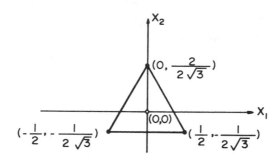

where only the three extreme points are used in the design.

The general n-dimensional design has $(n + 1)$ points and can be arrayed as a $(n + 1)$ by (n) matrix $\mathscr{X}$:

Point or treatment combination	Independent Variable (X)					
	1	2	3	4	...j...	n
1	$-\dfrac{1}{2}$	$-\dfrac{1}{2\sqrt{3}}$	$-\dfrac{1}{2\sqrt{6}}$	$-\dfrac{1}{2\sqrt{10}}$	$\cdots$	$-\dfrac{1}{\sqrt{2n(n+1)}}$
2	$\dfrac{1}{2}$	$-\dfrac{1}{2\sqrt{3}}$	$-\dfrac{1}{2\sqrt{6}}$	$-\dfrac{1}{2\sqrt{10}}$	$\cdots$	$-\dfrac{1}{\sqrt{2n(n+1)}}$
3	0	$\dfrac{2}{2\sqrt{3}}$	$-\dfrac{1}{2\sqrt{6}}$	$-\dfrac{1}{2\sqrt{10}}$	$\cdots$	$-\dfrac{1}{\sqrt{2n(n+1)}}$
4	0	0	$\dfrac{3}{2\sqrt{6}}$	$-\dfrac{1}{2\sqrt{10}}$	$\cdots$	$-\dfrac{1}{\sqrt{2n(n+1)}}$
. . .		.				
i	$\cdots$	$\cdots$	$\cdots$	$\cdots$	X_{ij}	$\cdots$
. . .	. . .	. . .	. . .	. . .	.	
$(n + 1)$	0	0	0	0	$\cdots$	$\dfrac{n}{\sqrt{2n(n+1)}}$

Notice that the two-dimensional design in the following tabulation as given previously in the upper left-hand corners of $\mathscr{X}$.

Point	Independent variable	
	1	2
1	$-\dfrac{1}{2} = X_{11}$	$-\left(\dfrac{1}{2\sqrt{3}}\right) = X_{12}$
2	$\dfrac{1}{2} = X_{21}$	$-\left(\dfrac{1}{2\sqrt{3}}\right) = X_{22}$
3	$0 = X_{31}$	$\left(\dfrac{2}{2\sqrt{3}}\right) = X_{32}$

Next we must show how one moves sequentially to the maximum yield or response. Let us assume that the point $[0, 2/(2\sqrt{3})]$ gives the minimum yield of the three points in the two-dimensional design. The sequential method is to delete that point and form another simplex with the remaining two points and a point on the opposite side. In this case the new point would be $[0, -4/(2\sqrt{3})]$. The algebraic method for finding the new point is to delete the minimum point from the array giving the remaining array as follows.

| | Independent variable | |
Point	1	2
1	$-\dfrac{1}{2}$	$-\dfrac{1}{2\sqrt{3}}$
2	$\dfrac{1}{2}$	$-\dfrac{1}{2\sqrt{3}}$

The next step is to add the levels of each factor and multiply each by $2/n$ (where n = the number of X's) or $2/2 = 1$ for this case, and subtract the level of the factor deleted. For variable 1 we have

$$\frac{2}{2} \left(- \frac{1}{2} + \frac{1}{2}\right) = 0$$

and subtract 0 giving 0. For variable 2

$$\frac{2}{2} \left(- \frac{1}{2\sqrt{3}} - \frac{1}{2\sqrt{3}}\right) = - \frac{2}{2\sqrt{3}}$$

and subtract $2/(2\sqrt{3})$ which gives $- 4/(2\sqrt{3})$.

Using the X_{ij} notation of the (n + 1) by n matrix given previously to denote the level of the jth variable of the ith treatment combination, we can find the new level X_{ij}^{*} by using

$$X_{ij}^{*} = [\frac{2}{n} \{X_{1j} + X_{2j} + \ldots + X_{i-1,j} + X_{i+1,j} + \ldots + X_{n+1,j}\} - X_{ij}]$$

The procedure to find the optimum region of the treatment combinations is to continue to delete the point or treatment

combination that has the lowest response in the simplex, replacing
this point by a point opposite to it as described above. This
procedure is followed until an optimum region is located. The
optimum region is indicated by several new points, say n, failing
to produce a response that is greater than that obtained at a
previous treatment combination. At this point the optimum
treatment combination should be repeated before experimentation is
stopped in order to assure that this point is truly optimum.

If in the process of stepping through the factor space with
successive simplexes, the yield of a new point is less than any
other point in the current simplex, do not go back to the point that
was previously vacated. In this case determine the next lowest
point and move opposite it. This makes up a different simplex which
is the basis for the next decision.

This concludes the design discussion. The analysis, in
addition to finding the optimum point, may include running a
regression analysis on those points near the optimum.

An example of the procedure is the following electrical
engineering problem in which the response y is output power and the
inputs are (X_1) voltage, (X_2) relative humidity, and (X_3)
temperature. The present system for X_1 is 110 V, X_2 is 43%, X_3 is
86°F with y = 58%. Hence the point

$$(0, 0, 0) = (110, 43, 86)$$

has a yield of 58. For X_1, X_2, and X_3 the experimenter felt that
increments of 10, 2, and 0.5, respectively, were equally important.
Using these increments to code about the centroid, we obtain the
four treatment combinations

$$
\begin{bmatrix}
1. & 105, & 42.4, & 85.8 \\
2. & 115, & 42.4, & 85.8 \\
3. & 110, & 44.2, & 85.8 \\
4. & 110, & 43.0, & 86.6
\end{bmatrix}
$$

from the following matrix:

$$\begin{bmatrix} -\dfrac{1}{2}, & -\dfrac{1}{2\sqrt{3}}, & -\dfrac{1}{2\sqrt{6}} \\[3ex] \dfrac{1}{2}, & -\dfrac{1}{2\sqrt{3}}, & -\dfrac{1}{2\sqrt{6}} \\[3ex] 0, & \dfrac{2}{2\sqrt{3}} & -\dfrac{1}{2\sqrt{6}} \\[3ex] 0, & 0, & \dfrac{3}{2\sqrt{6}} \end{bmatrix}$$

The yields for the four treatment combinations were 52, 62, 61, 57, respectively. Hence delete treatment combination 1 and replace it by the three levels:

$$\text{Level 1} = \frac{2}{3} (115 + 110 + 110) - 105 = 118.3$$

$$\text{Level 2} = \frac{2}{3} (42.4 + 44.2 + 43.0) - 42.4 = 44.0$$

$$\text{Level 3} = \frac{2}{3} (85.8 + 85.8 + 86.6) - 85.8 = 86.3$$

This new treatment combination (118.3, 44.0, 86.3), along with the remaining three treatment combinations, forms the new simplex. After running the new treatment combination, the sequential procedure continues as discussed previously.

The efficiency of sequential designs is measured by the number of points required to reach the maximum. For a given surface, one can determine the rate of advance to the maximum. Spendley, Hext, and Himsworth (1962) show that as the standard deviation (variation of the responses for the same treatment combination) increases the rate of advance decreases as {1/(standard deviation)}. For example, the rate of advance in an experiment with standard deviation of 3 is twice that of an experiment with standard deviation of 6.

On comparing achievement of sequential procedures to reach maximum, Spendley, Hext, and Himsworth (1962) indicate that the simplex design is second only to the steepest ascent (regression) procedure if there is no error and somewhat poorer (but not too bad) when error is present.

Box and Behnken (1960) describe a "simplex-sum" design, a type of second order design, that combines the vectors defining the points of the initial simplex in pairs, threes and so on. One can use the midpoints, or points off the midpoints, of the vectors joining the vertices to estimate the curvature. The concept of using intermediate points for estimation of curvature was given in Section 13.1.4.

> Problem 13.2.1. Consider the electrical engineering example in Section 13.2.2 in which y is output power and the inputs are (X_1) voltage, (X_2) relative humidity, and (X_3) temperature. If the new treatment combination gave a yield of 67, the next one gave a yield of 69, and the next 65, then 62, 68, 69, and finally 69.
> (a) Show the new treatment combinations for each new simplex.
> (b) What is the treatment combination that gives approximately the maximum yield?
> (c) Run a nonorthogonal regression analysis on these data, using the points you believe are applicable. Explain your model.

13.2.3 Computer Aided Designs

Given a set of conditions for a factorial type experiment with the lower and upper bounds on all factors established by the experimenter, the problem is to find out the specific order to run the possible treatment combination such that the most information for that number of sequential observations may be obtained over the factor space whenever the experimenter decides to stop the experiment. "Most information" in this case usually refers to minimum variance of the estimates of the parameters in the response surface equation over the defined factor space for the data obtained.

In this case Kennard and Stone (1969) use quantitative levels and linear associations to find the design points. This allows them to compute distances (actually squared distance is used) between points in the factor space using the computer. The technique is to choose two points that are farthest apart (these may not be unique so a criterion of choosing the smallest index is used for uniqueness) as the first two treatment combinations to be used in the design.

To obtain the next best (in the sense of giving most information using the distance criterion) point from the remaining (n - 2) points, one must find the point that is farthest from the first two. This point may not be unique either (as for the first two points), and the same criterion of taking the point with the smallest index is utilized.

This process of choosing succeeding points that are farthest from the ones already in the design is continued until the experimenter decides to stop or all points in the factor space are used. In general this design attempts to spread the observations evenly over the factor space with emphasis on using outermost points first and come into the center part next. Then go out from the center but not necessarily to the edge of the factor space and so on.

The basic idea that the observations may be taken at these points, an analysis run and a decision made as to whether or not more observations are needed allows much flexibility in this sequential type design.

Example of the Kennard and Stone Procedure

Consider a 5^2 equally spaced factorial. The points are numbered for the 25 possible combinations of X_1 and X_2 as follows:

		\(X_1\)				
		-2	-1	0	1	2
	+2	1	2	3	4	5
	+1	6	7	8	9	10
X_2	0	11	12	13	14	15
	-1	16	17	18	19	20
	-2	21	22	23	24	25

The pair of points (1, 25) and the pair (5, 21) are tied for being farthest apart. Since the index on the pair (1, 25) is smaller, namely 1, this pair is used in the design.

Now there are 23 points (n - 2) remaining. The squared distances to each of these points from 1 and 25 are computed. For example to point 3 from point 1 the squared distance is $(-2-0)^2$ or 4 and to 3 from 25 the squared distance using the Pythagorean theorem is $(2-0)^2 + (-2-2)^2 = 20$.

The same type of calculations are made for all 23 points and it is found that 5 and 21 are tied for being farthest from design points 1 and 25. Choosing the smaller index point, 5, the third design point is 5.

Continuing the process, design point four is 21 and design point five is 13. This is a basic "EVOP" design to be discussed later for a two factor design. That is, it is a two by two factorial with the points at the extreme vertices and one point in the center.

Continued application of the procedure gives a 3^2 for the first 9 design points since design point six is 3, design point seven is 11, design point eight is 15, and design point nine is 23.

One can continue this until all 25 points are in the design. This concludes the demonstration of the design.

It should be clear by now that this procedure allows a sequential design and a stopping point may be decided upon when the experimenter feels he has enough information. He may use a regression analysis at any time he has enough points to estimate the parameters of interest and has enough information on the experimental error.

This procedure is most flexible in that it is programmed to handle many factor factorials with varying levels, orthogonal and nonorthogonal, taking into account design points that are experimentally impossible to run and accounting for points already run.

Additional procedures for similar sequential type designs are discussed by Dykstra (1971) and Hebble and Mitchell (1972).

13.2.4 Steepest Ascent Method

The general mathematical concepts of utilizing regression techniques to examine yield surfaces has been known for many years. Hotelling (1941) discussed the problem using experimental procedures, but it was not until after the Box and Wilson (1951) paper that experimenters fully appreciated the usefulness of this tool. Other references are Myers (1971, Chapter 5), and John (1971, Chapter 10).

The general principle that is followed in the steepest ascent method is to establish an initial design, say a 2^n factorial, with a center point at some "guessed at" maximum in the factor space. A regression analysis is performed on the data of this initial design in order to determine which direction, if any, one should move from the center point of the design in order to increase the yield in the most efficient manner, i.e., along the path of steepest ascent. One usually follows along this path with equally spaced observations until a decline or leveling off occurs in the response variable. This region may then be used for a new initial design and the process repeated, if necessary. Another alternative at this point is to construct a second-order design, take the observations for this design, analyze the data using a regression analysis and finally use calculus techniques to determine an optimum treatment combination.

A numerical example given by Davies (1971, pp. 511-518) is· reproduced here for illustrative purposes. The experiment was a laboratory study of a chemical reaction of the type

$$A + B + C \longrightarrow D + \text{other products}$$

where the reaction was taking place in the presence of solvent E.

Object: To maximize the yield of D for a given amount of A (kept constant in the experiment), the most expensive material.

Given: About 45% of the theoretical yield (the yield which would be obtained if all of A were changed to D), but it was thought that a 75% yield was possible. The experimental error was expected to be 1% or less.

The five factors and their levels were as follows:

		Factor level	
		-1	+1
X_1,	Amount of E, cc.	200	250
X_2,	Amount of C, mole/mole of A	4.0	4.5
X_3,	Concentration of C, %	90	93
X_4,	Time of reaction, hour	1	2
X_5,	Amount of B, mole/mole of A	3.0	3.5

The center (00000) of the design was chosen in the best-known operating region. Since there seemed to be large possible gains, and interactions would probably not be as important as main effects, a quarter replicate was carried out assuming all two factor interactions except one, $Z_1 Z_3$, were negligible. The design allowed for this as is indicated by the following identity:

$$I = Z_1 Z_2 Z_3 Z_4 = -Z_2 Z_3 Z_5 = -Z_1 Z_4 Z_5$$

The block containing the following treatment combinations was used, and the corresponding yields obtained are as follows:

	Z_1	Z_2	Z_3	Z_4	Z_5	Treatment combination	Yield (y)
1.	-1	-1	-1	-1	-1	(1)	34.4
2.	-1	-1	1	1	1	cde	51.6
3.	-1	1	-1	1	1	bde	31.2
4.	-1	1	1	-1	-1	bc	45.1
5.	1	-1	-1	1	-1	ad	54.1
6.	1	-1	1	-1	1	ace	62.4
7.	1	1	-1	-1	1	abe	50.2
8.	1	1	1	1	-1	abcd	58.6

The following estimates of the regression coefficients were obtained using

$$\hat{\alpha} = \frac{\Sigma Zy}{\Sigma Z^2}$$

$$\hat{\alpha}_0 \rightarrow + \alpha_0 \ (+\alpha_{11} + \alpha_{22} + \alpha_{33} + \alpha_{44} + \alpha_{55}) = 48.5$$
$$\hat{\alpha}_1 \rightarrow + \alpha_1 \ (-\alpha_{45}) = 7.9$$
$$\hat{\alpha}_2 \rightarrow + \alpha_2 \ (-\alpha_{35}) = -2.2$$
$$\hat{\alpha}_3 \rightarrow + \alpha_3 \ (-\alpha_{25}) = 6.0$$
$$\hat{\alpha}_4 \rightarrow + \alpha_4 \ (-\alpha_{15}) = 0.4$$
$$-\hat{\alpha}_5 \rightarrow - \alpha_5 \ (+\alpha_{14} + \alpha_{23}) = -0.4$$
$$\hat{\alpha}_{13} \rightarrow + \alpha_{13} \ (+\alpha_{24}) = -1.8$$
$$\hat{\alpha}_{12} \rightarrow + \alpha_{12} \ (+\alpha_{34}) = 0.2$$

where $\rightarrow$ means estimates. Hence the model considered is:

$$y = 48.5 + 7.9Z_1 - 2.2Z_2 + 6.0Z_3 + 0.4Z_4 + 0.4Z_5 - 1.8Z_1Z_3 + 0.2Z_1Z_2$$

The approximate standard error of estimate of each α is

$$\sqrt{\frac{\sigma^2}{\Sigma Z^2}} = \sqrt{\frac{1}{8}} = \sqrt{0.125} \doteq .4$$

assuming $\sigma = 1$.

In general, the steepest ascent procedure is easy to calculate and visualize when only first order effects are used. Hence the $\hat{\alpha}_{13}$ term will be dropped out of the calculations even though it is significant. Note that this term was not as large as three of the first order effects (further support for omitting it). Ordinarily, the first order effects will be sufficient when one is relatively far from the optimum treatment combination. As the experimenter comes closer to optimum and second order terms become more important, he will need to change the method of surface exploration.

Coming back to the example on steepest ascent methodology, the calculation of the direction in which maximum gain would be expected is given in the following table:

Calculation of Path of Steepest Ascent and
Subsequent Trials on the Path*

	X_1	X_2	X_3	X_4	X_5
	cc	mole	%	hour	mole
(1) Base level	225	4.25	91.5	1.5	3.25
(2) Unit	25	0.25	1.5	0.5	0.25
(3) Estimated slope $\hat{\alpha}$ (Change in Yield Per Unit)	7.9	-2.2	6.0	0.4	+0.4
(4) Unit x $\hat{\alpha}$ [(2) x (3)]	197.5	-0.55	9.0	0.2	0.1
(5) Change in level per 10 cc change in X_1	10	-0.028	0.456	0.010	0.005
(6) Path of steepest	225	4.25	91.5	1.5	3.25
ascent represented	235	4.22	92.0	1.5	3.25
by a series of possible	245	4.19	92.4	1.5	3.26
trials on it	255	4.17	92.9	1.5	3.26
	265	4.14	93.3	1.5	3.27
Trial (9)	275	4.11	93.8	1.6	3.27
	285	4.08	94.2	1.6	3.28
Trial (10)	295	4.06	94.7	1.6	3.28
	305	4.03	95.1	1.6	3.29

*Data from Design and Analysis Industrial Experiments, edited
by Owen L. Davies, published by Oliver and Boyd, Hafner, New York,
1971, Table 11-22, p. 513.

In this calculation line (1) shows the base level for the
factors, line (2) shows the unit, i.e., the change in level for the
X-variable corresponding to the change from 0 to 1 for the
Z-variable, and line (3) shows the estimated slopes calculated on
the assumption that three and higher factor interactions are
negligible. The levels of the factors are then changed in proportion
to these slopes. For every 7.9 units increase in Z_1, Z_2 should be
decreased 2.2 units, Z_3 increased 6.0 units, and so on. Hence, for
each 25 x 7.9 cc. that X_1 is changed, X_2 should be changed by (.25)
x (-2.2) mole. These changes are given on line (4) and define the

direction of steepest ascent at the point given by the base levels
in line (1). The path leading from the origin in this direction
can be obtained by taking a convenient increment in one variable
(10 cc in X_1), and calculating the proportionate changes to be made
in the other variables. These are given in line (5). These
quantities may then be successively added to the base level to give
path (6).

At this point we deviate slightly from the example given in
Davies (1971). We recommend that the next observation be taken
along the path of steepest ascent slightly outside the limits of
the original design, say at (255, 4.17, 92.9, 1.5, 3.26). The next
observation would be taken one or two steps further along the path.
This would continue until no further increase in yield is observed.

Now, additional first order designs may be implemented to
arrive in the region of the maximum yield. Let us assume that this
region has been reached. The experimenter will then want to explore
the region near the maximum using a second order design in order to
investigate the curvature in the response surface.

The main weakness of the steepest ascent method is that it is
not invariant to scale changes (see Box and Draper, 1969, p. 166);
and Myers, 1971, p. 93).

13.2.5 Canonical Analysis

After the experimenter is satisfied that he has located the
region near the maximum response, he will usually want to explore
this region thoroughly before recommending the exact treatment
combination to be used in practice. Even if he knew the exact
optimum treatment combination, he will want to know the shape of
the response surface near that point so that he has the option of
picking alternative "near" optimum treatment combinations that may
be less expensive to use with little loss in yield.

In general, the design to explore the near maximum response region should be second order to allow for the investigation of curvature. Assume that a second order design has been used, and a quadratic regression equation evolved from the data. To interpret this response surface using the regression equation alone is sometimes most difficult. If one obtains the canonical form of the equation, however, interpretation of the results is much easier.

The following example gives the reader an approach to canonical analysis so that he may use the concept in practice. If the regression equation for two variables is

$$y = 6 + 3 X_1 + 5 X_2 - 4 X_1^2 - 3 X_2^2 - 2 X_1 X_2 \qquad (13.2.1)$$

one may find the contour for $y = 8$ by setting Eq. (13.2.1) equal to 8 and substituting a value for X_1. This makes the equation a quadratic in X_2 which allows one to solve for two points. Next, change the value of X_1 and solve for two new X_2 points. Continue this until one obtains the contour of 8 and the picture shown in Fig. 13.2.1.

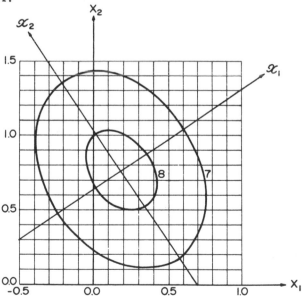

Fig. 13.2.1. Surface contours.

To find the canonical form of Eq. (13.2.1) differentiate with respect to X_1 and X_2. Then set the equations equal to zero to find the maximum point

$$\frac{\partial y}{\partial X_1} = 3 - 8 X_1 - 2 X_2 = 0$$

$$\frac{\partial y}{\partial X_2} = 5 - 2 X_1 - 6 X_2 = 0$$

Solve and obtain the point of maximum, $(X_1 = \frac{2}{11}, X_2 = \frac{17}{22})$, and refer to it as M. Substituting these solutions into Eq. (13.2.1), we obtain the maximum $y_M = 8.2$.

To determine the coefficients in the canonical form set up the determinant from Eq. (13.2.1)

$$\begin{vmatrix} b_{11} - \lambda & 1/2\ b_{12} \\ 1/2\ b_{12} & b_{22} - \lambda \end{vmatrix} = \begin{vmatrix} -4-\lambda & -1 \\ -1 & -3-\lambda \end{vmatrix}$$

and the characteristic equation becomes

$$\lambda^2 + 7\lambda + 11 = 0.$$

The characteristic roots of this equation are $\lambda = -4.62$ and -2.38.

The general form of the canonical equation is

$$y - y_m = B_{11}\mathcal{X}_1^2 + B_{22}\mathcal{X}_2^2 \qquad (13.2.2)$$

where B_{11} and B_{22} are the characteristic roots of the characteristic equation and are the coefficients of a new set of axes having the optimum treatment combination as their origin. The exact relationship between the old and new coordinate systems may be found by a procedure given in Myers (1971, p. 83).

In this example the canonical equation becomes

$$y - 8.2 = -4.62\mathcal{X}_1^2 - 2.38\mathcal{X}_2^2 \qquad (13.2.3)$$

where the $\mathcal{X}_1$ and $\mathcal{X}_2$ axes are labeled by observing that the contours drop off faster in the $\mathcal{X}_1$ direction than in the $\mathcal{X}_2$ direction.

The interpretation of Eq. (13.2.2) is that the change in yield on moving away from M to some point $(\mathscr{X}_1, \mathscr{X}_2)$ is given by the right-hand side of Eq. (13.2.2). In the example, Eq. (13.2.3), the coefficients B_{11} and B_{22} are both negative. Hence there is always a loss in yield whichever way one goes from M. This concludes the example.

If one found in his canonical analysis that B_{11} was negative and $B_{22} = 0$, there would be no unique center, and

$$y - y_M = B_{11}\mathscr{X}_1^2$$

This is a stationary ridge.

An example:

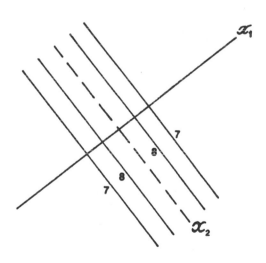

If

$$y - y_M = B_{11}\mathscr{X}_1^2 + B_2\mathscr{X}_2$$

where B_{11} is negative; B_2 is positive and measures the rate of increase in yield along the $\mathscr{X}_2$ axis, and M is on the $\mathscr{X}_2$ axis at infinity, one has a rising ridge. An example:

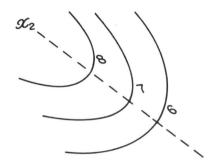

If B_{11} is negative and B_{22} is positive, the contours are hyperbolas. In the example, we have a saddle as

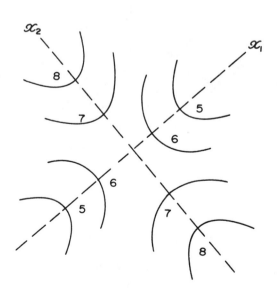

13.2.6 Evolutionary Operation (EVOP)

In the previous chapters and sections of this chapter various
designs have been presented and primarily intended for use by
experimenters associated with research laboratories, experimental
or developmental organizations. In this section, a design is
presented which is intended for use on a production line. This
concept is fairly new, having first been exposed in a paper by Box
(1957). This subject was later elaborated on by Box and Draper
(1969).

This method of experimentation is appealing to production
personnel because the basic idea is to select treatment combinations
in a relatively narrow range of each factor under consideration.
This type of selection of the factor levels has a tendency not to
upset the equilibrium of the production process but whenever the
complete design is replicated a sufficient number of times the
design will indicate conditions which will increase productivity.

Since each factor is constrained to a relatively narrow band
large differences in the response variable are not normally
expected among treatment combinations. For this reason a 2^n design
with a center point is usually employed. A 2^n design lends itself
to easy analysis; hence standard forms may be utilized to aid
production people to do the analysis. A form for this use is
illustrated in the following example. The term phase on this form
is used to record the design location in use, i.e., whenever the
data from the first phase indicates that the design location should
be shifted, then a new phase of the experiment is implemented.
Individual replications of the experiment in any one phase is
referred to as a cycle.

In the example that follows, three cycles are illustrated and
the method of calculation is that presented by Box and Hunter (1959)
in Technometrics. It should be pointed out that a more efficient
analysis may be carried out using the straight forward ANOVA
techniques presented earlier in this text.

Example 13.2:   The data from three cycles of an EVOP
application in a steel mill will be analyzed.   A method for
determining the location of the second phase will be shown.   The
location of design for Phase 1 and the data obtained from these
cycles is as follows:

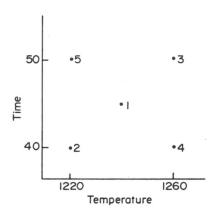

Operating condition	1	2	3	4	5
Cycle 1	27.0	28.6	25.1	26.0	26.8
Cycle 2	26.2	29.1	24.0	26.4	27.8
Cycle 3	26.5	29.2	24.2	25.6	27.0

The form for Cycle 1 has very little filled in since there was
no previous information and no estimate of the standard deviation
can be made with only one set of observations.   The calculations
shown for Cycle 2 in Table 13.2.2 shows the form much more complete.
All data shown under the heading of "Previous" have been transferred
from the form used for the calculations of Cycle 1.   The value of
the range in line (iv) is calculated by finding the range of the
five differences shown on that same line.   In order to find the
New s use the value of $f_{k,n}$ given in Table 13.2.4.   Note that the
two sigma error limits For New Effects is 0.89 which is smaller than
the temperature and time effects indicating that there is

opportunity for improving the location of the design. This fact is also supported by the relatively small change in mean effect, 0.10 as compared to the error limit of 0.79. Note that a significant change in mean effect would indicate curvature in the response surface over the region of design of the phase under consideration. In order to display the method the remainder of the results of Cycle 3 are given in Table 13.2.3. Again, notice that the temperature and time effects are significant indicating that a new phase should be implemented.

To locate a new set of operating conditions for Phase 2 one can apply the calculational procedure given in the steepest ascent method, Section 13.2.4. To do this denote the factor of temperature and time by X_1 and X_2, respectively; and

$$\hat{\alpha} = \frac{\Sigma zy}{\Sigma z^2} \text{ from Section 13.2.4, but in this example}$$

$$\hat{\alpha}_1 = \frac{\Sigma X_1 y}{\Sigma X_1^2} = \frac{(-5.8)}{4} = \frac{1}{2} \text{ (temperature effect).}$$

Using the data of Cycle 3 then, the calculations are the following:

$$\hat{\alpha}_1 = \frac{1}{2} \text{ (temperature effect)} = -1.45$$

$$\hat{\alpha}_2 = \frac{1}{2} \text{ (time effect)} = -0.85$$

	X_1	X_2
Base level	1240	45
Unit	20	5
Estimated slope	-1.45	-0.85
Unit x slope	-29.0	-4.25
Amount of change per 20°	-20.0	-2.93

TABLE 13.2.1

EVOP Calculation Form

Time: 5   3
       1
       2 Temp 4

Cycle: n = 1   Phase: 1

Response: Tensile strength (1000 lb)   Date: 5/20 - 5/22

Calculation of averages						Calculation of standard deviation s	
Operating Conditions	(1)	(2)	(3)	(4)	(5)		
(i) Previous cycle sum						Previous sum s	=
(ii) Previous cycle average						Previous average s	=
(iii) New observations	27.0	28.6	25.1	26.0	26.8	New s = range × $f_{k,n}$	=
(iv) Differences [(ii)-(iii)]						Range of (iv)	=
(v) New sums [(i)+(iii)]	27.0	28.6	25.1	26.0	26.8	New sum s	=
(vi) New average $\bar{y}_i$[(v)/n]	27.0	28.6	25.1	26.0	26.8	New average s = (New sum s)/(n-1)	=

Calculation of effects		Calculation of error limits	
Temperature effect = $\frac{1}{2}$ $(\bar{y}_3+\bar{y}_4-\bar{y}_2-\bar{y}_5)$	= -2.15		
Time effect = $\frac{1}{2}$ $(\bar{y}_3+\bar{y}_5-\bar{y}_2-\bar{y}_4)$	= -1.35	For new averages 2 s/√n	=
Temp × time interaction = $\frac{1}{2}$ $(\bar{y}_2+\bar{y}_3-\bar{y}_4-\bar{y}_5)$	= 0.45	For new effects 2 s/√n	=
Changes in mean effect = $\frac{1}{5}$ $(\bar{y}_1+\bar{y}_2+\bar{y}_3+\bar{y}_4+\bar{y}_5-5\,\bar{y}_1)$	= -0.30	For change in mean 1.78 s/√n	=

TABLE 13.2.2

EVOP Calculation Form

Time   5   3   Cycle: n = 2   Phase: 1

      1

      2 Temp 4   Response: Tensile strength (1000 lb)   Date: 5/23 - 5/25

Calculation of averages	(1)	(2)	(3)	(4)	(5)	Calculation of standard deviation s
Operating Conditions						
(i) Previous cycle sum	27.0	28.6	25.1	26.0	26.8	Previous sum s =
(ii) Previous cycle average	27.0	28.6	25.1	26.0	26.8	Previous average s =
(iii) New observations	26.2	29.1	24.0	26.4	27.8	New s = range $\times f_{k,n} = 2.1 \times .30 = 0.63$ (5,2)
(iv) Differences [(ii)-(iii)]	0.8	-0.5	1.1	-0.4	-1.0	Range of (iv) $= 1.1 - (-1.0) = 2.1$
(v) New sums [(i)+(iii)]	53.2	57.7	49.1	52.4	54.6	New sum s $= 0.63$
(vi) New averages $\bar{y}_i$ [(v)/n]	26.6	28.8	24.6	26.2	27.3	New average s = (New sum s)/(n-1) $= 0.63$

Calculation of effects		Calculation of error limits	
Temperature effect $= \frac{1}{2}(\bar{y}_3+\bar{y}_4-\bar{y}_2-\bar{y}_5)$	$= -2.65$	For new averages	$2\,s/\sqrt{n} = 0.89$
Time effect $= \frac{1}{2}(\bar{y}_3+\bar{y}_5-\bar{y}_2-\bar{y}_4)$	$= -1.55$	For new effects	$2\,s/\sqrt{n} = 0.89$
Temp x time interaction $= \frac{1}{2}(\bar{y}_2+\bar{y}_3-\bar{y}_4-\bar{y}_5)$	$= -0.05$	For change in mean	$1.78\,s/\sqrt{n} = 0.79$
Changes in mean effect $= \frac{1}{5}(\bar{y}_1+\bar{y}_2+\bar{y}_3+\bar{y}_4+\bar{y}_5-5\,\bar{y}_1)$	$= 0.10$		

TABLE 13.2.3

EVOP Calculation Form

Time 5 3
2 Temp 4
1 Cycle: n = 3
Response: Tensile strength (1000 lb)

Phase: 1
Date: 5/27 - 5/29

Calculation of averages						Calculation of standard deviation s	
Operating conditions	(1)	(2)	(3)	(4)	(5)		
(i) Previous cycle sum	53.2	57.7	49.1	52.4	54.6	Previous sum s	= 0.63
(ii) Previous cycle average	26.6	28.8	24.6	26.2	27.3	Previous average s	= 0.63
(iii) New observations	26.5	29.2	24.2	25.6	27.0	New s = range x $f_{k,n}$	= 0.35
(iv) Differences [(ii)-(iii)]	0.1	-0.4	0.4	0.6	0.3	Range of (iv)	= 1.0
(v) New sums [(i)+(iii)]	79.7	86.9	73.3	78.0	81.6	New sum s	= 0.98
(vi) New averages $\bar{y}_i$ [(v)/n]	26.6	29.0	24.4	26.0	27.2	New average s = (New sum s)/(n-1)	= 0.49

Calculation of effects			Calculation of error limits	
Temperature effect	$= \frac{1}{2} (\bar{y}_3+\bar{y}_4-\bar{y}_2-\bar{y}_5)$	= -2.90	For new averages $2\,s/\sqrt{n}$ =	0.56
Time effect	$= \frac{1}{2} (\bar{y}_3+\bar{y}_5-\bar{y}_2-\bar{y}_4)$	= -1.70	For new effects $2\,s/\sqrt{n}$ =	0.56
Temp x time interaction	$= \frac{1}{2} (\bar{y}_2+\bar{y}_3-\bar{y}_4-\bar{y}_5)$	= 0.10	For change in mean $1.78\,s/\sqrt{n}$ =	0.50
Changes in mean effect	$= \frac{1}{5} (\bar{y}_1+\bar{y}_2+\bar{y}_3+\bar{y}_4+\bar{y}_5-5\,\bar{y}_1)$ =	0.04		

Thus the new center point for Phase 2 could be

$$\text{Temperature } 1240 - 20 = 1220$$
$$\text{Time} \qquad 45 - 2.93 \approx \quad 42$$

Some of the mathematical rationale behind the method of calculation is given in Hicks (1973, Chapter 16) and Box (1957). The value of $f_{k,n}$ is shown to be

$$f_{k,n} = \sqrt{\frac{n-1}{n}} \; \frac{1}{d_2}$$

where d_2 is the quality control symbol representing the factor that must be divided into the range of a set of observations in order to estimate the standard deviation. Values of $f_{k,n}$ are given in Table 13.2.4.

TABLE 13.2.4

Values of $f_{k,n}$

k \ n	2	3	4	5	6	7	8	9	10
4	0.34	0.40	0.42	0.43	0.44	0.45	0.45	0.46	0.46
5	0.30	0.35	0.37	0.38	0.39	0.40	0.40	0.40	0.41
6	0.28	0.32	0.34	0.35	0.36	0.37	0.37	0.37	0.37
7	0.26	0.30	0.32	0.33	0.34	0.34	0.35	0.35	0.35
8	0.25	0.29	0.30	0.31	0.32	0.32	0.33	0.33	0.33
9	0.24	0.27	0.29	0.30	0.31	0.31	0.31	0.32	0.32
10	0.23	0.26	0.28	0.29	0.30	0.30	0.30	0.31	0.31

13.3 REFERENCES

Booth, K. H. V. and Cox, D. R. Technometrics 4:489 (1962).

Box, G. E. P. Applied Statistics 6:81 (1957).

Box, G. E. P. and Behnken, D. W. Ann. Math. Stat. 31:838 (1960).

Box, G. E. P. and Draper, N. R. Evolutionary Operation, Wiley, New York, 1969.

Box, G. E. P. and Hunter, J. S. Ann. Math. Stat. 28:195 (1957).

Box, G. E. P. and Hunter, J. S. Technometrics 1:77 (1959).

Box, G. E. P. and Wilson, K. B. J. Royal Stat. Soc. B, 13:1 (1951).

Brooks, S. H. Operations Research 6:244 (1958).

Brooks, S. H. Operations Research 7:430 (1959).

Cochran, W. G. and Cox, G. M. Experimental Designs, 2nd ed., Wiley, New York, 1957.

Davies, O. L. (Ed.), Design and Analysis of Industrial Experiments, 2nd ed., Oliver and Boyd, Edinburgh, 1971, Chapter 11.

Dykstra, O. Technometrics 13:682 (1971).

Gorman, J. W. and Hinman, J. E. Technometrics 4:463 (1962).

Hebble, T. L. and Mitchell, T. J. Technometrics 14:767 (1972).

Hicks, C. R. Fundamental Concepts in Design of Experiments, 2nd ed., Holt, Rinehart and Winston, New York, 1973, Chapter 16.

Hotelling, H. Ann. Math. Stat. 12:20 (1941).

John, P. W. M. Statistical Design and Analysis of Experiments, Macmillan, New York, 1971.

Kennard, R. W. and Stone, L. A. Technometrics 11:137 (1969).

McLean, R. A. and Anderson, V. L. Technometrics 8:447 (1966).

Myers, R. Response Surface Methodology, Allyn and Bacon, Boston, Massachusetts, 1971.

Plackett, R. L. and Burman, J. P. Biometrika 33:303 (1946).

Satterthwaite, F. E. "New Developments in Experimental Design." Talk given in Boston, (1956).

Satterthwaite, F. E. Technometrics 1:111 (1959).

Spendley, W., Hext, G. R., and Himsworth, F. R. Technometrics 4:441 (1962).

Scheffé, H. J. Roy. Stat. Soc. Series B, 20:344 (1958).

Scheffé, H. J. Roy. Stat. Soc. Series B, 25:235 (1963).

Snee, R. D. *Technometrics* 17:149 (1975)

Snee, R. D. and Marquardt, D. W. *Technometrics* 16:399 (1974).

Stapleton, J. H. *American Statistician*, 13:4 October, 1959.

Appendix 1

Random Numbers

	00-04	05-09	10-14	15-19	20-24	25-29	30-34	35-39	40-44	45-49
00	01826	72696	67261	13748	57834	27748	47472	43428	85524	19311
01	70731	12890	90395	45245	71282	15960	02749	86763	80564	02651
02	46616	84522	17249	78172	14197	84272	53226	96719	83462	05628
03	46384	26607	53444	68780	00458	27488	04949	30717	14304	60222
04	58645	90453	83413	15983	42345	27118	29425	42172	58412	31268
05	41874	83916	07454	42647	09410	12882	80001	98932	63277	94255
06	26032	68402	68176	87347	52572	79056	22703	83175	95807	60134
07	56142	74950	98878	36160	28616	77908	04908	63013	30555	51831
08	57893	57918	56290	00612	53884	76761	47934	20050	84977	11053
09	46729	47281	30473	94911	69061	87486	96782	90636	91405	01219
10	14107	46068	15859	99140	29872	70750	35757	25344	84845	95688
11	27713	13273	09015	42262	24580	30925	96900	56246	10613	17230
12	49886	62560	23023	09812	91948	92265	77407	93047	45352	61689
13	86651	47164	79270	45746	73141	67388	08454	68246	39046	57933
14	96133	52026	25837	64698	06911	21730	86390	19749	25859	80005
15	00926	46212	00204	58304	73907	97914	06786	47324	81225	50754
16	89036	39245	74371	87236	92131	38908	22146	15710	36858	35904
17	79592	66982	87984	30503	86953	56490	20414	07662	11122	95517
18	25084	83491	79984	64583	33924	36146	54692	44989	71603	31274
19	90261	79382	24483	75975	14765	49897	75468	66940	82187	68875
20	94937	36303	85514	76537	86864	88600	51843	18255	49660	87326
21	10178	16103	33283	38239	32402	02950	90428	24405	38580	61198
22	05983	63553	93546	66079	18389	57705	86746	62498	71642	60219
23	80832	36191	32042	76015	51331	40592	04887	50771	23810	64650
24	85889	08924	41542	04758	98753	52695	28165	87031	00479	95974
25	77032	15760	83026	92461	51806	65495	32148	40714	49107	05758
26	84080	67254	89239	13272	50218	17737	00242	15203	69060	39047
27	41022	70272	73710	17625	51325	69525	63464	89526	76738	80210
28	72398	03426	36476	62922	43624	98779	63289	34550	94270	01263
29	52224	04363	92979	26520	91076	13849	17740	25964	22169	23110
30	26223	68525	82483	15232	93800	78903	18831	19546	50469	74328
31	57872	36109	88383	83512	58763	57230	35952	75716	57094	53951
32	44529	90455	96666	31804	24979	51863	42983	13367	14111	38541
33	13586	36649	80041	54602	17281	22392	63074	53831	85782	94644
34	49084	08133	45510	21472	14644	38592	04490	52187	33945	78491
35	42309	57412	33314	07820	31345	45074	93547	30023	47632	93222
36	08174	49231	02588	36639	53978	40177	45572	44117	97946	44214
37	58679	57695	94704	16260	41928	27300	14053	93050	87103	34434
38	46847	17986	19347	74125	64945	97496	00465	65830	56564	46884
39	91053	64137	26104	29911	01242	58960	60582	83119	35911	68859
40	98796	24540	75578	79529	63199	78959	60734	73433	91995	81783
41	66824	16251	63616	91787	77519	14592	15333	63525	88097	79340
42	35492	57116	78056	72780	44502	97058	25342	53590	29196	06552
43	53794	57593	27187	99355	12652	77611	95229	41602	99946	94801
44	86332	05856	07786	78241	13584	29243	81609	55078	25937	90051
45	56530	97998	92028	18839	05879	43900	91769	87557	50750	05667
46	22966	67453	46806	57010	49278	99735	84996	74079	77917	48474
47	17068	92733	58021	81102	19871	17442	45375	27328	58055	20323
48	21570	52311	25131	76662	18619	91141	89382	41329	91702	39975
49	09699	95392	78594	40396	83605	01008	48694	85571	64427	23547

Appendix 2

Cumulative Normal Frequency Distribution

(Area under the standard normal curve from 0 to Z)

Z	0.00	0.01	0.02	0.03	0.04	0.05	0.06	0.07	0.08	0.09
0.0	0.0000	0.0040	0.0080	0.0120	0.0160	0.0199	0.0239	0.0279	0.0319	0.0359
0.1	.0398	.0438	.0478	.0517	.0557	.0596	.0636	.0675	0.714	.0753
0.2	.0793	.0832	.0871	.0910	.0948	.0987	.1026	.1064	.1103	.1141
0.3	.1179	.1217	.1255	.1293	.1331	.1368	.1406	.1443	.1480	.1517
0.4	.1554	.1591	.1628	.1664	.1700	.1736	.1772	.1808	.1844	.1879
0.5	.1915	.1950	.1985	.2019	.2054	.2088	.2123	.2157	.2190	.2224
0.6	.2257	.2291	.2324	.2357	.2389	.2422	.2454	.2486	.2517	.2549
0.7	.2580	.2611	.2642	.2673	.2704	.2734	.2764	.2794	.2823	.2852
0.8	.2881	.2910	.2939	.2967	.2995	.3023	.3051	.3078	.3106	.3133
0.9	.3159	.3186	.3212	.3238	.3264	.3289	.3315	.3340	.3365	.3389
1.0	.3413	.3438	.3461	.3485	.3508	.3531	.3554	.3577	.3599	.3621
1.1	.3643	.3665	.3686	.3708	.3729	.3749	.3770	.3790	.3810	.3830
1.2	.3849	.3869	.3888	.3907	.3925	.3944	.3962	.3980	.3997	.4015
1.3	.4032	.4049	.4066	.4082	.4099	.4115	.4131	.4147	.4162	.4177
1.4	.4192	.4207	.4222	.4236	.4251	.4265	.4279	.4292	.4306	.4319
1.5	.4332	.4345	.4357	.4370	.4382	.4394	.4406	.4418	.4429	.4441
1.6	.4452	.4463	.4474	.4484	.4495	.4505	.4515	.4525	.4535	.4545
1.7	.4554	.4564	.4573	.4582	.4591	.4599	.4608	.4616	.4625	.4633
1.8	.4641	.4649	.4656	.4664	.4671	.4678	.4686	.4693	.4699	.4706
1.9	.4713	.4719	.4726	.4732	.4738	.4744	.4750	.4756	.4761	.4767
2.0	.4772	.4778	.4783	.4788	.4793	.4798	.4803	.4808	.4812	.4817
2.1	.4821	.4826	.4830	.4834	.4838	.4842	.4846	.4850	.4854	.4857
2.2	.4861	.4864	.4868	.4871	.4875	.4878	.4881	.4884	.4887	.4890
2.3	.4893	.4896	.4898	.4901	.4904	.4906	.4909	.4911	.4913	.4916
2.4	.4918	.4920	.4922	.4925	.4927	.4929	.4931	.4932	.4934	.4936
2.5	.4938	.4940	.4941	.4943	.4945	.4946	.4948	.4949	.4951	.4952
2.6	.4953	.4955	.4956	.4957	.4959	.4960	.4961	.4962	.4963	.4964
2.7	.4965	.4966	.4967	.4968	.4969	.4970	.4971	.4972	.4973	.4974
2.8	.4974	.4975	.4976	.4977	.4977	.4978	.4979	.4979	.4980	.4981
2.9	.4981	.4982	.4982	.4983	.4984	.4984	.4985	.4985	.4986	.4986
3.0	.4987	.4987	.4987	.4988	.4988	.4989	.4989	.4989	.4990	.4990
3.1	.4990	.4991	.4991	.4991	.4992	.4992	.4992	.4992	.4993	.4993
3.2	.4993	.4993	.4994	.4994	.4994	.4994	.4994	.4995	.4995	.4995
3.3	.4995	.4995	.4995	.4996	.4996	.4996	.4996	.4996	.4996	.4997
3.4	.4997	.4997	.4997	.4997	.4997	.4997	.4997	.4997	.4997	.4998
3.6	.4998	.4998	.4999	.4999	.4999	.4999	.4999	.4999	.4999	.4999
3.9	.5000									

Percentage points of the t-distribution

v	Q=0.4	0.25	0.1	0.05	0.025	0.01	0.005	0.0025	0.001	0.0005
	2Q=0.8	0.5	0.2	0.1	0.05	0.02	0.01	0.005	0.002	0.001
1	0.325	1.000	3.078	6.314	12.706	31.821	63.657	127.32	318.31	636.62
2	.289	0.816	1.886	2.920	4.303	6.965	9.925	14.089	22.327	31.598
3	.277	.765	1.638	2.353	3.182	4.541	5.841	7.453	10.214	12.924
4	.271	.741	1.533	2.132	2.776	3.747	4.604	5.598	7.173	8.610
5	0.267	0.727	1.476	2.015	2.571	3.365	4.032	4.773	5.893	6.869
6	.265	.718	1.440	1.943	2.447	3.143	3.707	4.317	5.208	5.959
7	.263	.711	1.415	1.895	2.365	2.998	3.499	4.029	4.785	5.408
8	.262	.706	1.397	1.860	2.306	2.896	3.355	3.833	4.501	5.041
9	.261	.703	1.383	1.833	2.262	2.821	3.250	3.690	4.297	4.781
10	0.260	0.700	1.372	1.812	2.228	2.764	3.169	3.581	4.144	4.587
11	.260	.697	1.363	1.796	2.201	2.718	3.106	3.497	4.025	4.437
12	.259	.695	1.356	1.782	2.179	2.681	3.055	3.428	3.930	4.318
13	.259	.694	1.350	1.771	2.160	2.650	3.012	3.372	3.852	4.221
14	.258	.692	1.345	1.761	2.145	2.624	2.977	3.326	3.787	4.140
15	0.258	0.691	1.341	1.753	2.131	2.602	2.947	3.286	3.733	4.073
16	.258	.690	1.337	1.746	2.120	2.583	2.921	3.252	3.686	4.015
17	.257	.689	1.333	1.740	2.110	2.567	2.898	3.222	3.646	3.965
18	.257	.688	1.330	1.734	2.101	2.552	2.878	3.197	3.610	3.922
19	.257	.688	1.328	1.729	2.093	2.539	2.861	3.174	3.579	3.883
20	0.257	0.687	1.325	1.725	2.086	2.528	2.845	3.153	3.552	3.850
21	.257	.686	1.323	1.721	2.080	2.518	2.831	3.135	3.527	3.819
22	.256	.686	1.321	1.717	2.074	2.508	2.819	3.119	3.505	3.792
23	.256	.685	1.319	1.714	2.069	2.500	2.807	3.104	3.485	3.767
24	.256	.685	1.318	1.711	2.064	2.492	2.797	3.091	3.467	3.745
25	0.256	0.684	1.316	1.708	2.060	2.485	2.787	3.078	3.450	3.725
26	.256	.684	1.315	1.706	2.056	2.479	2.779	3.067	3.435	3.707
27	.256	.684	1.314	1.703	2.052	2.473	2.771	3.057	3.421	3.690
28	.256	.683	1.313	1.701	2.048	2.467	2.763	3.047	3.408	3.674
29	.256	.683	1.311	1.699	2.045	2.462	2.756	3.038	3.396	3.659
30	0.256	0.683	1.310	1.697	2.042	2.457	2.750	3.030	3.385	3.646
40	.255	.681	1.303	1.684	2.021	2.423	2.704	2.971	3.307	3.551
60	.254	.679	1.296	1.671	2.000	2.390	2.660	2.915	3.232	3.460
120	.254	.677	1.289	1.658	1.980	2.358	2.617	2.860	3.160	3.373
∞	.253	.674	1.282	1.645	1.960	2.326	2.576	2.807	3.090	3.291

$Q=1-P(t|v)$ is the upper-tail area of the distribution for v degrees of freedom, appropriate for use in a single-tail test. For a two-tail test, 2Q must be used.

Appendix 4

Number of Observations per Sample: Using t for difference of means

Column groups are by Single-Sided Test α (0.005, 0.01, 0.025, 0.05), which correspond to Double-Sided Test α (0.01, 0.02, 0.05, 0.1). Within each α group, sub-columns are values of β. Rows are the Value of $D = \delta/\sigma$.

$D = \delta/\sigma$	α=0.005 (β=0.01)	0.05	0.1	0.2	0.5	α=0.01 (β=0.01)	0.05	0.1	0.2	0.5	α=0.025 (β=0.01)	0.05	0.1	0.2	0.5	α=0.05 (β=0.05)	0.1	0.2	0.5
0.30										123					87				61
0.40					85					70				100	50		108	51	35
0.50				96	55			106	82	45		106	86	64	32	88	70	36	23
0.60		101	85	67	39		90	74	58	32	104	74	60	45	23	61	49	26	16
0.70	100	75	63	50	29	90	66	55	43	24	76	55	44	34	17	45	36	26	12
0.75	88	66	55	44	26	79	58	48	38	21	67	48	39	29	15	40	32	23	11
0.80	77	58	49	39	23	70	51	43	33	19	59	42	34	26	14	35	28	21	10
0.85	69	51	43	35	21	62	46	38	30	17	52	37	31	23	12	31	25	18	9
0.90	62	46	39	31	19	55	41	34	27	15	47	34	27	21	11	28	22	16	8
0.95	55	42	35	28	17	50	37	31	24	14	42	30	25	19	10	25	20	15	7
1.00	50	38	32	26	15	45	33	28	22	13	38	27	23	17	9	23	18	14	7
1.1	42	32	27	22	13	38	28	23	19	11	32	23	19	14	8	19	15	12	6
1.2	36	27	23	18	11	32	24	20	16	9	27	20	16	12	7	16	13	10	5
1.3	31	23	20	16	10	28	21	17	14	8	23	17	14	11	6	14	11	9	5
1.4	27	20	17	14	9	24	18	15	12	8	20	15	12	10	6	12	10	8	4
1.5	24	18	15	13	8	21	16	14	11	7	18	13	11	9	5	11	9	7	4
1.6	21	16	14	11	7	19	14	12	10	6	16	12	10	8	5	10	8	6	4
1.7	19	15	13	10	7	17	13	11	9	6	14	11	9	7	4	9	7	6	3
1.8	17	13	11	10	6	15	12	10	8	5	13	10	8	6	4	8	7	5	
1.9	16	12	11	9	6	14	11	9	8	5	12	9	7	6	4	7	6	5	
2.0	14	11	10	8	6	13	10	9	7	5	11	8	7	6	4	7	6	4	
2.5	10	8	7	6	4	9	7	6	5	4	8	6	5	4		5	4	3	
3.0	8	6	6	5	4	7	6	5	4	3	6	5	4	4		4	3		
3.5	6	5	5	4	3	6	5	4	4		5	4	4	3		3			
4.0	6	5	4	4		5	4	4	3		4	4	3						

Appendix 5
F-Distribution
Upper 25%

v_2 \ v_1	1	2	3	4	5	6	7	8	9	10	12	15	20	24	30	40	60	120	∞
1	5.83	7.50	8.20	8.58	8.82	8.98	9.10	9.19	9.26	9.32	9.41	9.49	9.58	9.63	9.67	9.71	9.76	9.80	9.85
2	2.57	3.00	3.15	3.23	3.28	3.31	3.34	3.35	3.37	3.38	3.39	3.41	3.43	3.43	3.44	3.45	3.46	3.47	3.48
3	2.02	2.28	2.36	2.39	2.41	2.42	2.43	2.44	2.44	2.44	2.45	2.46	2.46	2.46	2.47	2.47	2.47	2.47	2.47
4	1.81	2.00	2.05	2.06	2.07	2.08	2.08	2.08	2.08	2.08	2.08	2.08	2.08	2.08	2.08	2.08	2.08	2.08	2.08
5	1.69	1.85	1.88	1.89	1.89	1.89	1.89	1.89	1.89	1.89	1.89	1.89	1.88	1.88	1.88	1.88	1.87	1.87	1.87
6	1.62	1.76	1.78	1.79	1.79	1.78	1.78	1.78	1.77	1.77	1.77	1.76	1.76	1.75	1.75	1.75	1.74	1.74	1.74
7	1.57	1.70	1.72	1.72	1.71	1.71	1.70	1.70	1.69	1.69	1.68	1.68	1.67	1.67	1.66	1.66	1.65	1.65	1.65
8	1.54	1.66	1.67	1.66	1.66	1.65	1.64	1.64	1.63	1.63	1.62	1.62	1.61	1.60	1.60	1.59	1.59	1.58	1.58
9	1.51	1.62	1.63	1.63	1.62	1.61	1.60	1.60	1.59	1.59	1.58	1.57	1.56	1.56	1.55	1.54	1.54	1.53	1.53
10	1.49	1.60	1.60	1.59	1.59	1.58	1.57	1.56	1.56	1.55	1.54	1.53	1.52	1.52	1.51	1.51	1.50	1.49	1.48
11	1.47	1.58	1.58	1.57	1.56	1.55	1.54	1.53	1.53	1.52	1.51	1.50	1.49	1.49	1.48	1.47	1.47	1.46	1.45
12	1.46	1.56	1.56	1.55	1.54	1.53	1.52	1.51	1.51	1.50	1.49	1.48	1.47	1.46	1.45	1.45	1.44	1.43	1.42
13	1.45	1.55	1.55	1.53	1.52	1.51	1.50	1.49	1.49	1.48	1.47	1.46	1.45	1.44	1.43	1.42	1.42	1.41	1.40
14	1.44	1.53	1.53	1.52	1.51	1.50	1.49	1.48	1.47	1.46	1.45	1.44	1.43	1.42	1.41	1.41	1.40	1.39	1.38
15	1.43	1.52	1.52	1.51	1.49	1.48	1.47	1.46	1.46	1.45	1.44	1.43	1.41	1.41	1.40	1.39	1.38	1.37	1.36
16	1.42	1.51	1.51	1.50	1.48	1.47	1.46	1.45	1.44	1.44	1.43	1.41	1.40	1.39	1.38	1.37	1.36	1.35	1.34
17	1.42	1.51	1.50	1.49	1.47	1.46	1.45	1.44	1.43	1.43	1.41	1.40	1.39	1.38	1.37	1.36	1.35	1.34	1.33
18	1.41	1.50	1.49	1.48	1.46	1.45	1.44	1.43	1.42	1.42	1.40	1.39	1.38	1.37	1.36	1.35	1.34	1.33	1.32
19	1.41	1.49	1.49	1.47	1.46	1.44	1.43	1.42	1.41	1.41	1.40	1.38	1.37	1.36	1.35	1.34	1.33	1.32	1.30
20	1.40	1.49	1.48	1.47	1.45	1.44	1.43	1.42	1.41	1.40	1.39	1.37	1.36	1.35	1.34	1.33	1.32	1.31	1.29
21	1.40	1.48	1.48	1.46	1.44	1.43	1.42	1.41	1.40	1.39	1.38	1.37	1.35	1.34	1.33	1.32	1.31	1.30	1.28
22	1.40	1.48	1.47	1.45	1.44	1.42	1.41	1.40	1.39	1.39	1.37	1.36	1.34	1.33	1.32	1.31	1.30	1.29	1.28
23	1.39	1.47	1.47	1.45	1.43	1.42	1.41	1.40	1.39	1.38	1.37	1.35	1.34	1.33	1.32	1.31	1.30	1.28	1.27
24	1.39	1.47	1.46	1.44	1.43	1.41	1.40	1.39	1.38	1.38	1.36	1.35	1.33	1.32	1.31	1.30	1.29	1.28	1.26
25	1.39	1.47	1.46	1.44	1.42	1.41	1.40	1.39	1.38	1.37	1.36	1.34	1.33	1.32	1.31	1.29	1.28	1.27	1.25
26	1.38	1.46	1.45	1.44	1.42	1.41	1.39	1.38	1.37	1.37	1.35	1.34	1.32	1.31	1.30	1.29	1.28	1.26	1.25
27	1.38	1.46	1.45	1.43	1.42	1.40	1.39	1.38	1.37	1.36	1.35	1.33	1.32	1.31	1.30	1.28	1.27	1.26	1.24
28	1.38	1.46	1.45	1.43	1.41	1.40	1.39	1.38	1.37	1.36	1.34	1.33	1.31	1.30	1.29	1.28	1.27	1.25	1.24
29	1.38	1.45	1.45	1.43	1.41	1.40	1.38	1.37	1.36	1.35	1.34	1.32	1.31	1.30	1.29	1.27	1.26	1.25	1.23
30	1.38	1.45	1.44	1.42	1.41	1.39	1.38	1.37	1.36	1.35	1.34	1.32	1.30	1.29	1.28	1.27	1.26	1.24	1.23
40	1.36	1.44	1.42	1.40	1.39	1.37	1.36	1.35	1.34	1.33	1.31	1.30	1.28	1.26	1.25	1.24	1.22	1.21	1.19
60	1.35	1.42	1.41	1.38	1.37	1.35	1.33	1.32	1.31	1.30	1.29	1.27	1.25	1.24	1.22	1.21	1.19	1.17	1.15
120	1.34	1.40	1.39	1.37	1.35	1.33	1.31	1.30	1.29	1.28	1.26	1.24	1.22	1.21	1.19	1.18	1.16	1.13	1.10
∞	1.32	1.39	1.37	1.35	1.33	1.31	1.29	1.28	1.27	1.25	1.24	1.22	1.19	1.18	1.16	1.14	1.12	1.08	1.00

F-Distribution
Upper 10%

v_2 \ v_1	1	2	3	4	5	6	7	8	9	10	12	15	20	24	30	40	60	120	∞
1	39.86	49.50	53.59	55.83	57.24	58.20	58.91	59.44	59.86	60.19	60.71	61.22	61.74	62.00	62.26	62.53	62.79	63.06	63.33
2	8.53	9.00	9.16	9.24	9.29	9.33	9.35	9.37	9.38	9.39	9.41	9.42	9.44	9.45	9.46	9.47	9.47	9.48	9.49
3	5.54	5.46	5.39	5.34	5.31	5.28	5.27	5.25	5.24	5.23	5.22	5.20	5.18	5.18	5.17	5.16	5.15	5.14	5.13
4	4.54	4.32	4.19	4.11	4.05	4.01	3.98	3.95	3.94	3.92	3.90	3.87	3.84	3.83	3.82	3.80	3.79	3.78	3.76
5	4.06	3.78	3.62	3.52	3.45	3.40	3.37	3.34	3.32	3.30	3.27	3.24	3.21	3.19	3.17	3.16	3.14	3.12	3.10
6	3.78	3.46	3.29	3.18	3.11	3.05	3.01	2.98	2.96	2.94	2.90	2.87	2.84	2.82	2.80	2.78	2.76	2.74	2.72
7	3.59	3.26	3.07	2.96	2.88	2.83	2.78	2.75	2.72	2.70	2.67	2.63	2.59	2.58	2.56	2.54	2.51	2.49	2.47
8	3.46	3.11	2.92	2.81	2.73	2.67	2.62	2.59	2.56	2.54	2.50	2.46	2.42	2.40	2.38	2.36	2.34	2.32	2.29
9	3.36	3.01	2.81	2.69	2.61	2.55	2.51	2.47	2.44	2.42	2.38	2.34	2.30	2.28	2.25	2.23	2.21	2.18	2.16
10	3.29	2.92	2.73	2.61	2.52	2.46	2.41	2.38	2.35	2.32	2.28	2.24	2.20	2.18	2.16	2.13	2.11	2.08	2.06
11	3.23	2.86	2.66	2.54	2.45	2.39	2.34	2.30	2.27	2.25	2.21	2.17	2.12	2.10	2.08	2.05	2.03	2.00	1.97
12	3.18	2.81	2.61	2.48	2.39	2.33	2.28	2.24	2.21	2.19	2.15	2.10	2.06	2.04	2.01	1.99	1.96	1.93	1.90
13	3.14	2.76	2.56	2.43	2.35	2.28	2.23	2.20	2.16	2.14	2.10	2.05	2.01	1.98	1.96	1.93	1.90	1.88	1.85
14	3.10	2.73	2.52	2.39	2.31	2.24	2.19	2.15	2.12	2.10	2.05	2.01	1.96	1.94	1.91	1.89	1.86	1.83	1.80
15	3.07	2.70	2.49	2.36	2.27	2.21	2.16	2.12	2.09	2.06	2.02	1.97	1.92	1.90	1.87	1.85	1.82	1.79	1.76
16	3.05	2.67	2.46	2.33	2.24	2.18	2.13	2.09	2.06	2.03	1.99	1.94	1.89	1.87	1.84	1.81	1.78	1.75	1.72
17	3.03	2.64	2.44	2.31	2.22	2.15	2.10	2.06	2.03	2.00	1.96	1.91	1.86	1.84	1.81	1.78	1.75	1.72	1.69
18	3.01	2.62	2.42	2.29	2.20	2.13	2.08	2.04	2.00	1.98	1.93	1.89	1.84	1.81	1.78	1.75	1.72	1.69	1.66
19	2.99	2.61	2.40	2.27	2.18	2.11	2.06	2.02	1.98	1.96	1.91	1.86	1.81	1.79	1.76	1.73	1.70	1.67	1.63
20	2.97	2.59	2.38	2.25	2.16	2.09	2.04	2.00	1.96	1.94	1.89	1.84	1.79	1.77	1.74	1.71	1.68	1.64	1.61
21	2.96	2.57	2.36	2.23	2.14	2.08	2.02	1.98	1.95	1.92	1.87	1.83	1.78	1.75	1.72	1.69	1.66	1.62	1.59
22	2.95	2.56	2.35	2.22	2.13	2.06	2.01	1.97	1.93	1.90	1.86	1.81	1.76	1.73	1.70	1.67	1.64	1.60	1.57
23	2.94	2.55	2.34	2.21	2.11	2.05	1.99	1.95	1.92	1.89	1.84	1.80	1.74	1.72	1.69	1.66	1.62	1.59	1.55
24	2.93	2.54	2.33	2.19	2.10	2.04	1.98	1.94	1.91	1.88	1.83	1.78	1.73	1.70	1.67	1.64	1.61	1.57	1.53
25	2.92	2.53	2.32	2.18	2.09	2.02	1.97	1.93	1.89	1.87	1.82	1.77	1.72	1.69	1.66	1.63	1.59	1.56	1.52
26	2.91	2.52	2.31	2.17	2.08	2.01	1.96	1.92	1.88	1.86	1.81	1.76	1.71	1.68	1.65	1.61	1.58	1.54	1.50
27	2.90	2.51	2.30	2.17	2.07	2.00	1.95	1.91	1.87	1.85	1.80	1.75	1.70	1.67	1.64	1.60	1.57	1.53	1.49
28	2.89	2.50	2.29	2.16	2.06	2.00	1.94	1.90	1.87	1.84	1.79	1.74	1.69	1.66	1.63	1.59	1.56	1.52	1.48
29	2.89	2.50	2.28	2.15	2.06	1.99	1.93	1.89	1.86	1.83	1.78	1.73	1.68	1.65	1.62	1.58	1.55	1.51	1.47
30	2.88	2.49	2.28	2.14	2.05	1.98	1.93	1.88	1.85	1.82	1.77	1.72	1.67	1.64	1.61	1.57	1.54	1.50	1.46
40	2.84	2.44	2.23	2.09	2.00	1.93	1.87	1.83	1.79	1.76	1.71	1.66	1.61	1.57	1.54	1.51	1.47	1.42	1.38
60	2.79	2.39	2.18	2.04	1.95	1.87	1.82	1.77	1.74	1.71	1.66	1.60	1.54	1.51	1.48	1.44	1.40	1.35	1.29
120	2.75	2.35	2.13	1.99	1.90	1.82	1.77	1.72	1.68	1.65	1.60	1.55	1.48	1.45	1.41	1.37	1.32	1.26	1.19
∞	2.71	2.30	2.08	1.94	1.85	1.77	1.72	1.67	1.63	1.60	1.55	1.49	1.42	1.38	1.34	1.30	1.24	1.17	1.00

Appendix 5 (Continued)
F-Distribution
Upper 5%

v_2 \ v_1	1	2	3	4	5	6	7	8	9	10	12	15	20	24	30	40	60	120	∞
1	161.	200.	216.	225.	230.	234.	237.	239.	241.	242.	244.	246.	248.	249.	250.	251.	252.	253.	254.
2	18.51	19.00	19.16	19.25	19.30	19.33	19.35	19.37	19.38	19.40	19.41	19.43	19.45	19.45	19.46	19.47	19.48	19.49	19.50
3	10.13	9.55	9.28	9.12	9.01	8.94	8.89	8.85	8.81	8.79	8.74	8.70	8.66	8.64	8.62	8.59	8.57	8.55	8.53
4	7.71	6.94	6.59	6.39	6.26	6.16	6.09	6.04	6.00	5.96	5.91	5.86	5.80	5.77	5.75	5.72	5.69	5.66	5.63
5	6.61	5.79	5.41	5.19	5.05	4.95	4.88	4.82	4.77	4.74	4.68	4.62	4.56	4.53	4.50	4.46	4.43	4.40	4.36
6	5.99	5.14	4.76	4.53	4.39	4.28	4.21	4.15	4.10	4.06	4.00	3.94	3.87	3.84	3.81	3.77	3.74	3.70	3.67
7	5.59	4.74	4.35	4.12	3.97	3.87	3.79	3.73	3.68	3.64	3.57	3.51	3.44	3.41	3.38	3.34	3.30	3.27	3.23
8	5.32	4.46	4.07	3.84	3.69	3.58	3.50	3.44	3.39	3.35	3.28	3.22	3.15	3.12	3.08	3.04	3.01	2.97	2.93
9	5.12	4.26	3.86	3.63	3.48	3.37	3.29	3.23	3.18	3.14	3.07	3.01	2.94	2.90	2.86	2.83	2.79	2.75	2.71
10	4.96	4.10	3.71	3.48	3.33	3.22	3.14	3.07	3.02	2.98	2.91	2.85	2.77	2.74	2.70	2.66	2.62	2.58	2.54
11	4.84	3.98	3.59	3.36	3.20	3.09	3.01	2.95	2.90	2.85	2.79	2.72	2.65	2.61	2.57	2.53	2.49	2.45	2.40
12	4.75	3.89	3.49	3.26	3.11	3.00	2.91	2.85	2.80	2.75	2.69	2.62	2.54	2.51	2.47	2.43	2.38	2.34	2.30
13	4.67	3.81	3.41	3.18	3.03	2.92	2.83	2.77	2.71	2.67	2.60	2.53	2.46	2.42	2.38	2.34	2.30	2.25	2.21
14	4.60	3.74	3.34	3.11	2.96	2.85	2.76	2.70	2.65	2.60	2.53	2.46	2.39	2.35	2.31	2.27	2.22	2.18	2.13
15	4.54	3.68	3.29	3.06	2.90	2.79	2.71	2.64	2.59	2.54	2.48	2.40	2.33	2.29	2.25	2.20	2.16	2.11	2.07
16	4.49	3.63	3.24	3.01	2.85	2.74	2.66	2.59	2.54	2.49	2.42	2.35	2.28	2.24	2.19	2.15	2.11	2.06	2.01
17	4.45	3.59	3.20	2.96	2.81	2.70	2.61	2.55	2.49	2.45	2.38	2.31	2.23	2.19	2.15	2.10	2.06	2.01	1.96
18	4.41	3.55	3.16	2.93	2.77	2.66	2.58	2.51	2.46	2.41	2.34	2.27	2.19	2.15	2.11	2.06	2.02	1.97	1.92
19	4.38	3.52	3.13	2.90	2.74	2.63	2.54	2.48	2.42	2.38	2.31	2.23	2.16	2.11	2.07	2.03	1.98	1.93	1.88
20	4.35	3.49	3.10	2.87	2.71	2.60	2.51	2.45	2.39	2.35	2.28	2.20	2.12	2.08	2.04	1.99	1.95	1.90	1.84
21	4.32	3.47	3.07	2.84	2.68	2.57	2.49	2.42	2.37	2.32	2.25	2.18	2.10	2.05	2.01	1.96	1.92	1.87	1.81
22	4.30	3.44	3.05	2.82	2.66	2.55	2.46	2.40	2.34	2.30	2.23	2.15	2.07	2.03	1.98	1.94	1.89	1.84	1.78
23	4.28	3.42	3.03	2.80	2.64	2.53	2.44	2.37	2.32	2.27	2.20	2.13	2.05	2.01	1.96	1.91	1.86	1.81	1.76
24	4.26	3.40	3.01	2.78	2.62	2.51	2.42	2.36	2.30	2.25	2.18	2.11	2.03	1.98	1.94	1.89	1.84	1.79	1.73
25	4.24	3.39	2.99	2.76	2.60	2.49	2.40	2.34	2.28	2.24	2.16	2.09	2.01	1.96	1.92	1.87	1.82	1.77	1.71
26	4.23	3.37	2.98	2.74	2.59	2.47	2.39	2.32	2.27	2.22	2.15	2.07	1.99	1.95	1.90	1.85	1.80	1.75	1.69
27	4.21	3.35	2.96	2.73	2.57	2.46	2.37	2.31	2.25	2.20	2.13	2.06	1.97	1.93	1.88	1.84	1.79	1.73	1.67
28	4.20	3.34	2.95	2.71	2.56	2.45	2.36	2.29	2.24	2.19	2.12	2.04	1.96	1.91	1.87	1.82	1.77	1.71	1.65
29	4.18	3.33	2.93	2.70	2.55	2.43	2.35	2.28	2.22	2.18	2.10	2.03	1.94	1.90	1.85	1.81	1.75	1.70	1.64
30	4.17	3.32	2.92	2.69	2.53	2.42	2.33	2.27	2.21	2.16	2.09	2.01	1.93	1.89	1.84	1.79	1.74	1.68	1.62
40	4.08	3.23	2.84	2.61	2.45	2.34	2.25	2.18	2.12	2.08	2.00	1.92	1.84	1.79	1.74	1.69	1.64	1.58	1.51
60	4.00	3.15	2.76	2.53	2.37	2.25	2.17	2.10	2.04	1.99	1.92	1.84	1.75	1.70	1.65	1.59	1.53	1.47	1.39
120	3.92	3.07	2.68	2.45	2.29	2.17	2.09	2.02	1.96	1.91	1.83	1.75	1.66	1.61	1.55	1.50	1.43	1.35	1.25
∞	3.84	3.00	2.60	2.37	2.21	2.10	2.01	1.94	1.88	1.83	1.75	1.67	1.57	1.52	1.46	1.39	1.32	1.22	1.00

Appendix 5 (Continued)
F-Distribution
Upper 1%

v_2 \ v_1	1	2	3	4	5	6	7	8	9	10	12	15	20	24	30	40	60	120	∞
1	4052.	5000.	5403.	5625.	5764.	5859.	5928.	5982.	6022.	6056.	6106.	6157.	6209.	6235.	6261.	6287.	6313.	6339.	6366.
2	98.50	99.00	99.17	99.25	99.30	99.33	99.36	99.37	99.39	99.40	99.42	99.43	99.45	99.46	99.47	99.47	99.48	99.49	99.50
3	34.12	30.82	29.46	28.71	28.24	27.91	27.67	27.49	27.35	27.23	27.05	26.87	26.69	26.60	26.50	26.41	26.32	26.22	26.13
4	21.20	18.00	16.69	15.98	15.52	15.21	14.98	14.80	14.66	14.55	14.37	14.20	14.02	13.93	13.84	13.75	13.65	13.56	13.46
5	16.26	13.27	12.06	11.39	10.97	10.67	10.46	10.29	10.16	10.05	9.89	9.72	9.55	9.47	9.38	9.29	9.20	9.11	9.02
6	13.75	10.92	9.78	9.15	8.75	8.47	8.26	8.10	7.98	7.87	7.72	7.56	7.40	7.31	7.23	7.14	7.06	6.97	6.88
7	12.25	9.55	8.45	7.85	7.46	7.19	6.99	6.84	6.72	6.62	6.47	6.31	6.16	6.07	5.99	5.91	5.82	5.74	5.65
8	11.26	8.65	7.59	7.01	6.63	6.37	6.18	6.03	5.91	5.81	5.67	5.52	5.36	5.28	5.20	5.12	5.03	4.95	4.86
9	10.56	8.02	6.99	6.42	6.06	5.80	5.61	5.47	5.35	5.26	5.11	4.96	4.81	4.73	4.65	4.57	4.48	4.40	4.31
10	10.04	7.56	6.55	5.99	5.64	5.39	5.20	5.06	4.94	4.85	4.71	4.56	4.41	4.33	4.25	4.17	4.08	4.00	3.91
11	9.65	7.21	6.22	5.67	5.32	5.07	4.89	4.74	4.63	4.54	4.40	4.25	4.10	4.02	3.94	3.86	3.78	3.69	3.60
12	9.33	6.93	5.95	5.41	5.06	4.82	4.64	4.50	4.39	4.30	4.16	4.01	3.86	3.78	3.70	3.62	3.54	3.45	3.36
13	9.07	6.70	5.74	5.21	4.86	4.62	4.44	4.30	4.19	4.10	3.96	3.82	3.66	3.59	3.51	3.43	3.34	3.25	3.17
14	8.86	6.51	5.56	5.04	4.69	4.46	4.28	4.14	4.03	3.94	3.80	3.66	3.51	3.43	3.35	3.27	3.18	3.09	3.00
15	8.68	6.36	5.42	4.89	4.56	4.32	4.14	4.00	3.89	3.80	3.67	3.52	3.37	3.29	3.21	3.13	3.05	2.96	2.87
16	8.53	6.23	5.29	4.77	4.44	4.20	4.03	3.89	3.78	3.69	3.55	3.41	3.26	3.18	3.10	3.02	2.93	2.84	2.75
17	8.40	6.11	5.18	4.67	4.34	4.10	3.93	3.79	3.68	3.59	3.46	3.31	3.16	3.08	3.00	2.92	2.83	2.75	2.65
18	8.29	6.01	5.09	4.58	4.25	4.01	3.84	3.71	3.60	3.51	3.37	3.23	3.08	3.00	2.92	2.84	2.75	2.66	2.57
19	8.18	5.93	5.01	4.50	4.17	3.94	3.77	3.63	3.52	3.43	3.30	3.15	3.00	2.92	2.84	2.76	2.67	2.58	2.49
20	8.10	5.85	4.94	4.43	4.10	3.87	3.70	3.56	3.46	3.37	3.23	3.09	2.94	2.86	2.78	2.69	2.61	2.52	2.42
21	8.02	5.78	4.87	4.37	4.04	3.81	3.64	3.51	3.40	3.31	3.17	3.03	2.88	2.80	2.72	2.64	2.55	2.46	2.36
22	7.95	5.72	4.82	4.31	3.99	3.76	3.59	3.45	3.35	3.26	3.12	2.98	2.83	2.75	2.67	2.58	2.50	2.40	2.31
23	7.88	5.66	4.76	4.26	3.94	3.71	3.54	3.41	3.30	3.21	3.07	2.93	2.78	2.70	2.62	2.54	2.45	2.35	2.26
24	7.82	5.61	4.72	4.22	3.90	3.67	3.50	3.36	3.26	3.17	3.03	2.89	2.74	2.66	2.58	2.49	2.40	2.31	2.21
25	7.77	5.57	4.68	4.18	3.85	3.63	3.46	3.32	3.22	3.13	2.99	2.85	2.70	2.62	2.54	2.45	2.36	2.27	2.17
26	7.72	5.53	4.64	4.14	3.82	3.59	3.42	3.29	3.18	3.09	2.96	2.81	2.66	2.58	2.50	2.42	2.33	2.23	2.13
27	7.68	5.49	4.60	4.11	3.78	3.56	3.39	3.26	3.15	3.06	2.93	2.78	2.63	2.55	2.47	2.38	2.29	2.20	2.10
28	7.64	5.45	4.57	4.07	3.75	3.53	3.36	3.23	3.12	3.03	2.90	2.75	2.60	2.52	2.44	2.35	2.26	2.17	2.06
29	7.60	5.42	4.54	4.04	3.73	3.50	3.33	3.20	3.09	3.00	2.87	2.73	2.57	2.49	2.41	2.33	2.23	2.14	2.03
30	7.56	5.39	4.51	4.02	3.70	3.47	3.30	3.17	3.07	2.98	2.84	2.70	2.55	2.47	2.39	2.30	2.21	2.11	2.01
40	7.31	5.18	4.31	3.83	3.51	3.29	3.12	2.99	2.89	2.80	2.66	2.52	2.37	2.29	2.20	2.11	2.02	1.92	1.80
60	7.08	4.98	4.13	3.65	3.34	3.12	2.95	2.82	2.72	2.63	2.50	2.35	2.20	2.12	2.03	1.94	1.84	1.73	1.60
120	6.85	4.79	3.95	3.48	3.17	2.96	2.79	2.66	2.56	2.47	2.34	2.19	2.03	1.95	1.86	1.76	1.66	1.53	1.38
∞	6.63	4.61	3.78	3.32	3.02	2.80	2.64	2.51	2.41	2.32	2.18	2.04	1.88	1.79	1.70	1.59	1.47	1.32	1.00

Appendix 6
Upper Percentage Points of the Studentized Range
$\alpha = .10$

df/k	2	3	4	5	6	7	8	9	10	12	15	30	60	100
1	8.929	13.44	16.36	18.49	20.15	21.51	22.64	23.62	24.48	25.92	27.62	32.50	36.91	39.91
2	4.130	5.733	6.773	7.538	8.139	8.633	9.049	9.409	9.725	10.26	10.89	12.73	14.40	15.54
3	3.328	4.467	5.199	5.738	6.162	6.511	6.806	7.062	7.287	7.667	8.120	9.440	10.65	11.48
4	3.015	3.976	4.586	5.035	5.388	5.679	5.926	6.139	6.327	6.645	7.025	8.135	9.156	9.860
5	2.850	3.717	4.264	4.664	4.979	5.238	5.458	5.648	5.816	6.101	6.440	7.435	8.353	8.988
6	2.748	3.559	4.065	4.435	4.726	4.966	5.168	5.344	5.499	5.762	6.075	6.996	7.848	8.438
7	2.680	3.451	3.931	4.280	4.555	4.780	4.972	5.137	5.283	5.530	5.826	6.695	7.500	8.059
8	2.630	3.374	3.834	4.169	4.431	4.646	4.829	4.987	5.126	5.362	5.644	6.475	7.245	7.780
9	2.592	3.316	3.761	4.084	4.337	4.545	4.721	4.873	5.007	5.234	5.506	6.306	7.050	7.566
10	2.563	3.270	3.704	4.018	4.264	4.465	4.636	4.783	4.913	5.134	5.397	6.173	6.895	7.396
11	2.540	3.234	3.658	3.965	4.205	4.401	4.568	4.711	4.838	5.053	5.309	6.065	6.768	7.258
12	2.521	3.204	3.621	3.922	4.156	4.349	4.511	4.652	4.776	4.986	5.236	5.976	6.663	7.142
13	2.505	3.179	3.589	3.885	4.116	4.305	4.464	4.602	4.724	4.930	5.176	5.900	6.575	7.045
14	2.491	3.158	3.563	3.854	4.081	4.267	4.424	4.560	4.680	4.882	5.124	5.836	6.499	6.961
15	2.479	3.140	3.540	3.828	4.052	4.235	4.390	4.524	4.641	4.841	5.079	5.780	6.433	6.888
16	2.469	3.124	3.520	3.804	4.026	4.207	4.360	4.492	4.608	4.805	5.040	5.732	6.376	6.825
17	2.460	3.110	3.503	3.784	4.004	4.183	4.334	4.464	4.579	4.774	5.005	5.869	6.325	6.769
18	2.452	3.098	3.488	3.767	3.984	4.161	4.311	4.440	4.554	4.746	4.975	5.650	6.280	6.719
19	2.445	3.087	3.474	3.751	3.966	4.142	4.290	4.418	4.531	4.721	4.948	5.616	6.239	6.674
20	2.439	3.078	3.462	3.736	3.950	4.124	4.271	4.398	4.510	4.699	4.924	5.586	6.203	6.633
24	2.420	3.047	3.423	3.692	3.900	4.070	4.213	4.336	4.445	4.628	4.847	5.489	6.086	6.503
30	2.400	3.017	3.386	3.648	3.851	4.016	4.155	4.275	4.381	4.559	4.770	5.392	5.969	6.372
40	2.381	2.988	3.349	3.605	3.803	3.963	4.099	4.215	4.317	4.490	4.695	5.294	5.850	6.238
60	2.363	2.959	3.312	3.562	3.755	3.911	4.042	4.155	4.254	4.421	4.619	5.196	5.730	6.102
120	2.344	2.930	3.276	3.520	3.707	3.859	3.987	4.096	4.191	4.353	4.543	5.097	5.606	5.960
∞	2.326	2.902	3.240	3.478	3.661	3.808	3.931	4.037	4.129	4.285	4.468	4.997	5.480	5.812

Appendix 6 (Cont'd.)

Upper Percentage Points of the Studentized Range

α = .05

df/k	2	3	4	5	6	7	8	9	10	12	15	30	60	100
1	17.97	26.98	32.82	37.08	40.41	43.12	45.40	47.36	49.07	51.96	55.36	65.15	73.97	79.98
2	6.085	8.331	9.798	10.88	11.74	12.44	13.03	13.54	13.99	14.75	15.65	18.27	20.66	22.29
3	4.501	5.910	6.825	7.502	8.037	8.478	8.853	9.177	9.462	9.946	10.53	12.21	13.76	14.82
4	3.927	5.040	5.757	6.287	6.707	7.053	7.347	7.602	7.826	8.208	8.664	10.00	11.24	12.09
5	3.635	4.602	5.218	5.673	6.033	6.330	6.582	6.802	6.995	7.324	7.717	8.875	9.949	10.69
6	3.461	4.339	4.896	5.305	5.628	5.895	6.122	6.319	6.493	6.789	7.143	8.189	9.163	9.839
7	3.344	4.165	4.681	5.060	5.359	5.606	5.815	5.998	6.158	6.431	6.759	7.728	8.632	9.261
8	3.261	4.041	4.529	4.886	5.167	5.399	5.597	5.767	5.918	6.175	6.483	7.395	8.248	8.843
9	3.199	3.949	4.415	4.756	5.024	5.244	5.432	5.595	5.739	5.983	6.276	7.145	7.958	8.526
10	3.151	3.877	4.327	4.654	4.912	5.124	5.305	5.461	5.599	5.833	6.114	6.948	7.730	8.276
11	3.113	3.820	4.256	4.574	4.823	5.028	5.202	5.353	5.487	5.713	5.984	6.790	7.546	8.075
12	3.082	3.773	4.199	4.508	4.751	4.950	5.119	5.265	5.395	5.615	5.878	6.660	7.394	7.909
13	3.055	3.735	4.151	4.453	4.690	4.885	5.049	5.192	5.318	5.533	5.789	6.551	7.267	7.769
14	3.033	3.702	4.111	4.407	4.639	4.829	4.990	5.131	5.254	5.463	5.714	6.459	7.159	7.650
15	3.014	3.674	4.076	4.367	4.595	4.782	4.940	5.077	5.198	5.404	5.649	6.379	7.065	7.546
16	2.998	3.649	4.046	4.333	4.557	4.741	4.897	5.031	5.150	5.352	5.593	6.310	6.984	7.457
17	2.984	3.628	4.020	4.303	4.524	4.705	4.858	4.991	5.108	5.307	5.544	6.249	6.912	7.377
18	2.971	3.609	3.997	4.277	4.495	4.673	4.824	4.956	5.071	5.267	5.501	6.195	6.848	7.307
19	2.960	3.593	3.977	4.253	4.469	4.645	4.794	4.924	5.038	5.231	5.462	6.147	6.792	7.244
20	2.950	3.578	3.958	4.232	4.445	4.620	4.768	4.896	5.008	5.199	5.427	6.104	6.740	7.187
24	2.919	3.532	3.901	4.166	4.373	4.541	4.684	4.807	4.915	5.099	5.319	5.968	6.579	7.008
30	2.888	3.486	3.845	4.102	4.302	4.464	4.602	4.720	4.824	5.001	5.211	5.833	6.417	6.827
40	2.858	3.442	3.791	4.039	4.232	4.389	4.521	4.635	4.735	4.904	5.106	5.700	6.255	6.645
60	2.829	3.399	3.737	3.977	4.163	4.314	4.441	4.550	4.646	4.808	5.001	5.566	6.093	6.462
120	2.800	3.356	3.685	3.917	4.096	4.241	4.363	4.468	4.560	4.714	4.898	5.434	5.929	6.275
∞	2.772	3.314	3.633	3.858	4.030	4.170	4.286	4.387	4.474	4.622	4.796	5.301	5.764	6.085

Appendix 6 (Cont'd.)

Upper Percentage Points of the Studentized Range

α = .01

df/k	2	3	4	5	6	7	8	9	10	12	15	30	60	100
1	90.03	135.0	164.3	185.6	202.2	215.8	227.2	237.0	245.6	260.0	277.0	326.0	370.1	400.1
2	14.04	19.02	22.29	24.72	26.63	28.20	29.53	30.68	31.69	33.40	35.43	41.32	46.70	50.38
3	8.261	10.62	12.17	13.33	14.24	15.00	15.64	16.20	16.69	17.53	18.52	21.44	24.13	25.99
4	6.512	8.120	9.173	9.958	10.58	11.10	11.55	11.93	12.27	12.84	13.53	15.57	17.46	18.77
5	5.702	6.976	7.804	8.421	8.913	9.321	9.669	9.972	10.24	10.70	11.24	12.87	14.39	15.45
6	5.243	6.331	7.033	7.556	7.973	8.318	8.613	8.869	9.097	9.485	9.951	11.34	12.65	13.55
7	4.949	5.919	6.543	7.005	7.373	7.679	7.939	8.166	8.368	8.711	9.124	10.36	11.52	12.34
8	4.746	5.635	6.204	6.625	6.960	7.237	7.474	7.681	7.863	8.176	8.552	9.678	10.75	11.49
9	4.596	5.428	5.957	6.348	6.658	6.915	7.134	7.325	7.495	7.784	8.132	9.177	10.17	10.87
10	4.482	5.270	5.769	6.136	6.428	6.669	6.875	7.055	7.213	7.485	7.812	8.794	9.726	10.39
11	4.392	5.146	5.621	5.970	6.247	6.476	6.672	6.842	6.992	7.250	7.560	8.491	9.377	10.00
12	4.320	5.046	5.502	5.836	6.101	6.321	6.507	6.670	6.814	7.060	7.356	8.246	9.094	9.693
13	4.260	4.964	5.404	5.727	5.981	6.192	6.372	6.528	6.667	6.903	7.188	8.043	8.859	9.436
14	4.210	4.895	5.322	5.634	5.881	6.085	6.258	6.409	6.543	6.772	7.047	7.873	8.661	9.219
15	4.168	4.836	5.252	5.556	5.796	5.994	6.162	6.309	6.439	6.660	6.927	7.728	8.492	9.035
16	4.131	4.786	5.192	5.489	5.722	5.915	6.079	6.222	6.349	6.564	6.823	7.602	8.347	8.874
17	4.099	4.742	5.140	5.430	5.659	5.847	6.007	6.147	6.270	6.480	6.734	7.493	8.219	8.735
18	4.071	4.703	5.094	5.379	5.603	5.788	5.944	6.081	6.201	6.407	6.655	7.398	8.107	8.611
19	4.046	4.670	5.054	5.334	5.554	5.735	5.889	6.022	6.141	6.342	6.585	7.313	8.008	8.502
20	4.024	4.639	5.018	5.294	5.510	5.688	5.839	5.970	6.087	6.285	6.523	7.237	7.919	8.404
24	3.956	4.546	4.907	5.168	5.374	5.542	5.685	5.809	5.919	6.106	6.330	7.001	7.642	8.097
30	3.889	4.455	4.799	5.048	5.242	5.401	5.536	5.653	5.756	5.932	6.143	6.772	7.370	7.796
40	3.825	4.367	4.696	4.931	5.114	5.265	5.392	5.502	5.599	5.764	5.961	6.547	7.104	7.500
60	3.762	4.282	4.595	4.818	4.991	5.133	5.253	5.356	5.447	5.601	5.785	6.330	6.843	7.207
120	3.702	4.200	4.497	4.709	4.872	5.005	5.118	5.214	5.299	5.443	5.614	6.117	6.588	6.919
∞	3.643	4.120	4.403	4.603	4.757	4.882	4.987	5.078	5.157	5.290	5.448	5.911	6.338	6.636

Appendix 6 (Cont'd.)

Upper Percentage Points of the Studentized Range

α = .001

df/k	2	3	4	5	6	7	8	9	10	12	15	30	60	100
1	900.3	1351.	1643.	1856.	2022.	2158.	2272.	2370.	2455.	2600.	2770.	3260.	3701.	4002.
2	44.69	60.42	70.77	78.43	84.49	89.46	93.67	97.30	100.5	105.9	112.3	131.0	148.0	159.7
3	18.28	23.32	26.65	29.13	31.11	32.74	34.12	35.33	36.39	38.20	40.35	46.68	52.51	56.53
4	12.18	14.99	16.84	18.23	19.34	20.26	21.04	21.73	22.33	23.36	24.59	28.24	31.65	34.00
5	9.714	11.67	12.96	13.93	14.71	15.35	15.90	16.38	16.81	17.53	18.41	21.01	23.45	25.15
6	8.427	9.960	10.97	11.72	12.32	12.83	13.26	13.63	13.97	14.54	15.22	17.28	19.22	20.58
7	7.648	8.930	9.768	10.40	10.90	11.32	11.68	11.99	12.27	12.74	13.32	15.05	16.69	17.85
8	7.130	8.250	8.978	9.522	9.958	10.32	10.64	10.91	11.15	11.56	12.06	13.57	15.01	16.02
9	6.762	7.768	8.419	8.906	9.295	9.619	9.897	10.14	10.36	10.73	11.18	12.53	13.82	14.74
10	6.487	7.411	8.006	8.450	8.804	9.099	9.352	9.573	9.769	10.11	10.52	11.75	12.94	13.78
11	6.275	7.136	7.687	8.098	8.426	8.699	8.933	9.138	9.319	9.630	10.01	11.16	12.25	13.04
12	6.106	6.917	7.436	7.821	8.127	8.383	8.601	8.793	8.962	9.254	9.606	10.68	11.71	12.45
13	5.970	6.740	7.231	7.595	7.885	8.126	8.333	8.513	8.673	8.948	9.281	10.29	11.27	11.97
14	5.856	6.594	7.062	7.409	7.685	7.915	8.110	8.282	8.434	8.696	9.012	9.972	10.91	11.57
15	5.760	6.470	6.920	7.252	7.517	7.736	7.925	8.088	8.234	8.483	8.786	9.703	10.59	11.23
16	5.678	6.365	6.799	7.119	7.374	7.585	7.766	7.923	8.063	8.303	8.593	9.475	10.34	10.95
17	5.608	6.275	6.695	7.005	7.250	7.454	7.629	7.781	7.916	8.148	8.427	9.277	10.10	10.70
18	5.546	6.196	6.604	6.905	7.143	7.341	7.510	7.657	7.788	8.012	8.283	9.106	9.904	10.48
19	5.492	6.127	6.525	6.817	7.049	7.242	7.405	7.549	7.676	7.893	8.156	8.955	9.730	10.29
20	5.444	6.065	6.454	6.740	6.966	7.154	7.313	7.453	7.577	7.788	8.044	8.821	9.575	10.12
24	5.297	5.877	6.238	6.503	6.712	6.884	7.031	7.159	7.272	7.467	7.701	8.411	9.100	9.596
30	5.156	5.698	6.033	6.278	6.470	6.628	6.763	6.880	6.984	7.162	7.375	8.021	8.647	9.096
40	5.022	5.528	5.838	6.063	6.240	6.386	6.509	6.616	6.711	6.872	7.067	7.651	7.214	8.618
60	4.894	5.365	5.653	5.860	6.022	6.155	6.268	6.366	6.451	6.598	6.774	7.299	7.802	8.161
120	4.771	5.211	5.476	5.667	5.815	5.937	6.039	6.128	6.206	6.339	6.496	6.966	7.411	7.726
∞	4.654	5.063	5.309	5.484	5.619	5.730	5.823	5.903	5.973	6.092	6.234	6.651	7.041	7.314

Cumulative Distribution of Chi-Square

Degrees of Freedom	Probability of a Greater Value													
	0.995	0.990	0.975	0.950	0.900	0.750	0.500	0.250	0.100	0.050	0.025	0.010	0.005	0.001
1					0.02	0.10	0.45	1.32	2.71	3.84	5.02	6.63	7.88	10.83
2	0.01	0.02	0.05	0.10	0.21	0.58	1.39	2.77	4.61	5.99	7.38	9.21	10.60	13.82
3	0.07	0.11	0.22	0.35	0.58	1.21	2.37	4.11	6.25	7.81	9.35	11.34	12.84	16.27
4	0.21	0.30	0.48	0.71	1.06	1.92	3.36	5.39	7.78	9.49	11.14	13.28	14.86	18.47
5	0.41	0.55	0.83	1.15	1.61	2.67	4.35	6.63	9.24	11.07	12.83	15.09	16.75	20.52
6	0.68	0.87	1.24	1.64	2.20	3.45	5.35	7.84	10.64	12.59	14.45	16.81	18.55	22.46
7	0.99	1.24	1.69	2.17	2.83	4.25	6.35	9.04	12.02	14.07	16.01	18.48	20.28	24.32
8	1.34	1.65	2.18	2.73	3.49	5.07	7.34	10.22	13.36	15.51	17.53	20.09	21.96	26.12
9	1.73	2.09	2.70	3.33	4.17	5.90	8.34	11.39	14.68	16.92	19.02	21.67	23.59	27.88
10	2.16	2.56	3.25	3.94	4.87	6.74	9.34	12.55	15.99	18.31	20.48	23.21	25.19	29.59
11	2.60	3.05	3.82	4.57	5.58	7.58	10.34	13.70	17.28	19.68	21.92	24.72	26.76	31.26
12	3.07	3.57	4.40	5.23	6.30	8.44	11.34	14.85	18.55	21.03	23.34	26.22	28.30	32.91
13	3.57	4.11	5.01	5.89	7.04	9.30	12.34	15.98	19.81	22.36	24.74	27.69	29.82	34.53
14	4.07	4.66	5.63	6.57	7.79	10.17	13.34	17.12	21.06	23.68	26.12	29.14	31.32	36.12
15	4.60	5.23	6.26	7.26	8.55	11.04	14.34	18.25	22.31	25.00	27.49	30.58	32.80	37.70
16	5.14	5.81	6.91	7.96	9.31	11.91	15.34	19.37	23.54	26.30	28.85	32.00	34.27	39.25
17	5.70	6.41	7.56	8.67	10.09	12.79	16.34	20.49	24.77	27.59	30.19	33.41	35.73	40.79
18	6.26	7.01	8.23	9.39	10.86	13.68	17.34	21.60	25.99	28.87	31.53	34.81	37.16	42.31
19	6.84	7.63	8.91	10.12	11.65	14.56	18.34	22.72	27.20	30.14	32.85	36.19	38.58	43.82
20	7.43	8.26	9.59	10.85	12.44	15.45	19.34	23.83	28.41	31.41	34.17	37.57	40.00	45.32
21	8.03	8.90	10.28	11.59	13.24	16.34	20.34	24.93	29.62	32.67	35.48	38.93	41.40	46.80
22	8.64	9.54	10.98	12.34	14.04	17.24	21.34	26.04	30.81	33.92	36.78	40.29	42.80	48.27
23	9.26	10.20	11.69	13.09	14.85	18.14	22.34	27.14	32.01	35.17	38.08	41.64	44.18	49.73
24	9.89	10.86	12.40	13.85	15.66	19.04	23.34	28.24	33.20	36.42	39.36	42.98	45.56	51.18
25	10.52	11.52	13.12	14.61	16.47	19.94	24.34	29.34	34.38	37.65	40.65	44.31	46.93	52.62
30	13.79	14.95	16.79	18.49	20.60	24.48	29.34	34.80	40.26	43.77	46.98	50.89	53.67	59.70
40	20.71	22.16	24.43	26.51	29.05	33.66	39.34	45.62	51.80	55.76	59.34	63.69	66.77	73.40
50	27.99	29.71	32.36	34.76	37.69	42.94	49.33	56.33	63.17	67.50	71.42	76.15	79.49	86.66
60	35.53	37.48	40.48	43.19	46.46	52.29	59.33	66.98	74.40	79.08	83.30	88.38	91.95	99.61
70	43.28	45.44	48.76	51.74	55.33	61.70	69.33	77.58	85.53	90.53	95.02	100.42	104.22	112.32
80	51.17	53.54	57.15	60.39	64.28	71.14	79.33	88.13	96.58	101.88	106.63	112.33	116.32	124.84
90	59.20	61.75	65.65	69.13	73.29	80.62	89.33	98.64	107.56	113.14	118.14	124.12	128.30	137.21
100	67.33	70.06	74.22	77.93	82.36	90.13	99.33	109.14	118.50	124.34	129.56	135.81	140.17	149.45

PERCENTILE POINTS FOR Q-TEST, FOR EQUAL DEGREES OF FREEDOM ν, AND FOR p SAMPLES

p	$\nu = 1$		$\nu = 2$		$\nu = 3$		$\nu = 4$	
	.99	.999	.99	.999	.99	.999	.99	.999
3	*	*	.863	*	.757	.919	.684	.828
4	.920	*	.720	.898	.605	.754	.549	.675
5	.828	*	.608	.773	.512	.644	.443	.552
6	.744	.949	.539	.690	.430	.546	.369	.461
7	.671	.865	.469	.606	.372	.471	.318	.394
8	.609	.793	.412	.537	.325	.411	.276	.342
9	.576	.750	.371	.481	.287	.363	.244	.300
10	.528	.694	.333	.433	.257	.324	.218	.267
12	.448	.598	.276	.358	.211	.265	.179	.217
14	.391	.522	.234	.303	.178	.222	.151	.181
15	.365	.490	.217	.280	.165	.205	.140	.167
16	.343	.460	.202	.261	.154	.190	.130	.155
18	.304	.409	.178	.228	.135	.165	.114	.135
20	.273	.367	.158	.202	.120	.146	.101	.119
22	.246	.332	.142	.180	.108	.130	.090	.106
24	.224	.302	.129	.162	.098	.117	.082	.096
26	.206	.276	.118	.148	.090	.107	.075	.087
28	.190	.254	.108	.135	.082	.098	.069	.080
30	.176	.234	.100	.124	.075	.090	.064	.074
32	.163	.218	.093	.115	.070	.083	.060	.068
36	.143	.189	.082	.100	.062	.072	.052	.060
40	.127	.167	.072	.088	.055	.064	.047	.053
45	.111	.145	.063	.076	.048	.055	.041	.046
50	.098	.127	.056	.067	.043	.049	.037	.041
60	.080	.102	.045	.053	.035	.039	.030	.033
64	.074	.094	.042	.049	.033	.037	.028	.031

*These entries exceeded 1 using the approximating distribution. Since $Q \geq 1$, they are omitted.

p	$\nu = 5$ .99	.999	$\nu = 6$ .99	.999	$\nu = 8$ .99	.999	$\nu = 10$ .99	.999
3	.631	.760	.593	.708	.539	.633	.512	.596
4	.498	.608	.461	.558	.413	.490	.383	.446
5	.399	.490	.368	.446	.328	.388	.303	.351
6	.334	.407	.307	.368	.271	.318	.250	.288
7	.284	.345	.261	.311	.230	.268	.212	.242
8	.246	.298	.226	.268	.199	.231	.184	.209
9	.217	.261	.199	.235	.176	.202	.162	.183
10	.194	.232	.178	.208	.157	.179	.145	.163
15	.123	.145	.113	.131	.101	.113	.094	.103
20	.090	.104	.083	.094	.074	.082	.069	.075
30	.058	.065	.053	.059	.048	.052	.045	.048
40	.042	.047	.039	.043	.035	.038	.033	.035
50	.033	.036	.031	.033	.028	.030	.026	.028
60	.027	.029	.025	.027	.023	.024	.022	.023

p	$\nu = 12$ .99	.999	$\nu = 14$ .99	.999	$\nu = 16$ .99	.999	$\nu = 20$ .99	.999
3	.486	.558	.466	.530	.451	.508	.429	.476
4	.362	.415	.347	.393	.335	.375	.319	.351
5	.287	.326	.275	.308	.265	.295	.252	.276
6	.236	.267	.227	.253	.219	.242	.209	.226
7	.201	.225	.192	.213	.186	.204	.178	.191
8	.174	.194	.167	.184	.162	.176	.154	.166
9	.154	.170	.148	.162	.143	.155	.136	.146
10	.137	.152	.132	.144	.128	.138	.122	.130
15	.089	.097	.086	.092	.083	.089	.080	.084
20	.066	.070	.063	.067	.062	.065	.059	.062
30	.043	.045	.042	.043	.040	.042	.039	.040
40	.032	.033	.031	.032	.030	.031	.029	.030
50	.025	.026	.024	.025	.024	.025	.023	.024
60	.021	.022	.020	.021	0.20	.020	.019	.020

For $\nu > 60$, calculate $p\nu(pq-1)$ and compare with χ^2 with $(p-1)$ degrees of freedom in Appendix 7.

Coefficients $\{a_{n-i+1}\}$ for the W test for normality,

for n = 2(1)50

i \ n	2	3	4	5	6	7	8	9	10
1	0.7071	0.7071	0.6872	0.6646	0.6431	0.6233	0.6052	0.5888	0.5739
2		.0000	.1677	.2413	.2806	.3031	.3164	.3244	.3291
3				.0000	.0875	.1401	.1743	.1976	.2141
4						.0000	.0561	.0947	.1224
5								.0000	.0399

i \ n	11	12	13	14	15	16	17	18	19	20
1	0.5601	0.5475	0.5359	0.5251	0.5150	0.5056	0.4968	0.4886	0.4808	0.4734
2	.3315	.3325	.3325	.3318	.3306	.3290	.3273	.3253	.3232	.3211
3	.2260	.2347	.2412	.2460	.2495	.2521	.2540	.2553	.2561	.2565
4	.1429	.1586	.1707	.1802	.1878	.1939	.1988	.2027	.2059	.2085
5	.0695	.0922	.1099	.1240	.1353	.1447	.1524	.1587	.1641	.1686
6	0.0000	0.0303	0.0539	0.0727	0.0880	0.1005	0.1109	0.1197	0.1271	0.1334
7			.0000	.0240	.0433	.0593	.0725	.0837	.0932	.1013
8					.0000	.0196	.0359	.0496	.0612	.0711
9							.0000	.0163	.0303	.0422
10									.0000	.0140

i \ n	21	22	23	24	25	26	27	28	29	30
1	0.4643	0.4590	0.4542	0.4493	0.4450	0.4407	0.4366	0.4328	0.4291	0.4254
2	.3185	.3156	.3126	.3098	.3069	.3043	.3018	.2992	.2968	.2944
3	.2578	.2571	.2563	.2554	.2543	.2533	.2522	.2510	.2499	.2487
4	.2119	.2131	.2139	.2145	.2148	.2151	.2152	.2151	.2150	.2148
5	.1736	.1764	.1787	.1807	.1822	.1836	.1848	.1857	.1864	.1870
6	0.1399	0.1443	0.1480	0.1512	0.1539	0.1563	0.1584	0.1601	0.1616	0.1630
7	.1092	.1150	.1201	.1245	.1283	.1316	.1346	.1372	.1395	.1415
8	.0804	.0878	.0941	.0997	.1046	.1089	.1128	.1162	.1192	.1219
9	.0530	.0618	.0696	.0764	.0823	.0876	.0923	.0965	.1002	.1036
10	.0263	.0368	.0459	.0539	.0610	.0672	.0728	.0778	.0822	.0862
11	0.0000	0.0122	0.0228	0.0321	0.0403	0.0476	0.0540	0.0598	0.0650	0.0697
12			.0000	.0107	.0200	.0284	.0358	.0424	.0483	.0537
13					.0000	.0094	.0178	.0253	.0320	.0381
14							.0000	.0084	.0159	.0227
15									.0000	.0076

Coefficients {a_{n-i+1}} for the W test for normality,

for n = 2(1)50 (cont.)

i \ n	31	32	33	34	35	36	37	38	39	40
1	0.4220	0.4188	0.4156	0.4127	0.4096	0.4068	0.4040	0.4015	0.3989	0.3964
2	.2921	.2898	.2876	.2854	.2834	.2813	.2794	.2774	.2755	.2737
3	.2475	.2463	.2451	.2439	.2427	.2415	.2403	.2391	.2380	.2368
4	.2145	.2141	.2137	.2132	.2127	.2121	.2116	.2110	.2104	.2098
5	.1874	.1878	.1880	.1882	.1883	.1883	.1883	.1881	.1880	.1878
6	0.1641	0.1651	0.1660	0.1667	0.1673	0.1678	0.1683	0.1686	0.1689	0.1691
7	.1433	.1449	.1463	.1475	.1487	.1496	.1505	.1513	.1520	.1526
8	.1243	.1265	.1284	.1301	.1317	.1331	.1344	.1356	.1366	.1376
9	.1066	.1093	.1118	.1140	.1160	.1179	.1196	.1211	.1225	.1237
10	.0899	.0931	.0961	.0988	.1013	.1036	.1056	.1075	.1092	.1108
11	0.0739	0.0777	0.0812	0.0844	0.0873	0.0900	0.0924	0.0947	0.0967	0.0986
12	.0585	.0629	.0669	.0706	.0739	.0770	.0798	.0824	.0848	.0870
13	.0435	.0485	.0530	.0572	.0610	.0645	.0677	.0706	.0733	.0759
14	.0289	.0344	.0395	.0441	.0484	.0523	.0559	.0592	.0622	.0651
15	.0144	.0206	.0262	.0314	.0361	.0404	.0444	.0481	.0515	.0546
16	0.0000	0.0068	0.0131	0.0187	0.0239	0.0287	0.0331	0.0372	0.0409	0.0444
17			.0000	.0062	.0119	.0172	.0220	.0264	.0305	.0343
18					.0000	.0057	.0110	.0158	.0203	.0244
19							.0000	.0053	.0101	.0146
20									.0000	.0049

i \ n	41	42	43	44	45	46	47	48	49	50
1	0.3940	0.3917	0.3894	0.3872	0.3850	0.3830	0.3808	0.3789	0.3770	0.3751
2	.2719	.2701	.2684	.2667	.2651	.2635	.2620	.2604	.2589	.2574
3	.2357	.2345	.2334	.2323	.2313	.2302	.2291	.2281	.2271	.2260
4	.2091	.2085	.2078	.2072	.2065	.2058	.2052	.2045	.2038	.2032
5	.1876	.1874	.1871	.1868	.1865	.1862	.1859	.1855	.1851	.1847
6	0.1693	0.1694	0.1695	0.1695	0.1695	0.1695	0.1695	0.1693	0.1692	0.1691
7	.1531	.1535	.1539	.1542	.1545	.1548	.1550	.1551	.1553	.1554
8	.1384	.1392	.1398	.1405	.1410	.1415	.1420	.1423	.1427	.1430
9	.1249	.1259	.1269	.1278	.1286	.1293	.1300	.1306	.1312	.1317
10	.1123	.1136	.1149	.1160	.1170	.1180	.1189	.1197	.1205	.1212
11	0.1004	0.1020	0.1035	0.1049	0.1062	0.1073	0.1085	0.1095	0.1105	0.1113
12	.0891	.0909	.0927	.0943	.0959	.0972	.0986	.0998	.1010	.1020
13	.0782	.0804	.0824	.0842	.0860	.0876	.0892	.0906	.0919	.0932
14	.0677	.0701	.0724	.0745	.0765	.0783	.0801	.0817	.0832	.0846
15	.0575	.0602	.0628	.0651	.0673	.0694	.0713	.0731	.0748	.0764
16	0.0476	0.0506	0.0534	0.0560	0.0584	0.0607	0.0628	0.0648	0.0667	0.0685
17	.0379	.0411	.0442	.0471	.0497	.0522	.0546	.0568	.0588	.0608
18	.0283	.0318	.0352	.0383	.0412	.0439	.0465	.0489	.0511	.0532
19	.0188	.0227	.0263	.0296	.0328	.0357	.0385	.0411	.0436	.0459
20	.0094	.0136	.0175	.0211	.0245	.0277	.0307	.0335	.0361	.0386
21	0.0000	0.0045	0.0087	0.0126	0.0163	0.0197	0.0229	0.0259	0.0288	0.0314
22		.0000	.0042	.0081	.0118	.0153	.0185	.0215	.0244
23				.0000	.0039	.0076	.0111	.0143	.0174
24						.0000	.0037	.0071	.0104
25								.0000	.0035

Appendix 10

Percentage points of the W test for n = 3(1)50

n	0.01	0.02	0.05	0.10	Level 0.50	0.90	0.95	0.98	0.99
3	0.753	0.756	0.767	0.789	0.959	0.998	0.999	1.000	1.000
4	.687	.707	.748	.792	.935	.987	.992	.996	.997
5	.686	.715	.762	.806	.927	.979	.986	.991	.993
6	0.713	0.743	0.788	0.826	0.927	0.974	0.981	0.986	0.989
7	.730	.760	.803	.838	.928	.972	.979	.985	.988
8	.749	.778	.818	.851	.932	.972	.978	.984	.987
9	.764	.791	.829	.859	.935	.972	.978	.984	.986
10	.781	.806	.842	.869	.938	.972	.978	.983	.986
11	0.792	0.817	0.850	0.876	0.940	0.973	0.979	0.984	0.986
12	.805	.828	.859	.883	.943	.973	.979	.984	.986
13	.814	.837	.866	.889	.945	.974	.979	.984	.986
14	.825	.846	.874	.895	.947	.975	.980	.984	.986
15	.835	.855	.881	.901	.950	.975	.980	.984	.987
16	0.844	0.863	0.887	0.906	0.952	0.976	0.981	0.985	0.987
17	.851	.869	.892	.910	.954	.977	.981	.985	.987
18	.858	.874	.897	.914	.956	.978	.982	.986	.988
19	.863	.879	.901	.917	.957	.978	.982	.986	.988
20	.868	.884	.905	.920	.959	.979	.983	.986	.988
21	0.873	0.888	0.908	0.923	0.960	0.980	0.983	0.987	0.989
22	.878	.892	.911	.926	.961	.980	.984	.987	.989
23	.881	.895	.914	.928	.962	.981	.984	.987	.989
24	.884	.898	.916	.930	.963	.981	.984	.987	.989
25	.888	.901	.918	.931	.964	.981	.985	.988	.989
26	0.891	0.904	0.920	0.933	0.965	0.982	0.985	0.988	0.989
27	.894	.906	.923	.935	.965	.982	.985	.988	.990
28	.896	.908	.924	.936	.966	.982	.985	.988	.990
29	.898	.910	.926	.937	.966	.982	.985	.988	.990
30	.900	.912	.927	.939	.967	.983	.985	.988	.900
31	0.902	0.914	0.929	0.940	0.967	0.983	0.986	0.988	0.990
32	.904	.915	.930	.941	.968	.983	.986	.988	.990
33	.906	.917	.931	.942	.968	.983	.986	.989	.990
34	.908	.919	.933	.943	.969	.983	.986	.989	.990
35	.910	.920	.934	.944	.969	.984	.986	.989	.990
36	0.912	0.922	0.935	0.945	0.970	0.984	0.986	0.989	0.990
37	.914	.924	.936	.946	.970	.984	.987	.989	.990
38	.916	.925	.938	.947	.971	.984	.987	.989	.990
39	.917	.927	.939	.948	.971	.984	.987	.989	.991
40	.919	.928	.940	.949	.972	.985	.987	.989	.991
41	0.920	0.929	0.941	0.950	0.972	0.985	0.987	0.989	0.991
42	.922	.930	.942	.951	.972	.985	.987	.989	.991
43	.923	.932	.943	.951	.973	.985	.987	.990	.991
44	.924	.933	.944	.952	.973	.985	.987	.990	.991
45	.926	.934	.945	.953	.973	.985	.988	.990	.991
46	0.927	0.935	0.945	0.953	0.974	0.985	0.988	0.990	0.991
47	.928	.936	.946	.954	.974	.985	.988	.990	.991
48	.929	.937	.947	.954	.974	.985	.988	.990	.991
49	.929	.937	.947	.955	.974	.985	.988	.990	.991
50	.930	.938	.947	.955	.974	.985	.988	.990	.991

Appendix 11

Coefficients of Orthogonal Polynomials

n	Polynomial	X=1	2	3	4	5	6	7	8	9	10	ΣZ^2	λ
2	Linear	-1	1									2	2
3	Linear	-1	0	1								2	1
	Quadratic	1	-2	1								6	3
4	Linear	-3	-1	1	3							20	2
	Quadratic	1	-1	-1	1							4	1
	Cubic	-1	3	-3	1							20	10/3
5	Linear	-2	-1	0	1	2						10	1
	Quadratic	2	-1	-2	-1	2						14	1
	Cubic	-1	2	0	-2	1						10	5/6
	Quartic	1	-4	6	-4	1						70	35/12
6	Linear	-5	-3	-1	1	3	5					70	2
	Quadratic	5	-1	-4	-4	-1	5					84	3/2
	Cubic	-5	7	4	-4	-7	5					180	5/3
	Quartic	1	-3	2	2	-3	1					28	7/12
7	Linear	-3	-2	-1	0	1	2	3				28	1
	Quadratic	5	0	-3	-4	-3	0	5				84	1
	Cubic	-1	1	1	0	-1	-1	1				6	1/6
	Quartic	3	-7	1	6	1	-7	3				154	7/12
8	Linear	-7	-5	-3	-1	1	3	5	7			168	2
	Quadratic	7	1	-3	-5	-5	-3	1	7			168	1
	Cubic	-7	5	7	3	-3	-7	-5	7			264	2/3
	Quartic	7	-13	-3	9	9	-3	-13	7			616	7/12
	Quintic	-7	23	-17	-15	15	17	-23	7			2184	7/10
9	Linear	-4	-3	-2	-1	0	1	2	3	4		60	1
	Quadratic	28	7	-8	-17	-20	-17	-8	7	28		2772	3
	Cubic	-14	7	13	9	0	-9	-13	-7	14		990	5/6
	Quartic	14	-21	-11	9	18	9	-11	-21	14		2002	7/12
	Quintic	-4	11	-4	-9	0	9	4	-11	4		468	3/20
10	Linear	-9	-7	-5	-3	-1	1	3	5	7	9	330	2
	Quadratic	6	2	-1	-3	-4	-4	-3	-1	2	6	132	1/2
	Cubic	-42	14	35	31	12	-12	-31	-35	-14	42	8580	5/3
	Quartic	18	-22	-17	3	18	18	3	-17	-22	18	2860	5/12
	Quintic	-6	14	-1	-11	-6	6	11	1	-14	6	780	1/10

Appendix 12

Variance of Means and Contrasts

for

Mixed Models

Assuming the correlations among all the components in the models are zero, the following derivations are appropriate [Anderson and Bancroft (1952), p. 340-344]:

1. <u>Variance of the column mean for a mixed two way classified analysis</u>

   Consider the layout

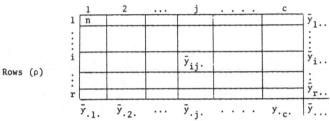

in which there is a two way classification with more than one observation per cell, n. If γ is fixed and ρ is random in the following model:

$$Y_{ijk} = \mu + \rho_i + \gamma_j + (\rho\gamma)_{ij} + \varepsilon_{ijk},$$

let ρ_i be $(0,\sigma_R^2)$, $(\rho\gamma)_{ij}$ be $(0,\sigma_{RC}^2)$ and ε_{ijk} be $(0,\sigma^2)$.

The $E(\gamma_j) = \gamma_j$ since γ_j is fixed and $E(\mu) = \mu$. We wish to find the variance of the column (fixed) mean. From the equation above, the jth column mean is:

$$\overline{Y}_{\cdot j \cdot} = \frac{\sum\limits_{i=1}^{r} \sum\limits_{k=1}^{n} Y_{ijk}}{rn} = \mu + \frac{n \sum\limits_{i} \rho_i}{rn} + \frac{rn\gamma_j}{rn}$$

$$+ \frac{n \sum\limits_{i=1}^{r} (\rho\gamma)_{ij}}{rn} + \frac{\sum\limits_{i} \sum\limits_{k=1}^{n} \varepsilon_{ijk}}{rn}$$

Since $E(\gamma_j) = \gamma_j$, $E(\mu) = \mu$, and E (all random variables in the model) = 0

$$E(CP) = 0$$

$$E(\overline{Y}\cdot j\cdot) = \mu + \gamma_j$$

The variance of the estimate of the jth column mean is:

$$V(\bar{Y} \cdot j \cdot) = E[\bar{Y}_{\cdot j} \cdot - E(\bar{Y} \cdot j \cdot)]^2 = E\left\{ \frac{\sum_i \rho_i}{r} + \frac{\sum_i (\rho\gamma)_{ij}}{r} + \frac{\sum_i \sum_k \epsilon_{ijk}}{rn} \right\}^2$$

but $E(\rho_i^2) = \sigma_R^2$, $E(\rho\gamma)_{ij}^2 = \sigma_{RC}^2$, and $E(\epsilon_{ijk})^2 = \sigma^2$.

Hence

$$V(\bar{Y}_{\cdot j}) = E\left\{ \left(\frac{\rho_1^2 + \dots + \rho_r^2 + \overset{(CP1)}{\text{Cross products 1}}}{r^2} \right) \right.$$

$$+ \left(\frac{(\rho\gamma)_{1j}^2 + \dots + (\rho\gamma)_{rj}^2 + CP_2}{r^2} \right)$$

$$\left. + \left(\frac{\epsilon_{1j1}^2 + \dots + \epsilon_{rjn}^2 + CP_3}{r^2 n^2} \right) + CP_4 \right\}$$

and $V(\bar{Y}_{\cdot j}) = \left(\frac{r\sigma_R^2}{r^2} + \frac{r\sigma_{RC}^2}{r^2} + \frac{r \, n \, \sigma^2}{r^2 n^2} \right) = \frac{\sigma_R^2}{r} + \frac{\sigma_{RC}^2}{r} + \frac{\sigma^2}{rn}$.

2. Variance of the mean inside the table.

Next we want to find the variance of the mean inside the table (with the assumption that $(\rho\gamma)_{ij}$ is random, the value of this mean is questionable):

$$\bar{Y}_{ij} \cdot = \frac{\sum_{k=1}^{n} Y_{ijk}}{n} = \mu + \frac{n\rho_i}{n} + \frac{n\gamma_j}{n} + \frac{n(\rho\gamma)_{ij}}{n} + \frac{\sum_{k=1}^{n} \epsilon_{ijk}}{n}$$

$$E(\bar{Y}_{ij} \cdot) = \mu + \gamma_j$$

and

$$V(\bar{Y}_{ij} \cdot) = E[\bar{Y}_{ij} \cdot - E(\bar{Y}_{ij} \cdot)]^2 = E\left\{ \rho_i + (\rho\gamma)_{ij} + \frac{\sum_{k=1}^{n} \epsilon_{ijk}}{n} \right\}^2$$

$$= E\left\{ \rho_i^2 + (\rho\gamma)_{ij}^2 + \frac{\epsilon_{ij1}^2 + \dots + \epsilon_{ijn}^2 + CP_5}{n^2} + CP_6 \right\}$$

$$= \sigma_R^2 + \sigma_{RC}^2 + \frac{n\sigma^2}{n^2} .$$

Finally

$$V(\bar{Y}_{ij} \cdot) = \sigma_R^2 + \sigma_{RC}^2 + \frac{\sigma^2}{n} .$$

3. Variance of a contrast.

Consider the comparison involving only two column means $\overline{Y} \cdot p \cdot$ and $\overline{Y} \cdot q \cdot$, then the difference is really the contrast:

$$(\overline{Y}_{\cdot p \cdot} - \overline{Y}_{\cdot q \cdot}) = \left\{ \mu + \frac{\sum\limits_i \rho_i}{r} + \gamma_p + \frac{\sum\limits_i (\rho\gamma)_{ip}}{r} + \frac{\sum\limits_i \sum\limits_k \epsilon_{ipk}}{rn} \right.$$

$$\left. - \mu - \frac{\sum\limits_i \rho_i}{r} - \gamma_q - \frac{\sum\limits_i (\rho\gamma)_{iq}}{r} - \frac{\sum\limits_i \sum\limits_k \epsilon_{iqk}}{rn} \right\}$$

$$(\overline{Y}_{\cdot p \cdot} - \overline{Y}_{\cdot q \cdot}) = \left\{ (\gamma_p - \gamma_q) + \frac{\sum\limits_i (\rho\gamma)_{ip}}{r} - \frac{\sum\limits_i (\rho\gamma)_{iq}}{r} \right.$$

$$\left. + \frac{\sum\limits_i \sum\limits_k \epsilon_{ipk}}{rn} - \frac{\sum\limits_i \sum\limits_k \epsilon_{iqk}}{rn} \right\} .$$

The variance of this contrast is

$$V(\overline{Y}_{\cdot p \cdot} - \overline{Y}_{\cdot q \cdot}) = E[(\overline{Y}_{\cdot p \cdot} - \overline{Y}_{\cdot q \cdot}) - E(\overline{Y}_{\cdot p \cdot} - \overline{Y}_{\cdot q \cdot})]^2 .$$

Now

$$E(\overline{Y}_{\cdot p \cdot} - \overline{Y}_{\cdot q \cdot}) = (\gamma_p - \gamma_q)$$

and

$$V(\overline{Y}_{\cdot p \cdot} - \overline{Y}_{\cdot q \cdot}) = E\left\{ \left(\frac{\sum\limits_i (\rho\gamma)_{ip}}{r} - \frac{\sum\limits_i (\rho\gamma)_{iq}}{r} \right) + \left(\frac{\sum\limits_i \sum\limits_k \epsilon_{ipk}}{rn} - \frac{\sum\limits_i \sum\limits_k \epsilon_{iqk}}{rn} \right) \right\}^2$$

$$= E\left\{ \frac{\sum\limits_i (\rho\gamma)_{ip}^2}{r^2} + \frac{\sum\limits_i (\rho\gamma)_{iq}^2}{r^2} + \frac{\sum\limits_i \sum\limits_k \epsilon_{ipk}^2}{r^2 n^2} + \frac{\sum\limits_i \sum\limits_k \epsilon_{iqk}^2}{r^2 n^2} + CP \right\} .$$

It follows that

$$V(\overline{Y}_{\cdot p \cdot} - \overline{Y}_{\cdot q \cdot}) = \frac{r\sigma_{RC}^2}{r^2} + \frac{r\sigma_{RC}^2}{r^2} + \frac{rn\sigma^2}{r^2 n^2} + \frac{rn\sigma^2}{r^2 n^2}$$

$$= \frac{2\sigma_{RC}^2}{r} + \frac{2\sigma^2}{rn} .$$

Knowing that E.M.S. (R.C.) $= n\sigma_{RC}^2 + \sigma^2$, the estimate of this variance is

$$\hat{V}(\overline{Y}_{\cdot p \cdot} - \overline{Y}_{\cdot q \cdot}) = \frac{2(\text{mean square RC})}{rn} .$$

The experimenter can always handle the fixed model problem because the standard error $\frac{s}{\sqrt{n}}$ is the basis for all tests. This is not true for mixed models, and depending upon what is fixed and what is random, the standard error may actually be quite different from the variance of the contrast. To be safe for mixed models, the experimenter must derive the appropriate standard error and variance of the contrast from the model.

Of course there is no interest in means for the random model; consequently there is no problem for the completely random case.

When investigating individual comparisons in a mixed model, there results are especially applicable. Consider the repeated t-tests

$$|\bar{y}_{.p.} - \bar{y}_{.q.}| > t_\alpha(\sqrt{2}\, s_{\bar{y}}) : \text{(for a fixed model)},$$

$(\sqrt{2}\, s_{\bar{y}})$ is really the standard error of the contrast and we merely replace this component by

$$\sqrt{\frac{2(\text{mean square RC})}{rn}}$$

for the mixed model. Of course the experimenter will thoroughly investigate the interaction before examining and eventually interpreting this contrast among the treatment means.

A similar procedure may be used on the various other tests. For example the Newman-Keuls test uses

$$R_k = q_\alpha(k,\, df)(s_{\bar{y}}) : \text{(for a fixed model)}.$$

But the standard error for the contrast is

$$\sqrt{2}\, s_{\bar{y}}.$$

Hence

$$R_k = \frac{q_\alpha(k, df)}{\sqrt{2}}\, (\sqrt{2}\, s_{\bar{y}}),$$

(Note that when k = 2, $\dfrac{q_\alpha(k,\, df)}{\sqrt{2}} = t_\alpha$) and, for the mixed model, substituting

$$\sqrt{\frac{2}{rn}\,(\text{mean square RC})}$$

for $\sqrt{2}\, s_{\bar{y}}$, we get

$$R_k = \frac{q_\alpha(k, df)}{\sqrt{2}}\, \sqrt{\frac{2}{rn}\,(\text{mean square RC})}$$

or

$$R_k = q_\alpha(k,df) \sqrt{\frac{\text{mean square RC}}{rn}}$$

and $\sqrt{\dfrac{\text{mean square RC}}{rn}}$ can be used

directly to replace $s_{\bar{y}}$. (Note this is not the usual $s_{\bar{y}}$ for a fixed model and results only because of the contrast causing σ_R^2 to drop out).

4. Nested Factorial for Subsection 6.2.3.

Derivation of estimated standard error of ration means from Equation (6.2.3):

$$y_{ijk} = \mu + R_i + H_{(i)j} + \delta_{(ij)} + W_k + RW_{ik} + HW_{(i)jk} + \epsilon_{(ijk)}$$

where: R_i and W_k are fixed and $H_{(i)j}$ are random,

$$\bar{y}_{i..} = \left[\mu + R_i + \frac{\sum_{j=1}^{8} H_{(i)j}}{8} + \frac{\sum_{j=1}^{8} \delta_{(ij)}}{8} \right.$$
$$\left. + \underbrace{\frac{\sum_{k=1}^{5} W_k}{5} + \frac{\sum_{k=1}^{5} RW_{ik}}{5} + \frac{\sum_{j=1}^{8}\sum_{k=1}^{5} HW_{(i)jk}}{40}}_{= 0} + \frac{\sum_{j=1}^{8}\sum_{k=1}^{5} \epsilon_{(ijk)}}{40} \right]$$

Since weeks (W_k) are fixed and hens $[H_{(i)j}]$ are random, the summation of the mixed interaction over the subscript k is defined as zero.

Thus $(\frac{1}{40}) \left(\sum_{j=1}^{8}\sum_{k=1}^{5} HW_{(i)jk} \right) = 0$.

$$E(\bar{y}_{i..}) = \mu + R_i$$
$$V(\bar{y}_{i..}) = E[\bar{y}_{i..} - E(\bar{y}_{i..})]^2 = E\left[\frac{\sum_{j=1}^{8} H_{(i)j}}{8} + \frac{\sum_{j=1}^{8} \delta_{(ij)}}{8} + \frac{\sum_{j=1}^{8}\sum_{k=1}^{5} \epsilon_{(ijk)}}{40} \right]^2$$

or

$$V(\bar{y}_{i..}) = \left(\frac{\hat{\sigma}_H^2}{8} + \frac{\hat{\sigma}_\delta^2}{8} + \frac{\hat{\sigma}^2}{40} \right)$$

as $s_{\bar{y}_{i..}} = \sqrt{\hat{V}(\bar{y}_{i..})} = \sqrt{\frac{\hat{\sigma}_H^2}{8} + \frac{\hat{\sigma}_\delta^2}{8} + \frac{\hat{\sigma}^2}{40}}$.

5. Extensions.

This approach of finding the variance of means and contrasts can be used for all fixed or mixed ANOVA models if the assumption of no correlation is appropriate for all of the components in all of the ANOVA equations.

SUBJECT INDEX